Energy Resources

Theodore I. Erski

Energy Resources

Cover design: Theodore Erski.

Cover credit/courtesy:

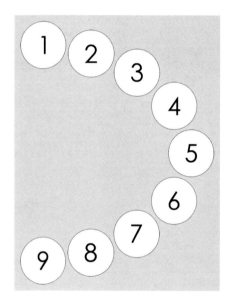

①Peabody Energy, ②BP, ③Siemens, ④United States Nuclear Regulatory Commission, ⑤USDA FSA, Farm Service Agency, ⑥ Voith Hydro, ⑦National Renewable Energy Laboratory, ⑧ Bureau of Land Management, ⑨ Siemens.

ISBN: 978-1-105-79089-8

Acknowledgements

Acknowledgements come first in any book because without the selfless assistance of others, the rest is impossible.

First, my sincere thanks to Paul Stahmann for tirelessly critiquing every chapter and offering salient points that broadened my prospective audience. Second, a thank you Steve Socol, because you did not pull any punches with your constructive criticism, especially with the uranium chapter. Third, to my numerous math and science colleagues who reviewed and critiqued various chapters within this textbook, including Steven Burks, Sharon Button, Bev Dow and Deb Firak. Thank you also to Kate Kramer for offering assistance with the petroleum chapter, and to Paul Hamill, who introduced me to Lulu. Thank you Meg Gallagher and the professionals at Peabody Energy for you constructive

and balanced review. Your comments strengthened the text's content and prose, and *Energy Resources* is a better book because of you. Finally, thank you Kelly Fallon, Joan Flanagan, Marla Garrison, Scotty Nath, Anne Martinique, Amy Maxeiner, Angela Sass and Rob Smith for encouraging me to complete this textbook.

Also, thank you to the dozens of individuals and organizations that contributed pictures and graphics to this textbook. Your contributions set this work apart from all the others and make the topic of energy resources accessible and interesting for many readers. As the fossil fuel age progresses deeper into the 21st century, we all need to become more informed energy consumers, and you help make that possible.

Contents

Longwall Miners, *Credit/Courtesy of the Beamish Museum*

Trans-Alaska Pipeline, *Credit/Courtesy of the Bureau of Land Management*

Nuclear Waste Storage Casks, *Credit/Courtesy of the United States Nuclear Regulatory Commission*

Part 3: Renewable Energy Sources

Wind Turbines in New Zealand, *Credit/Courtesy of Siemens*

Part of an 800 kV High-Voltage Direct Current (HVDC)
System, *Credit/Courtesy of Siemens*

Chapter 1
Resources & Electricity

Energy and Resources

The world is going to run out of oil. Clean coal technology provides pollution-free energy that is also cheap. Electricity comes from the outlet in my wall. Nuclear power is dangerous. Dams ruin the environment. Wind turbines are inefficient because the wind doesn't blow all the time. Solar power is great as long as the sun is shining. OPEC has too much power. Carbon dioxide is warming up the planet.

As the twenty-first century matures, the above statements are increasingly heard in homes, on the news and in political forums. Are they true? Since we are all energy consumers, it is important to know where our energy resources come from, how much exist, and how to responsibly use these resources to ensure access, market stability, and environmental quality. This book is designed to inform you about today's energy resources, and thereby empower you to make the best possible decisions about the resources you use, and the policies you support. The comprehensive scope and unbiased perspective of this work is designed to cut through the confusion, the hyperbole, the misinformation we are confronted with every day.

All the work herein is based on the science behind our energy resources. It details important virtues and vices of *every* energy resource. It recognizes that there are competing policies about energy resources and, perhaps most importantly, that there is no perfect solution to satisfying our ever-increasing demand for energy. While based on science, this book is written with the non-scientist in mind—the energy consumer—no matter the profession, educational background, or political inclination. The data, graphs, charts and illustrations were chosen and written about to appeal to the broadest possible audience while satisfying technical, scientific questions about resource extraction and marketing, environmental impacts, and policy decisions that we as energy consumers confront every day. The objective of this work is simple—to empower you with the knowledge to become an informed energy consumer.

Energy is the ability to do work. For instance, a car moving down the road works to transport people. A freezer works to chill food. A light bulb works to illuminate a room. A furnace works to heat a building. These mechanical devices ultimately rely on various forms of energy to accomplish their jobs, and the end result is an increase in our quality of life.

Sources of energy that ultimately power so many of the mechanical devices we rely upon are broadly or-ganized into two categories called kinetic and potential. **Kinetic energy is motion.** For example, electricity *moves* through power lines. Sunlight *streams* through the earth's atmosphere. Wind *blows* across the landscape. Water *flows* down a slope. **Potential energy is stored.** For example, coal *can be* ignited, but until then it simply remains a sedimentary rock. Petroleum *can be* refined into useful products, but until then it simply remains a viscous fluid. Natural gas *can* be ignited, but until then is simply an invisible and odorless gas. Uranium *can be* split, thereby producing heat, but until then it is simply one of many elements in the earth's crust. Finally, wood *can be* burned, but until then it is simply part of the living biomass on the earth's surface. We capture both kinetic and potential forms of energy in order to cook food, heat homes, and of course to power all the useful mechanical devices in the modern world.

Much of the energy that we use is created from non-renewable sources. **Non-renewable energy sources cannot be replenished in a short time span.** These sources include coal, petroleum, natural gas (CH_4) and uranium (U). **In 2010, non-renewable energy sources constituted about 92% of the total energy used in the United States** (United States Energy Information Administration, 2011). Time is of course relative, so in this context "short time span" can mean a year, twenty years, a human lifetime, or even several generations. Coal, for instance, cannot be replenished in a short time span. Approximately 300 million years must pass before specific kinds of plant matter, in specific kinds of environments, can change into a unique, carbon-based, combustible rock. Uranium too, cannot be replenished in a short time span. A star must die, specifically in a supernova event, in order to create uranium. That uranium dust must then coalesce with other matter to create a planet. Another way to think about this is to consider that all the uranium that exists on earth was created in a single cosmic, cataclysmic event over 4.6 billion years ago.

A small but growing portion of the energy we use is created from renewable sources. **Renewable energy sources can be replenished in a short time span.** These include biomass, hydropower, wind, geothermal and sunlight. **In 2010, renewable energy sources constituted about 8% of the total energy used in the United States** (United States Energy Information Administration, 2011). Consider the following renewable energy examples: Biomass is simply plant matter that is burned for energy, including the wood in a fireplace and the ethanol in a gas tank. Biomass is replenished every time a crop is harvested. Captured with a turbine, wind is nothing more than the nitrogen and oxygen gas molecules in our atmosphere moving from high to low pressure areas because of sunlight intensity variations across the earth's surface. Wind is continuously replenished because gas molecules constantly move. Sunlight too, is continuously replenished because the fusion process on our sun constantly creates electromagnetic radiation that is captured using active techniques (such as photovoltaic cells) or passive techniques (such as hot water heaters installed on rooftops).

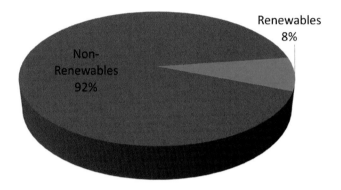

Figure 1-1, Role of Non-Renewable and Renewable Energy Sources in the United States, 2010

We use the terms "resource" and "reserve" when discussing energy. Of these two terms, "resource" is the most general and therefore least useful. **A resource is an existing energy source.** Existing energy resources covered in this book, for instance, are coal, oil, natural gas, uranium, biomass, water, wind, geothermal energy and sunlight. The energy industry and policy-makers are more concerned with the second term, "reserves". This is because the term "resource" does not consider if extracting (or capturing) the energy source is economically feasible. Thinking broadly, economically feasibility may include issues such as a break-even point, contracts, and regulatory red-tape. For example, if the cost of extracting a barrel of oil from a particular well is greater than what the market will pay, an oil drilling company will fail to cover its costs—it will fail to "break even"—and therefore decide not to work that particular well.

The term "reserve" on the other hand, is more specific and therefore more useful to the energy industry and to policy-makers. A "reserve" may be either "proven" or "probable". **A proven reserve is a resource that *is* economically feasible for extraction.** An example of this is the current oil field in Saudi Arabia named Ghawar, where about 5 million barrels of oil are extracted every day. **A probable reserve is a resource that *may become* economically feasible for extraction.** For example, the Arctic National Wildlife Refuge (ANWR) in northeast Alaska contains 7.7 billion barrels of oil (United States Geological Survey, 2001, p. 4). This resource is currently not being extracted, however, due to environmental and aesthetic concerns. The "regulatory red-tape" one could argue, renders the resource in ANWR a probable reserve.

Proven reserves are used to calculate the frequently misunderstood statistic of how many years of a particular resource are left, either in a region or on earth. This is called the reserve production (R/P) ratio. **The reserve production ratio is the remaining amount of a non-renewable resource, expressed in years.** This is an important statistic, as one might imagine, because it is used to forecast the future availability of oil, coal and natural gas. Energy companies, as well as policy-makers, need to anticipate incomes, employment and energy-security. They do so with the aid of an R/P ratio. The "R" of the formula is a resource's proven reserves. The "P" of the formula is the current production of that resource.

Concerning oil, for example, there are 1,341.572 billion (1,341,572,000,000) barrels of proven oil reserves on earth today. We produce—that is, extract from the ground—31.7 billion (31,700,000,000) barrels every year (United States Energy Information Administration, 2011b). Calculating the earth's petroleum R/P ratio is demonstrating in Calculation 1-1, and produces an answer of 42 years.

Sounds scary—only 42 years of oil left on earth! As stated above, however, this number is frequently misunderstood. R/P ratios are often, unfortunately, casually tossed around by the media and by policy-makers

---

Calculating Earth's Petroleum R/P Ratio

A) 1,342 billion barrels of oil reserves existed in 2009 (1,342 billion = 1.3×10^{12})
B) 31.7 billion barrels of oil were produced in 2009 31.7 billion = 3.17×10^{10})
C) Reserves / Production = $1.3 \times 10^{12}/3.17 \times 10^{10}$ = $.41 \times 10^2$ = 41 years.

Calculation 1-1, Calculating Earth's Petroleum R/P Ratio

Fuel	Unit of Measure	Proven Reserves	Production	R/P
Oil	Billions of barrels	1,341.572	31.7	42 years
Coal	Billion short tons	948	7	135 years
Natural Gas	Trillions of cubic feet	6,289.147	106	59 years

Table 1-1, Reserve Production Ratios for Worldwide Oil, Coal and Natural Gas, 2010

as they seek to advocate a particular agenda. What is missing in the hyperbole is the following important fact: **Any R/P ratio is applicable only to the particular moment it is calculated**. The numbers of years left of any particular resource constantly changes as proven reserves (the "R") and/or production (the "P") changes. For instance, if a new proven reserve of oil enters into the marketplace, the numerator of the formula—the "R'"—must necessarily change. Similarly, if oil production drops due to lack of demand, or grows due to expanding populations and their desire to drive a vehicle, the denominator of the formula—the "P"—must also be changed. Any change in "R" or the "P" necessarily changes the number of years left of any particular proven reserve.

Electricity

Coal produces the majority of the world's electricity. **Electricity is a secondary energy source, obtained by converting non-renewable and renewable energy resources.** Electricity is a tricky thing to get one's brain around because it is frequently discussed in terms of a "flow," and as such we imagine a physical thing—an electron—moving within a solid wire. This is only partially correct.

Copper is an especially useful medium through which an electric charge travels because the copper atom has a weakly-bonded electron in its outermost shell. When many copper atoms are joined to form copper wire, all those weakly-bonded electrons float freely within the metal's crystalline lattice. These free electrons (there are quintillions in any geometric plane of typical household wire) are readily compelled into regimented order whenever an electric force (this is called "voltage"—more about this a bit later) is placed across a wire (Livingston, 1996, pp. 65-68). The copper's free electrons are thus current carriers. Electricity travels at near light speed through a wire, and does so by using the metal's regimented electrons as the road on which to travel. These current-carrying electrons become regimented all at once, like marching, in-step soldiers moving as one entity. They never disappear, drain-away, or accumulate somewhere due to the presence of an electric current. If the electrical charge is turned off, they simple return to their wandering within the metal's crystalline lattice.

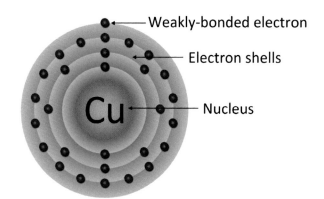

Weakly-bonded electron

Electron shells

Nucleus

Figure 1-2, A Copper Atom

An electrical force needs to be created—generated—in order to flow through a wire. **The most common electrical generation method in the world is the steam turbine** (United States Energy Information Administration, 2011c). Other common generation methods include internal-combustion engines, gas-combustion engines, water and wind turbines and photovoltaics. With the exception of photovoltaics, all the methods have one simple objective—to induce spinning. To generate electricity all one must do is spin a series of copper coils within a magnetic field, or spin a series of magnets in proximity to a collection of copper coils (these devices are called generators or alternators, depending on how they are constructed). However it is built, when a generator spins it forces otherwise free electrons into regimented order and thereby induces an electromagnetic current that can then flow through a wire.

Electricity generation and consumption is measured in Watts. The name "Watt" is in honor of the eighteenth century Scottish engineer named James Watt. Watt improved the efficiency of early steam engines, thereby helping launch the Industrial Revolution which

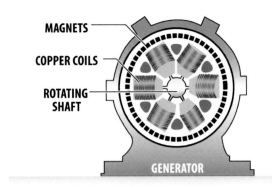

The copper coils spin inside a ring of magnets. This creates an electric field, producting electricity.

Figure 1-3, Model of a Generator, *Credit/Courtesy of National Energy Education Development Project*

radically transformed human society. Today the world honors Watt by using his name to measure electricity production and consumption because, as stated above, most of the world's electricity is generated by steam turbines. Knowing the background of the name, however, is of limited use—after all, what is a Watt? **A Watt is one joule of energy transfer per second.** This of course begs another question—what is a joule? A joule is a measure of work performed. While it is usually defined mathematically, consider this straightforward (and hopefully more digestible) example. Take a small apple in your hand and lift it one meter straight up—this effort takes about one joule. If you accomplish this task in one second, you've just performed one Watt. For another example, consider that "one joule can be defined as the kinetic energy of a 2-kilogram mass traveling at 1 meter per second" (Cox & Forshaw, 2009, p. 140).

We are of course most familiar with the term Watt when we change the light bulbs in our home. A common one is the 60-Watt incandescent bulb ("incandescent" simply means something that emits light because it is hot). When we turn on a 60-Watt bulb, it performs 60 joules of work each second. We experience this work as light and heat. If we decide to replace the 60-Watt bulb with a 100-Watt bulb, it in turn performs 100 joules of work each second. We would probably notice that the 100-Watt bulb is brighter and hotter than the 60-Watt bulb. This is because each second the 100-Watt bulb performs 40 more joules of work than the 60-Watt bulb. Since it is doing more work, more electricity is required, which we ultimately experience as a higher electricity bill.

Consumers pay their electric bill each month. Our electricity consumption is tallied by a local utility com-

pany and reported not in Watts, but in something called kilowatt-hours, which is abbreviated "kWh" (notice the "k" and the "h" are lower case, while the "W" is upper case—again in honor of James Watt). To understand kilowatt-hours we have to first consider a "kilowatt" and then tackle the "hours" part. The word "kilo" is a prefix multiplier. It is used in front of another word, such as gram, meter or Watt, in order to express a larger unit. **Kilo means 1,000.** For example, 1 kilogram is 1,000 grams, 1 kilometer is 1,000 meters, and of course 1 kilowatt is 1,000 Watts. Prefix multipliers are used because they simplify numerical discussions and make errors less likely when calculations are required.

Imagine, then, turning on some lights in your home, perhaps an appliance or two, and maybe a radio and a television. All these devices consume electricity in varying amounts. If, in 1 hour, 1,000 Watts have been consumed by these devices, you'll owe the local utility company whatever it costs for 1 kilowatt-hour (kWh) of electricity. Cost per kWh varies by location (and is different for residential, commercial and industrial consumers), but in 2009 a single kWh of electricity cost residential consumers about 11.5 cents (United States Energy Information Administration, 2011d, p. 61).

If, during that hour, you used less than 1,000 Watts (say, .87 kWh), you'll owe the local utility company whatever it costs for a fraction of a kilowatt-hour (if the rate is 11.5 cents/kWh, you'll owe about 1 dime because 11.5 x .87 = 10.005 cents). What happens, if more than 1,000 Watts (say, 1.35 kWh) are consumed by all those appliances in 1 hour? Then, for that single hour of time, you will owe the local utility company whatever it costs for more than 1 kWh (if the rate is 11.5 cents/kWh, you'll owe about 16 cents because 11.5 x 1.35 = 15.52 cents). In this manner, even though there are 24 hours in a day, a typical home may in fact consume more than 24 kWh of electricity in a day. Similarly, even though there are 720 hours in a month of 30 days, a typical household may receive a bill showing that 908 kWh were consumed in that month. This was in fact the average monthly consumption for an American home in 2009 (United States Energy Information Administration, 2011e).

Average monthly kWh consumption varies by state and census division. In 2009, for example, households in Maine consumed an average of 521 kWh per month, while households in Louisiana consumed an average of 1,273kWh per month (United States Energy Information Administration, 2011f). Since electricity cost an average of 11.5 cents per kWh, then a typical household in Maine would owe an average of $59.92 for electricity during any particular month (11.5 x 521 = 5,991.5 cents / 100 = $59.915), while a typical household in Louisiana would owe an average of $146.40 for electricity during any particular month (11.5 x 1,273 = 14,639.5 cents / 100 = $146.395).

If you have ever paid an electric bill, however, you know that consumers are charged for more than electricity with each bill. Flat charges, as well as variable charges based on monthly consumption, can include the following: customer charge, meter charge, distribution charge, transmission charge, municipal tax and state tax. In all, approximately 60% of any electric bill is for the electricity, and approximately 40% is for other related charges.

Converting Energy Values

Comparing energy production and consumption across people and places is simplified by converting different values into a common measurement standard. In this manner, one can more accurately compare a household that, for example, cooks with natural gas, to another household that cooks using a stove with electric burners. Which household uses more energy? By converting both the natural gas use (measured in hundreds of cubic feet) and the electricity use (measured in kWh) to a common measurement standard, we can determine the answer. That common measurement standard varies, but a very common one is called the British thermal unit (Btu). **A British thermal unit is the amount of energy needed to raise 1 pound of water 1 °Fahrenheit.** By knowing the total amount (and quality) of natural gas the first household used, we can convert that value to yield a total number of Btu (1 ft^3 of CH$_4$ = 1028 Btu). Similarly, by knowing the total amount of kWh the second household used, we can convert that value to yield a total number of Btu (1 kWh = 3412 Btu). With two Btu values, the two households can be accurately compared even though they use two different energy resources.

Energy production and consumption is also compared across people and places using a common measure-

ment standard called barrel of oil equivalent (BOE). **Barrel of oil equivalent is the amount of energy contained in one barrel of crude oil.** For example, about 5,642 cubic feet of natural gas contains the same amount of energy as one barrel of crude oil (5,642 ft^3 of CH$_4$ = 1 BOE). We'll find in later chapters, however, that this is a rough measurement because both natural gas and crude oil resources vary in energy content. Another important issue to remember about crude oil is that 1 barrel is 42 gallons. This is an odd number (i.e., why not 40, or even 50 gallons?). 42 gallons is the inherited, standard size due to late nineteenth century Pennsylvanian coopers and the size of barrels they made at the beginning of the oil boom. Yet another important and potentially confusing issue is the abbreviation of the word "barrel" and "barrels". **"Barrel" is abbreviated Bbl. "Barrels" is abbreviated Bbls.** This is unfortunate because, of course, there are two "b"s in each abbreviation. Wouldn't "Bl" and "Bls" be more logical? Yes! However inconvenient, this is today's abbreviation standard, which is also inherited from sometime in the past. Be careful *not* to insert "billions" in place of the first (capital) "B" in the above two abbreviations. If one wants to discuss "billions of barrels of oil" the standard abbreviation is Bbbls.

The size of energy production and consumption is enormous, especially when considering the topic from a global perspective. Very large units must be used to accurately and reliably express energy reserves and production measurements. We already examined the prefix multiplier "kilo" (kilo = 1,000). In the energy industry, however, units many times greater than 1,000 are common. Discussions and calculations comprising millions, billions and even trillions of units (be they barrels of oil, Watts or therms) are commonplace. For example, a nuclear power plant may have a nameplate capacity of 2,000 megawatts (2,000 MW). **Mega means 1,000,000.** So, 2,000 MW is the same number as 2,000,000,000 Watts (2,000 x 1,000,000 = 2,000,000,000 Watts). Notice that 2,000,000,000 Watts is an enormous number—a 2 with nine zeros after it. Calculations using so many zeros are unwieldy and error-prone, so an even larger unit of measure, called a giga, is often employed. **Giga means 1,000,000,000.** Rather than writing "2,000,000,000 Watts," it's faster

Table 1-2, Prefix Multipliers

Prefix	Abbreviation	Value	Name
kilo	k	1,000	thousand
mega	M	1,000,000	million
giga	G	1,000,000,000	billion
tera	T	1,000,000,000,000	trillion

and less error-prone to simply write "2 GW". Running with this idea then, that nuclear power plant's capacity is correctly expressed as 2,000,000,000 Watts, 2,000,000 kW, 2,000 MW, or 2 GW.

An efficient and reliable manner in which to express and calculate very large units of energy reserves and production is with scientific notation. Admittedly, this system at first glance looks cumbersome and perhaps even intimidating. Keep in mind, however, that it exists to *simplify* numbers, thereby making our work more reliable and accurate. For example, one could write the number one-hundred-twenty-three trillion as 123,000,000,000,000. Notice that this number is huge, with the 123 followed by twelve zeros. Using such a large number, in this form, is awkward and therefore error-prone. A better form is this: 1.23×10^{14}. It expresses the exact same number but does so without having to write or enter all those unwieldy zeros. Sometimes it may be expressed as 1.23E+14, or even as 1.23 x 10^14. All three expressions are scientific notation, and all three communicate the same number.

There are two rules for writing numbers using scientific notation. Rule #1 is that the first number (called the "coefficient") must be greater than or equal to 1, but less than 10. Rule #2 is that the second number (called the "base") must be 10, and also contain an exponent. In the above example, 1.23 is the coefficient, correctly written because it is greater than 1 but less than 10, and 10^{14} is the base, also correctly written because it is the number 10 and contains an exponent (number 14 is the exponent). In the above example, the decimal point was simply moved from the end of the number (the last zero) to the left until Rule #1 was obtained.

Expressing any number using scientific notation requires moving the decimal point to the left (for large numbers) or to the right (for small numbers) until Rule #1 is obtained. The number of places the decimal point moves becomes the exponent of the base. For example, Canada exported about 2,302,000 barrels of oil per day to the United States in 2010 (United States Energy Information Administration, 2011g). This means that in 2010 the United States received a total of about 840,230,000 barrels of Canadian oil. This number can also be expressed (with some rounding error) as 8.4×10^8 barrels of oil. Similarly, one would not likely use the number 60,644,000,000,000 to express the total cubic feet of proven shale gas reserves in the United States in 2009 (United States Energy Information Administration, 2010). Instead, 6×10^{13} cubic feet is a much easier and less error prone expression of essentially the same number.

Energy Consumption & Development

Humans harness energy resources and put them to productive use via increasingly complicated mechanical devices. This ability spurs rising levels of social and economic growth. Consider how far we've come since first learning to harness the energy contained in wood (by burning it to light a fire). Today we harness a viscous fluid from the ground (petroleum) and create, among other things, gasoline, plastics and fertilizer. Similarly, we harness the very air we breathe (wind) to create electricity, which in turn can power millions of different kinds of useful machines, from simple toasters to the most complicated supercomputers.

Economic growth is often coupled with increasing levels of energy consumption, especially for developing countries. Access to inexpensive and reliable electricity, for example, promotes economic growth by increasing a country's productivity and competitiveness. As these variables grow, they in turn spur increasing levels of energy consumption. The relationship is linear because, for example, a 3% increase in Gross Domestic Product (GDP) requires a concurrent 3% increase in energy consumption. Since energy enables development, ensuring access is critical if the world's impoverished billions are to ever hope of an improved life. The need is critical—the International Energy Agency (IEA) estimates that in 2009, 1.3 billion people did not have access to electricity, and 2.7 billion relied on the traditional use of biomass for cooking (International Energy Agency, 2011, p. 10).

Economic growth may become decoupled from energy consumption as a country's economic development matures, but only a select few countries have demonstrated this, and only with substantial policies aimed at reducing energy consumption (Johnson & Lambe, 2009, p. 2). For example, a 3% increase in GDP may only require a concurrent 1.5% increase in energy consumption. This non-linear trend can happen as increasing efficiencies are adopted and a society becomes able to squeeze more value from a single unit of energy. Societies can thus get "more bang for their buck" by using energy resources more efficiently. This positive trend is called "dropping energy intensity". **Energy intensity measures how much an economy is dependent on energy per unit of GDP.** In 1973, for example, 15,400 Btu were required to create one dollar of GDP in the United States. By 2009, only 7,400 Btu were required to create that same dollar, and by 2035 it is projected that only 4,400 Btu will be required (United States Energy Information Administration, 2011h, p. 59).

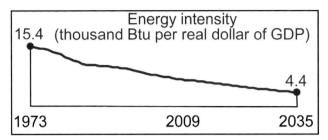

Figure 1-4, Dropping Energy Intensity, *Credit/Courtesy of United States Energy Information Administration*

It is important to note that dropping energy intensity does not necessarily mean a corresponding slowdown in energy consumption. In fact, since the late twentieth century, the opposite pattern exists. Ever increasing energy consumption is the norm, while more value is squeezed from each unit of energy as each year passes. **Energy intensity is dropping while energy consumption is rising.** This is illustrated by tracking overall energy consumption, including the renewables. Energy consumption is ever-rising because 1) the human population keeps growing, 2) people aspire to live an energy-saturated life similar to those in North America, Europe, Japan and Australia, and 3) energy consumers use ever increasing numbers of electronic appliances as each year passes.

By 2035 humanity is likely to increase its total energy consumption by nearly 50%. In 2008 humanity consumed approximately 493 quadrillion Btu (United States Energy Information Administration, 2011i). **A quadrillion is 1,000,000,000,000,000.** This massive energy consumption is therefore expressed as 493,000,000,000,000,000 Btu, or, 4.93×10^{17}, or even simpler, as 493 quads. By 2035 many analysts expect humanity to consume 739 quads (United States Energy Information Administration, 2010b, p. 9).

This future energy consumption will be unequally distributed around the planet. Developed countries, such as those members of the Organization for Economic Cooperation and Development (OECD), will likely strive to maintain their current energy consumption hegemony, using whatever political, economic and perhaps military means at their disposal. These are comparatively wealthy countries, such as Canada, the United States, members of the European Union, Australia, Japan and South Korea. They currently hold about 20% of the world's population, and in 2008 consumed approximately 200 million Btu per capita. By contrast, non-OECD countries (many of them are often called "developing" countries) currently hold about 80% of the world's population, and in 2008 consumed only about 46 million Btu per capita (United States Energy Information Administration, 2011j).

This difference, of about 154 million Btu per capita between energy consumers in OECD and non-OECD countries, makes a big difference in productivity and therefore quality of life. Access to cheap and reliable energy resources, as discussed previously, increases social and economic growth. Note, however, that people living in non-OECD places, such as Central America, South America, Southwest Asia, Africa, South Asia and China, have quickly expanding workforces comprised of comparatively young populations. These workers want high a quality of life and are therefore working hard to achieve this goal. In doing so, humanity's overall energy consumption will rise. We can therefore expect an environment of political and economic tension, to say the least, as regions and countries strive to secure essential energy resources.

Figure 1-5, Total Primary Energy Consumption (Quadrillion Btu)

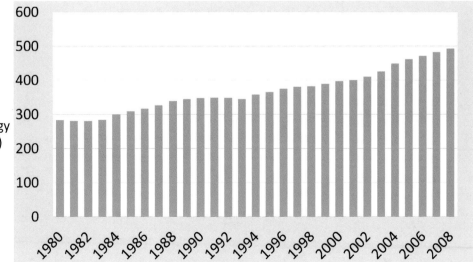

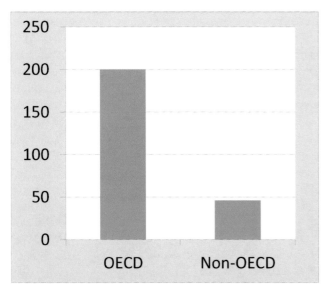

Figure1- 6, Total Primary Energy Consumption per Capita (Million Btu/capita), OECD & Non-OECD Countries

Electricity Generation and the Grid

Electricity is a resource (albeit a secondary one), and can be examined in a number of different ways. For instance, one might study how it is generated and its corresponding consumption in a state or country. Another examination might consider how many megawatt-hours are produced in a year by a single utility company. Still another inquiry might consider the "end users" of electricity—for instance, residential or industrial consumers—and their electrical needs over a set time period. Whatever the particular discussion, the term "capacity" will probably arise. **Capacity is the maximum output that a generating system can supply.** Depending on a conversation's context, a "generating system" can mean a single generator, a single power plant (which might have several generators), a state, a country, a region, or even the entire planet. In most discussions, however, the term is associated with power plants and is used to compare their relative maximum outputs.

Power plants are, of course, places that convert resources (non-renewable and/or renewable) into electricity. Often the term "nameplate capacity" is used, which literally refers to a metal plate physically attached to a generator that states the unit's maximum output. Energy analysts often consider the nameplate capacity of a particular power plant, or all the power plants in a state, or even all the power plants in a country. For example, the Diablo Canyon nuclear power plant in California has a nameplate capacity of 2,240 MW, but is just one of many dozens of power plants within the state, all of which add up to a 2009 state-wide capacity of 65,948 MW (United States Energy Information Administration, 2011k).

In 2009 the United States had 1,025,400 MW (or 1,025 GW) of electricity generating capacity (United States Energy Information Administration, 2010c). This was the maximum output, at any instant, if every power plant in the country operated at 100%. This, as you might expect, never actually happens. Power plants have down-times when maintenance is performed, fuel is renewed, the sun does not shine or the wind does not blow. Imagine yourself on a bicycle. You could pedal at your maximum capacity, but eventually you would need to stop for maintenance and to refuel with food and water. A power plant operates in much the same manner. This leads us to an essential term called "capacity factor". **Capacity factor is the percent of time a generating system operates at maximum capacity.**

For an example of capacity factor, consider the largest nuclear power plant in the United States, the Palo Verde Nuclear Generating Station located in Wintersberg, Arizona. The three reactors located at Palo Verde have the following capacity: Palo Verde-1 = 1,311 MW, Palo Verde-2 = 1,314 MW, and Palo Verde-3 = 1,317 MW. Adding all three reactors' capacities nets a station capacity of 3,942 MW. This means that 94,608 MWh would be produced if all the reactors operated at 100% for a single day (3,942 x 24 = 94,608). Correspondingly, 34,531,920 MWh would be produced if all the reactors operated at 100% for an entire year (94,608 x 365 = 34,531,920). In 2009, however, Palo Verde's three generators produced 30,661,851 MWh (United States Energy Information Administration, 2011l). Thus Palo Verde had a capacity factor in 2009 of about 89% (30,661,851 / 34,531,920 = .8879 x 100 = 88.8%).

Capacity factor is relatively high (>80%) for fossil fuel and nuclear power plants because an uninterrupted fuel supply possible. Nonstop coal deliveries occur by train, natural gas and oil via pipeline, and uranium by truck or train. Conversely, capacity factor is relatively low (<40%) for some renewable energy sources because the "fuel" may be intermittent. For example, because wind is fickle and sometimes blows at very high speeds while occasionally dropping off to zero, wind turbines have a relatively low capacity factor.

In 2009, the United States generated 3,950,331 thousand megawatthours (MWh) of electricity (United States Energy Information Administration, 2010d). Do not get confused by the fact that there is a massive difference between the *capacity* and *generation*—the two terms are not interchangeable. This is because, as stated earlier, more than 24 kWh can be consumed in a single 24 hour day. Since there are 8760 hours in a year, the United States generated an average of 451 GW every hour of 2009 (3,950,331,000 MWh/8760 hours in a year = 450,951 MW per hour, or 451 GW per hour). **When discussing capacity, *only* wattage is considered, but when discussing generation and/ or consumption, wattage *and* hours must be considered.** We'll find in later chapters that this average is in fact a pretty rough measurement because electricity demand fluctuates depending upon the time of day, location and season. For example, in Texas during July, electricity demand may peak at perhaps 2:00 PM, and may ebb to its lowest demand level at 3:00 AM.

Coal generates the most electricity in the United States, followed by natural gas and uranium. These are non-renewable energy resources that are acquired both domestically and internationally. They create heat (through combustion or fission), which we do our best to capture, often by boiling water to create steam. The steam is then used to spin a turbine, which in turn spins a generator, which (finally) forces an electromagnetic current through a wire. We are utterly dependent upon this protracted logistical operation. As if all these steps were not challenging enough, the barefaced inefficiency of the process should make us all stand up and pay attention.

The average efficiency of the coal-fired power plant fleet in the United States is about 33% (Eisenhauer & Scheer, 2009, p. 1). Another way to consider this is that most ageing coal-fired power plants only get about 33% of the energy out of any ton of coal. The rest—about 67% of the energy in that ton—is wasted. Engineers are of course working hard to improve this number, and innovations such as ultrasupercritical (USC) steam turbines are improving efficiencies of the latest generation of coal-fired plants. USC systems operate at temperature of about 1,100 °F and at pressure higher than 3,500 pounds per square inch (psi). The next best systems, conventional supercritical units, operate at a maximum of 1,050 °F and 3500 psi. Such differences allow USC systems to obtain efficiency ratings as high as 48%. This efficiency leap not only stretches a country's coal supply, it reduces the amount of carbon dioxide (and other emissions) produced per MWh, from 1.06 tons/MWh to .99 tons/MWh (Klotz, Davis, & Pickering, 2009). Modernizing the coal-fired power plant fleet in the United States is therefore an essential policy issue that is advocated by many often unrelated stakeholders.

The inefficiency of steam-driven power plants is partially due to something called conversion losses. Conversion losses occur because the chemical energy of the fuel is converted into heat and motion inside a power plant. Heat and motion are essential for generation, but since neither are electricity, they are consumed in the creation of the "product" of the power plant. For example, consider the steam line in a typical power plant. The water used in steam turbines is

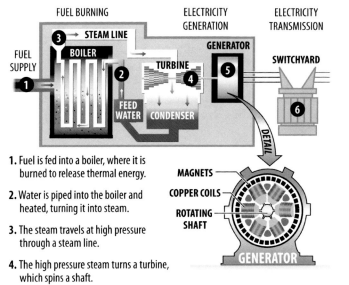

Figure 1-7, Model of a Steam-Driven Power Plant, *Credit/Courtesy of National Energy Education Development Project*

1. Fuel is fed into a boiler, where it is burned to release thermal energy.

2. Water is piped into the boiler and heated, turning it into steam.

3. The steam travels at high pressure through a steam line.

4. The high pressure steam turns a turbine, which spins a shaft.

5. Inside the generator, the shaft spins coils of copper wire inside a ring of magnets. This creates an electric field, producing electricity.

6. Electricity is sent to a switchyard, where a transformer increases the voltage, allowing it to travel through the electric grid.

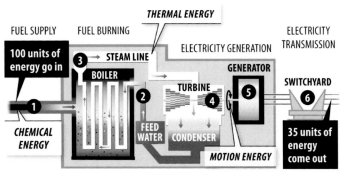

1. Fuel is fed into a boiler, where it is burned to release thermal energy.

2. Water is piped into the boiler and heated, turning it into steam.

3. The steam travels at high pressure through a steam line.

4. The high pressure steam turns a turbine, which spins a shaft.

5. Inside the generator, the shaft spins coils of copper wire inside a ring of magnets. This creates an electric field, producing electricity.

6. Electricity is sent to a switchyard, where a transformer increases the voltage, allowing it to travel through the electric grid.

Figure 1-8, Model of Efficiency Losses in Ageing Steam-Driven Power Plants, *Credit/Courtesy of National Energy Education Development Project*

continuously boiled to make steam, routed through a turbine, condensed back into a liquid, only to be re-boiled again. **After running through a turbine, steam is deliberately condensed back into a liquid.** Changing liquid into steam requires a lot of energy. Consider, for example, how long a pot of water needs to boil on your stovetop before it all evaporates. When water is boiled, it absorbs energy, when it is condensed, it sheds energy. So where does this energy go after the steam is condensed? Much of it goes up the smoke-stacks and/or into a cooling pond or cooling towers.

Consider that even on the coldest winter day, the cooling pond outside a major power plant will not completely freeze over. This is because heat from the power plant's condenser is cycled into the pond, which keeps the water warm.

If steam is what is needed to spin a turbine, why would a power plant *deliberately* condense the steam it *just made*, only to have to re-heat that same water to make more steam? The steam entering a turbine is hot and under very high pressure. Its energy is con-

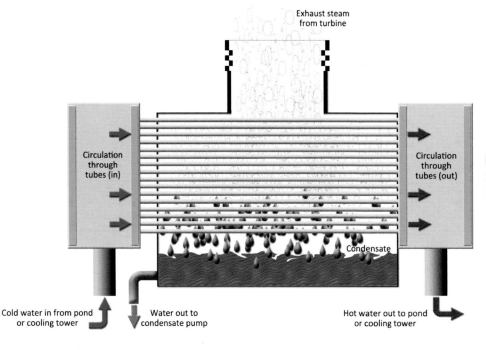

Figure 1-9, Model of a Power Plant's Condenser Unit

Figure 1-10,
Rotor Blades of a Steam
Turbine, *Credit/Courtesy
of Siemens*

verted into mechanical energy (spin) via the turbine's blades. Engineers have long known, however, that more of the steam's energy will convert into mechanical energy if a low pressure steam outlet is situated in the turbine. A good way to squeeze more mechanical energy from the incoming steam is to condense the turbine's outlet steam into liquid. Thus, after moving past the turbine's blades, the high-pressure steam hits a condenser, which turns it back into a liquid. This state change creates a low-pressure area—a vacuum. The greater vacuum between the inlet and outlets, the more mechanical energy is obtained (Heat Exchange Institute, 2005).

When a generator spins as a result of the mechanical energy produced by a turbine, an electric current is forced through the wires leaving the power plant. Sometimes this electricity must travel very long distances before it reaches customers. To do this economically, power companies increase the electricity's voltage. **Voltage is the force that pushes an electric current through a wire.** We see how electricity is transported whenever spotting high-voltage transmission lines marching across the landscape. When approaching consumers, this high-voltage is then "stepped-down" to a useful voltage. Stepping-up and stepping-down voltage is accomplished with transformers. **Transformers increase or decrease electricity's voltage.** For example, high-voltage transmission lines might have a voltage of 500 kilovolts (500 kV). 500 kV is not useful for a home because it is far too much force. To put this in perspective, imagine filling a glass with water under your sink's tap. Try the same thing using a fire hose. Water from the fire hose would not only fail to fill your glass, its force would destroy the glass itself. So too would high-voltage destroy your home's appliances and electronic devices. A typical home needs only 120 volts. Changing 500,000 volts to 120 volts requires step-down transformers.

The network of power plants, transmission lines, distribution lines, and consumers, is called the grid. In the United States, the largest capacity power plant is the Grand Coulee Dam with over 7,000 MW. Many thousands of other smaller power plants also serve the grid. Consider, for example, the coal-burning power plant in Waukegan, Illinois, a city north of Chicago, with a capacity of 689 MW. Or the workhorse

Figure 1-11, The SST-900 Steam Turbine, *Credit/ Courtesy of Siemens*

of the wind industry, a single 1.5 MW wind turbine. Or perhaps the PH Robinson natural gas power plant, southeast of Houston, Texas, with a nameplate capacity of 1,198 MW. These and many, many others serve electricity into the grid and thereby satisfy the power requirements of North America.

Regardless of their respective capacities, the volume of electricity flowing from power plants must be continually monitored in order to balance supply with demand. Bulk electricity cannot be stored (to any appreciable extent), so there is never any inventory and as such many dozens of dispatch centers scattered across North America must constantly match the needs of consumers with supplies from producers (North American Electric Reliability Corporation, 2011). Coordinating all this, and establishing and enforcing grid reliability standards, is the job of the North American

Electric Reliability Corporation (NERC), which was established in 1965. NERC develops the planning and operating standards of all electric utilities throughout the United States and Canada. It also audits utilities to ensure compliance and thereby works to ensure an uninterrupted electricity flow throughout the continent.

Important Ideas:

- Energy is the ability to do work.

- inetic energy is motion.

- Potential energy is stored.

- Non-renewable energy sources cannot be replenished in a short time span.

- In 2010, non-renewable energy sources constituted about 92% of the total energy used in the United States.

- Renewable energy sources can be replenished in a short time span.

- In 2010, renewable energy sources constituted about 8% of the total energy used in the United States.

- A resource is an existing energy source.

- A proven reserve is a resource that is economically feasible for extraction.

- A probable reserve is a resource that may become economically feasible for extraction.

- The reserve production ratio is the remaining amount of a non-renewable resource, expressed in years.

- Any R/P ratio is applicable only to the particular moment it is calculated.

- Electricity is a secondary energy source, obtained by converting non-renewable and renewable energy resources.

- The most common electrical generation method in the world is the steam turbine.

- A Watt is one joule of energy transfer per second.

- Kilo means 1,000.

- A British thermal unit is the amount of energy needed to raise 1 pound of water 1 °Fahrenheit.

- Barrel of oil equivalent is the amount of energy contained in one barrel of crude oil.

- "Barrel" is abbreviated Bbl. "Barrels" is abbreviated Bbls.

- Mega means 1,000,000.

- Giga means 1,000,000,000.

- Energy intensity measures how much an economy is dependent on energy per unit of GDP.

- Energy intensity is dropping while energy consumption is rising.

- A quadrillion is 1,000,000,000,000,000.

- Capacity is the maximum output that a generating system can supply.

- In 2009 the United States had 1,025,400 MW (or 1,025 GW) of electricity generating capacity.

- Capacity factor is the percent of time a generating system operates at maximum capacity.

- In 2009, the United States generated 3,950,331 thousand megawatthours (MWh) of electricity.

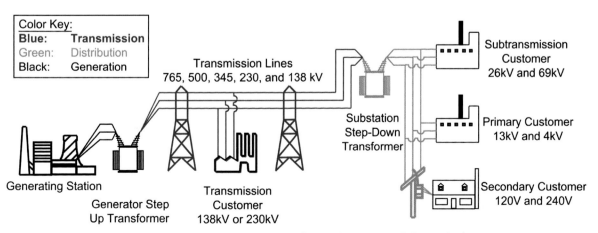

Figure 1-12, Electricity Generation, Transmission and Distribution Model, *Credit/ Courtesy of U.S. Canada Power System Outage Task Force*

- When discussing capacity, *only* wattage is considered, but when discussing generation and/or consumption, wattage *and* hours must be considered.

- Coal generates the most electricity in the United States, followed by natural gas and uranium.

- The average efficiency of the coal-fired power plant fleet in the United States is about 33%.

- After running through a turbine, steam is deliberately condensed back into a liquid.

- Voltage is the force that pushes an electric current through a wire.

- Transformers increase or decrease electricity's voltage.

- The network of power plants, transmission lines, distribution lines, and consumers, is called the grid.

Works Cited

Cox, B., & Forshaw, J. (2009). *Why Does E=mc2?: (And Why Should We Care?) [Kindle Edition]*. Cambridge, MA: Da Capo Press.

Eisenhauer, J., & Scheer, R. (2009, July 15-16). *Opportunities to Improve the Efficiency of Existing Coal-fired Power Plants*. Retrieved September 14, 2011, from National Energy Technology Laboratory, Improving Power Plant Efficiency, Technical Workshop Report: http://www.netl.doe.gov/energy-analyses/pubs/NETL%20Power%20Plant%20Efficiency%20Workshop%20Report%20Final.pdf

Heat Exchange Institute. (2005, February 11). *Tech Sheet #113*. Retrieved September 14, 2011, from Condenser Basics: http://www.heat-exchange.org/pub/pdf/edu/Tech%20Sheet%20113.pdf

International Energy Agency. (2011, October). *World Energy Outlook 2011*. Retrieved November 21, 2011, from Energy for All: Financing Access for the Poor, Special Early Excerpt of the World Energy Outlook 2011: http://www.iea.org/Papers/2011/weo2011_energy_for_all.pdf

Johnson, F. X., & Lambe, F. (2009, May). *Comission on Climate Change and Development*. Retrieved November 21, 2011, from Energy Access, Climate and Development: http://www.ccdcommission.org/Filer/commissioners/Energy.pdf

Klotz, H., Davis, K., & Pickering, E. (2009, July 1). *Designing an Ultrasupercritical Steam Turbine*. Retrieved November 18, 2011, from Power Magazine: http://www.powermag.com/issues/features/Designing-an-Ultrasupercritical-Steam-Turbine_1993.html

Livingston, J. D. (1996). *Driving Force: The Natural Magic of Magnets*. Cambridge: Harvard University Press.

North American Electric Reliability Corporation. (2011). *About NERC*. Retrieved September 14, 2011, from Understanding the Grid: http://www.nerc.com/page.php?cid=1|15

United States Energy Information Administration. (2010, December 30). *Independent Statistics and Analysis*. Retrieved September 13, 2011, from Natural Gas, Shale Gas, Proved Reserves, as of December 31, 2009: http://www.eia.gov/dnav/ng/ng_enr_shalegas_a_EPG0_R5301_Bcf_a.htm

United States Energy Information Administration. (2010b, July). *Office of Integrated Analysis and Forecasting*. Retrieved September 14, 2011, from International Energy Outlook 2010, DOE/EIA-0484(2010): http://www.eia.gov/oiaf/ieo/pdf/0484(2010).pdf

United States Energy Information Administration. (2010c, November 23). *Independent Statistics and Analysis*. Retrieved September 14, 2011, from Electric Power Annual, Existing Net Summer Capacity by Energy Source and Producer Type, 1998 through 2009 (megawatts): http://www.eia.gov/cneaf/electricity/epa/epat1p1.html

United States Energy Information Administration. (2010d, November 23). *Independent Statistics and Analysis*. Retrieved September 14, 2011, from Electric Power Annual, Net Generation by Energy Source by Type of Producer, 1998 through 2009 (Thousand Megawatthours): http://www.eia.gov/cneaf/electricity/epa/epat2p1.html

United States Energy Information Administration. (2011, July 14). *Independent Statistics and Analysis*. Retrieved September 13, 2011, from What Role Does Renewable Energy Play in the United States?: http://www.eia.doe.gov/energyexplained/index.cfm?page=renewable_home

United States Energy Information Administration. (2011b). *Independent Statistics and Analysis, International Energy Statistics*. Retrieved September 13, 2011, from Proven Reserves of Crude Oil: http://tonto.eia.doe.gov/cfapps/ipdbproject/iedindex3.cfm?tid=5&pid =57&aid=6&cid=&syid=2009&eyid=2009&unit=BB

United States Energy Information Administration. (2011c, September 13). *Independent Statistics and Analysis*. Retrieved September 13, 2011, from International Energy Statistics, Electrcitiy Generation, Total Conventional Thermal Electricity Net Generation (Billion Kilowatthours) and Total Electricity Net Generation (Billion Kilowatthours) : http://www.eia.gov/cfapps/ipdbproject/iedindex3.cf m?tid=2&pid=2&aid=12&cid=ww,&syid=2008&eyid=2008&unit=BKWH

United States Energy Information Administration. (2011d, April 13). *Independent Statistics and Analysis, Electric Power Annual 2009*. Retrieved September 13, 2011, from Table 7.4. Average Retail Price of Electricity to Ultimate Customers by End-Use Sector, 1998 through 2009 (Cents per kilowatthour): http://205.254.135.24/cneaf/electricity/epa/epa.pdf

United States Energy Information Administration. (2011e, June 1). *Independent Statistics and Analysis*. Retrieved September 13, 2011, from How much electricity does an American home use?: http://www.eia.gov/tools/faqs/faq.cfm?id=97&t=3

United States Energy Information Administration. (2011f, September 13). *Independent Statistics and Analysis*. Retrieved September 13, 2011, from Table 5A. Residential Average Monthly Bill by Census Division, and State 2009: http://www.eia.gov/cneaf/electricity/ esr/table5_a.xls

United States Energy Information Administration. (2011g, July 28). *Independent Statistics and Analysis*. Retrieved September 14, 2011, from Petroleum and Other liquids, U.S. Net Imports by Country: http://www.eia.gov/dnav/pet/pet_move_neti_a_ep00_IMN_ mbblpd_a.htm

United States Energy Information Administration. (2011h, April). *Independent Statistcis and Analysis*. Retrieved September 14, 2011, from Annual Energy Outlook 2011: http://www.eia.gov/forecasts/aeo/pdf/0383(2011).pdf

United States Energy Information Administration. (2011i). *Independent Statistics and Analysis*. Retrieved September 14, 2011, from International Energy Statistics, Total Primary Energy Consumption (Quadrillion Btu), World: http://www.eia.gov/cfapps/ipdbproject/ iedindex3.cfm?tid=44&pid=44&aid=2&cid=ww,&syid=2008&eyid=2008&unit=QBTU

United States Energy Information Administration. (2011j). *Independent Statistics and Analysis*. Retrieved September 14, 2011, from International Energy Statistics, Total Primary Energy Consumption per Capita (Million Btu per Person), 2008: http://www.eia.gov/ cfapps/ipdbproject/iedindex3.cfm?tid=44&pid=45&aid=2&cid=CG6,CG5,&syid=2008&eyid=2008&unit=MBTUPP

United States Energy Information Administration. (2011k, August 25). *Independent Statistsics and Analysis*. Retrieved September 14, 2011, from US States, California, Data, Reserves & Supply: http://www.eia.gov/state/state-energy-profiles-data.cfm?sid=CA

United States Energy Information Administration. (2011l, January). *Independent Statistics and Analysis*. Retrieved September 14, 2011, from U.S. Nuclear Statistics, Table 1. Nuclear Reactor, State, Type, Net Capacity, Generation, and Capacity Factor: http://www. eia.gov/cneaf/nuclear/page/operation/statoperation.html

United States Geological Survey. (2001, April). *Arctic National Wildlife Refuge, 1002 Area, Petroleum Assessment, 1998, Including Economic Analysis*. Retrieved September 13, 2011, from United States Geological Survey: http://pubs.usgs.gov/fs/fs-0028-01/fs-0028-01.pdf

A Dragline Excavating Coal at the North Antelope Rochelle Mine (NARM) in Wyoming, *Credit/Courtesy of Peabody Energy*

Chapter 2
Coal

Coal and the Carboniferous Period

Imagine picking up a rock and lighting it on fire. While burning, the rock releases much more heat, and also burns longer, than a comparable amount of wood. Now imagine that this special type of rock is widely available throughout the world and relatively easy to extract, thereby making it inexpensive. This rock is of course coal. **Coal is a combustible sedimentary rock of biological origin.** It is plentiful, packs a lot of Btu in each pound, and is largely responsible for fueling the industrial age beginning around the middle to late nineteenth century. The twenty-first century not only stands on the shoulders of coal, it continues to be heavily dependent upon it, especially as a means of generating electricity.

Coal, like petroleum and natural gas, is a fossil fuel. **Fossil fuels are energy resources that come from the transformed remains of plants and animals that lived millions of years ago.** Fossil fuels can be solid (such as coal), liquid (such as petroleum) and gas (such as methane). They all combust—that is, we can light them on fire to produce heat and light—because they all contain hydrogen (H) and carbon (C). Because fossil fuels contain both hydrogen and carbon, they are frequently referred to as "hydrocarbons." The term "fossil" in "fossil fuels" is a bit of a misnomer. While there may in fact be actual fossils in a bed of coal, it is not itself a fossil and neither are petroleum or natural gas. We use the term "fossil" because, like actual fossils, these energy resources are very, very old.

The plants that eventually became coal lived during the Carboniferous geologic period. Carboniferous means coal-bearing, which is appropriate since a lot of the organic sediment deposited during this time became the coal we use today. **The Carboniferous period lasted 60 million years.** It began 359 million years ago and ended 299 million years ago (the abbreviation "Ma" is frequently used to express millions of years ago—"M" for million and "a" for annum). This was not, as is commonly mistaken, the time when dinosaurs roamed the earth. Dinosaurs lived long *after* the Carboniferous period, beginning 230 Ma and ending 65 Ma. When we burn coal for electricity, we are *not* using the transformed remains of tyrannosaurus, brontosaurus, or any other dinosaurs.

During the Carboniferous period, sea level was lower than it is today because of the Karoo Ice Age. This ice age began at the same time as the Carboniferous period, but lasted 40 million years longer. The effect of cooler global temperatures during the ice age was to lock vast quantities of earth's water into ice, especially at higher latitudes (near the poles). Ice sheets (these exist on land) effectively grew at the expense of sea level—as land-based ice grew, sea level dropped. Lower global sea levels encouraged habitat expansion of low-lying swamps and forests, especially at lower latitudes (near the equator). Physical characteristics of the vegetation living in these swamps and forests made them especially resistant to decay. When this vegetation died, and was then buried by other generations of vegetation, the natural decay process was

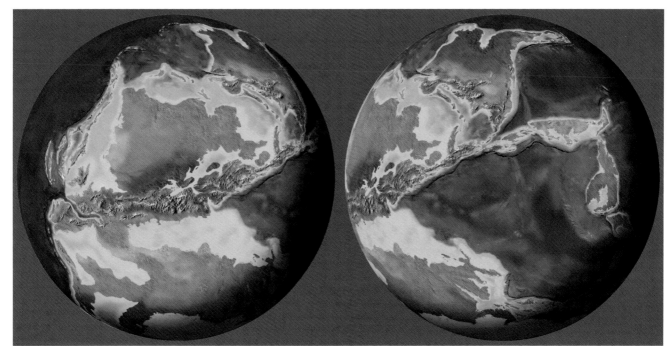

Figure 2-1, Earth's Continents During the Carboniferous Period, *Credit/Courtesy of Ron Blakey, NAU Geology*

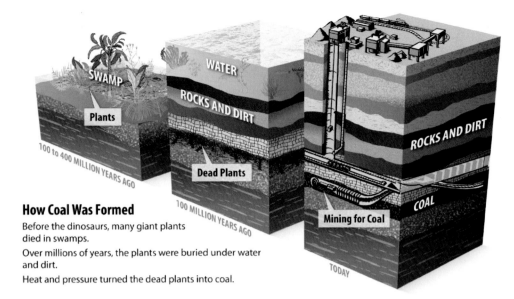

How Coal Was Formed

Before the dinosaurs, many giant plants died in swamps.

Over millions of years, the plants were buried under water and dirt.

Heat and pressure turned the dead plants into coal.

Figure 2-2, Model of Coal Formation, *Credit/Courtesy of National Energy Education Development Project*

substantially slowed. Eventually time, pressure and heat transformed the vegetation into coal. When burning coal (or other fossil fuels) today, we release the energy—and the carbon—that the vegetation acquired during its lifetime millions of years ago. In this sense one can correctly argue that we are using stored solar energy—and stored carbon—when we burn any fossil fuel.

Coal Ranks & Combustion

All coal is not the same. Two general categories of coal—soft and hard—dominate energy discussions. Soft coal is more plentiful than hard coal, but contains comparatively less energy per pound. **"Lignite" and "sub-bituminous" are two ranks of soft coal.** Lignite contains about 6,500 Btu per pound, while sub-bituminous holds about 8,750 Btu per pound. Both types of soft coal are used primarily as "thermal" coals, that is, they are burned to generate heat in steam-driven power plants. Hard coal, which is globally less plentiful than soft coal, contains comparatively more energy per pound. **"Bituminous" and "anthracite" are two ranks of hard coal.** Bituminous coal (the most plentiful type in the United States) contains about 12,000 Btu per pound, while Anthracite holds approximately 12,500 Btu per pound (United States Energy Information Administration, 2011). Hard coal is used both as a thermal coal, as well as to make coke (coke is used in the steel industry as a fuel).

Figure 2-3, Exposed Coal Beds from a Road Cut, *Credit/Courtesy of Royce Bair*

Figure 2-4,
Lignite, Bituminous and
Anthracite Types of Coal

As one might imagine, coal with higher Btu per pound is generally more valuable than that with lower Btu per pound. This is dramatically illustrated by examining the price per ton of different types of coal. The last week of August, 2011, for example, coal from central Appalachia, containing 12,500 Btu per pound, sold for $80.15 per ton. That same day, coal from the Powder River Basin (in Wyoming) containing 8,800 Btu per pound, sold for only $14.10 per ton (United States Energy Information Administration, 2011b).

When power plants buy high Btu coal, comparatively less must be transported in order to generate the same amount of electricity from a larger quantity of lower Btu coal. Mining companies too, of course want to maximize their efforts, and will therefore seek to extract the highest Btu coal, if at all possible. But Btu is just one of several variables in coal pricing and does not wholly determine market activity. For example, due to environmental regulations in the United States, the low-sulfur and low-nitrogen coals from the Powder River Basin (PRB) are currently in high demand even though this coal has a relatively low Btu content when compared to those from other coal-producing areas. Using PRB coal, or mixing it with higher Btu coal, has "proven extremely effective in meeting state and federal emission requirements" for coal-fired power plants (Labbe, 2009).

The United States Geologic Survey (USGS) also weighs in with its judgments about coal resources. According to the USGS, soft coal beds must be ≥ (greater than or equal to) 76.2 centimeters thick in order to be considered candidates for mining operations. Hard coal beds, on the other hand, need only be ≥ 36.6 centimeters thick in order to attract mining operations (United States Geological Survey, 2011). Another way to consider this is that hard coal beds need only be about 48% as thick as soft coal beds in order to be economically feasible for extraction.

Newly mined (raw) coal must be processed at a coal preparation plant prior to shipping. These plants clean and grade the coal, thereby removing impurities such as rocks and clay, and producing a cleaner burning product of uniform size and Btu content. A common cleaning method involves placing the raw coal into a bed of water and using pulsing currents to separate the materials. Since coal is less dense than the refuse, it rises in the bed and is thus easily removed. Thermal dryers evaporate any remaining water before the cleaned coal is placed into a holding silo for future transportation via train or barge.

At a power plant, coal is pulverized (milled) into a fine powder prior to combustion. Crushing the coal increases its surface area, thereby making it more ex-

Week Ended	Central Appalachia 12,500 Btu	Northern Appalachia 13,000 Btu	Illinois Basin 11,800 Btu	Powder River Basin 8,800 Btu	Uinta Basin 11,700 Btu
15-July-11	$79.65	$78.20	$47.80	$14.60	$41.00
22-July-11	$80.15	$78.20	$47.80	$14.70	$41.00
29-July-11	$80.15	$78.20	$47.80	$14.60	$41.00
05-July-111	$80.15	$78.20	$47.80	$14.55	$41.00
12-Aug-11	$80.15	$78.20	$47.80	$14.35	$41.00
19-Aug-11	$80.15	$76.00	$47.80	$14.25	$41.00
26-Aug-11	$80.15	$76.00	$47.80	$14.10	$41.00

Table 2-1, Average Weekly Coal Prices, Dollars per Short Ton (2000 lbs)

plosive and allowing it to burn quickly and at a high temperature (World Coal Association, 2012). The pulverized coal is then blown into a combustion chamber where burners direct flames at the explosive powder, thereby igniting the coal and producing heat. Hollow tubing situated on the sides and at the top of the boiler routes water through the combustion chamber. After absorbing enough heat, the water transitions into vapor (steam) which is then routed to a steam turbine (International Energy Agency, 2007, p. 27). The entire process continues non-stop and thereby produces baseload power so long as a supply of coal is continuously delivered to the power plant.

Coal Production in the United States

Coal production in the United States experienced substantial growth through the 1970s and 1980s. In 1974, for example, the United States produced 610 million tons (610,000,000 tons). In 1990, by contrast, the United States produced over 1 billion tons (1,029,100,000 tons to be precise). **The United States produced about 69% more coal in 1990 than in 1974.** The year 1990 is an important marker because prior to that year United States annual coal production never exceeded 1 billion tons. In addition, since 1994, United States annual coal production has never dropped below 1 billion tons (United States Energy Information Administration, 2010).

So what happened in the early 1970s to spur coal production in the United States? In October of 1973, members of the Organization of Arab Petroleum Exporting Countries (OAPEC) declared an oil embargo (an intention to limit or stop trade) against any coun-

try that supported Israel during the Yom Kippur War. This had long-term, complex economic and political repercussions, some of which we undoubtedly continue experiencing today. The immediate effect, however, was to create massive petroleum shortages and thus equally impressive price increases. We cannot, of course, put coal into our gasoline tanks, so why would a *petroleum* shortage provoke an increase in *coal* production? As prices climbed and gasoline shortages appeared, the Nixon administration announced Project Independence, an initiative to make the United States self-sufficient in energy by 1980. Annual domestic coal production began climbing the following year as structural changes within the energy generation industry took place across the country.

In 2009, the United States had 1,375 operational coal mines. About 94% of these mines were located east of the Mississippi River and heavily concentrated throughout the Appalachian region. Though numerous, these mines produced only about 42% of the coal. **The bulk of United States coal production currently occurs west of the Mississippi River, specifically in the Powder River and Unita Basins.** Here, a total of 85 mines (only about 6% of the total coal mines in the United Sates) produced 58% of all the coal in 2009 (United States Energy Information Administration, 2011c). Clearly, mining operations west of the Mississippi River are larger and more productive. This is due to the relative age of eastern mining areas, efficient and relatively inexpensive western mining techniques, and of course the size of coal bed deposits.

The 87,592 employees (United States Energy Information Administration, 2011d) in the United States coal mining industry in 2009 work in a modern and heavily regulated industry. No longer are great num-

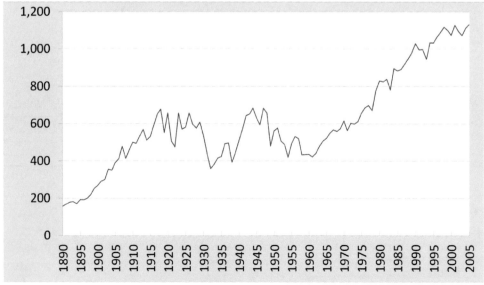

Figure 2-5, Coal Production in the United States, 1890-2005 (million short tons)

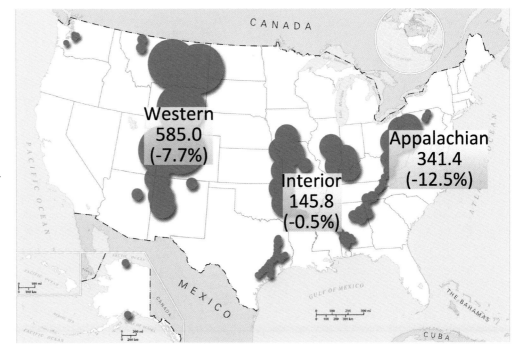

Figure 2-6,
Coal Production by Coal-
Producing Region, 2009

bers of men descending underground with pickaxes, randomly searching for elusive coal seams that may never appear. Instead, continuous mining machines are used underground, while massive draglines make short work of coal deposits at surface mines. Both methods are employed only after geologists conduct a thorough analysis of a potential site and judge it economically feasible for extraction operations. Both underground and surface mining techniques are more efficient and safer than in the past, allowing steadily fewer employees to extract ever greater quantities of this resource. Regulating worker safety in the United States is the responsibility of the Mine Safety and Health Administration (MSHA). At a minimum, this organization must inspect surface mines twice a year, and underground mines four times a year, in order to ensure compliance with existing safety and health standards (United States Department of Labor, 2011). In practice, MSHA inspects many surface mines more than twice a year, and many underground mines more than four times a year.

Coal Mining

Mining coal underground is relatively dangerous and inefficient compared to surface mining. Efficient surface extraction techniques, in addition to the availability of massive, easily-exploitable western surface deposits, have gradually eroded the percentage share of coal produced from underground mines over the past 40 years. **The majority of coal produced in the United States comes from surface mines.** In 1973, 50% of the

coal produced in the United States came from underground mines. By 2003, that share dropped to 33%, with surface mines picking up the difference (United States Energy Information Administration, 2006, p. 5). This pattern of relative decline continues today, with underground mines accounting for only 31% of all coal produced in the United States in 2009.

Underground mining extracts coal via conventional and longwall techniques. **Conventional mining creates a "room and pillar" layout in the midst of a coal bed.** "Rooms" are opened underground with a series of drilling and blasting measures, after which the broken coal is transported out of the mine via conveyor. Massive "pillars" are deliberately left standing throughout the mine in order to support the roof and prevent collapse. These pillars are themselves coal, and it is thus tempting (although fraught with danger) to whittle away at these in an attempt to extract more coal from an existing underground mine (Lind, 2005).

A second underground mining technique is called longwall mining. **Longwall mining extracts entire panels of coal, leaving no space or support pillars.** Longwall mining is possible only if a coal bed is large, horizontal (most are), and of uniform dimensions. The typical minimum size of a coal panel "face" (this is where the active cutting occurs) is about 900 feet wide, while a typical panel's width is greater than 8,000 feet long (remember, a single 1 mile is 5,280 feet) (United States Energy Information Administration, 2011e). During longwall mining, an automated "shearer" travels back and forth against the panel's face, progressively shearing away many tons of coal

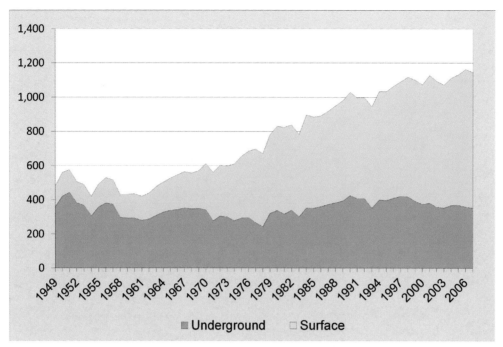

Figure 2-7, US Coal Production by Mine Type (million short tons)

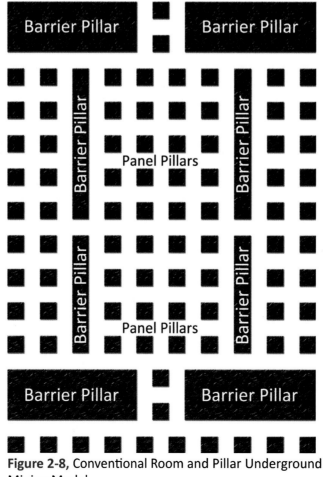

Figure 2-8, Conventional Room and Pillar Underground Mining Model

Figure 2-9, A Continuous Mining Machine in Conventional
Room and Pillar Mining, *Credit/Courtesy of Peabody Energy*

Figure 2-10, A Continuous Mining Machine in
Wearmouth Colliery, United Kingdom (closed in
1993), *Credit/Courtesy of hoggy03*

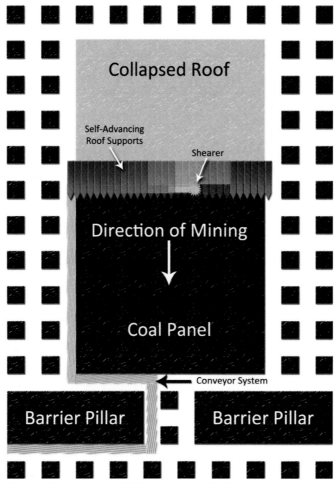

Figure 2-11, Longwall Underground Mining Model

Figure 2-12, Longwall Mining, *Credit/Courtesy of Peabody Energy*

Figure 2-13, Longwall Mining Shearer at the Deutsches Bergbau-Museum (DBM) Mining museumin Bochum, Germany, *Credit/Courtesy of Helen Simonsson*

with each pass. During the process, broken coal continuously falls onto a conveyor and is transported up to the surface. As you might imagine, the shearer is a large and very "toothy" piece of spinning machinery. It is housed within a linear, mobile, steel roofing support structure that advances as the shearer opens space underground. The roof of the mine deliberately collapses behind the longwall roofing structure as it advances down the panel.

There are four methods of surface mining: contour mining, highwall mining, area mining and mountaintop mining (MTM). **Contour mining, as the name suggests, occurs along "contours" in sloping terrain.** In this context, a contour is a line connecting points of equal elevation. In hilly terrain, as one moves up or down a slope, various horizontal sedimentary rock beds may be accessible as outcrops. A coal bed out-

crop might exist between, say, elevation 1,015 and 1,020 feet above sea level. This 5 foot thick bed of coal can be mined by extracting the coal along the slope's contour. Debris is cast downslope or used as fill in an adjacent, previously mined area. Operations progress as deeply as possible into the slope until the overburden (this is anything that lies on top of the bed of coal) becomes too great, and thus too costly, to remove.

Highwall mining often occurs as a final extraction method when overburden becomes too thick and thus too costly for removal, especially in contour mining operations. Imagine a steep, vertical face of overburden, with a bed of coal at its bottom—this this entire face is called a "highwall". One could spend the time, fuel, labor and opportunity costs removing the unwanted overburden in order to gain access to the

Figure 2-14, Contour Mining, *Credit/Courtesy of Kentuckians for The Commonwealth, www. kftc.org*

Figure 2-15, Highwall Mining Machine & Rectangular Bore-holes, *Credit/Courtesy of ADDCAR*

Figure 2-16, Producing Coal with a Highwall Miner, *Credit/Courtesy of ADDCAR*

Figure 2-17, Workers Inside a Highwall Mining Machine, *Credit/Courtesy of ADDCAR*

Figure 2-18, Rectangular Boreholes from Highwall Mining, *Credit/Courtesy of National Institute for Occupational Safety and Health (NIOSH) Mining Division*

coal, but this may be a money-losing option given the price of the resource. A more cost-effective technique is to simply leave the overburden in place and burrow into the exposed coal bed using a highwall miner. These massive machines push a rotating shearer (called a "cutterhead boom") into the exposed coal face at the bottom of a highwall, and progressively extract coal from a steadily lengthening, rectangular bore-hole. These holes can progress horizontally up to about 1,000 feet, thereby economically extracting many tons of additional coal from an otherwise inaccessible deposit.

Area mining occurs in places with relatively flat terrain and proceeds by removing vast, sectional strips of overburden. Topsoil is first removed and piled into a mound, which is often seeded to prevent erosion. Subsoil and additional overburden is removed and become the fill of a previously mined strip. Eventually the top of the coal bed appears. Extraction proceeds with dragline excavators scooping up the coal and transferring it into a continuous procession of enormous mining trucks. When all the recoverable coal is removed, the subsoil and topsoil is then spread back over the strip, seeded, and the land is returned to a vegetated state. The surface coal mines in the Powder River Basin of Wyoming are area-mines. For an example, consider the most productive area-mine in North America—the North Antelope Rochelle Mine,

Figure 2-19, Area Mining with Dragline Excavator, *Credit/Courtesy of Peabody Energy*

Figure 2-20, Dragline Excavator Loading Coal, *Credit/ Courtesy of Peabody Energy*

located about 65 miles south of Gillette, Wyoming, and operated by Peabody Energy. In 2010 this single mine produced record 106,000,000 short tons of coal (a "short" ton is 2,000 pounds) (Gallagher, 2011).

Mountaintop mining (MTM) is a form of surface mining that involves removing approximately 400 vertical feet of mountain peak to access numerous beds of coal. This mining technique extracts thin seams of

coal at varying depths as each layer of mountaintop overburden is removed and cast into adjacent valleys. A single mountaintop may contain approximately 7 different horizontally-situated coal seams, some of which are especially high-rank. All these seams, however, are too slender (perhaps only a foot thick) to economically extract using any other mining technique. MTM currently occurs in Virginia, West Virginia, Kentucky and Tennessee, and began in the mid-1990s as

Figure 2-21, Caballo Surface Mine, Powder River Basin, Wyoming, (Coal Seams Average 68 Feet Thick), *Credit/Courtesy of Peabody Energy*

Figure 2-22, Mountain Top Mining, *Credit/Courtesy of Vivian Stockman/www.ohvec.org, Flyover courtesy SouthWings.org*

demand for low-sulfur coal rose, coupled with the development of massive draglines capable of moving 100 cubic yards of material in a single scoop (Copeland, 2008).

Controversy surrounds MTM for several reasons. The most significant concerns aesthetics, and with little surprise considering the impact this practice levies upon the visible environment. Even long-time coal mining communities can be stunned when seeing nearby mountain peaks steadily disappear, literally flattened amidst persistent blasting and dragline scooping. Re-grading the blasted peaks post-mining is an option for coal companies, but a more economical alternative is to close their MTM operations by simply covering the flattened plane with topsoil and seeding it with native grasses. While technically "reclaimed", the table-top plane is too artificial a landscape for some area residents, who then raise public objections to the practice. Another controversial issue surrounding MTM is that peak overburden fills adjacent valleys and thereby interrupts natural watersheds and stream-flow patterns. In addition to concerns about habitat loss, water quality issues also arise because the rocky debris often contains heavy metals and

other toxins that pollute local waterways (McQuaid, 2009).

Surface mining, especially MTM, can leave an especially visible impact on the landscape. **The Surface Mining Control and Reclamation Act of 1977 (United States Public Law 95-87) states that society and the environment be protected from the adverse effects of surface mining** (United States Government, 1993). Helping to apply this law is the United States Office of Surface Mining Reclamation and Enforcement (OSM), which develops surface mining oversight programs that are most often instituted through state-specific mining regulation agencies. As a result of United States Public Law 95-87 and OSM guidelines, land reclamation procedures are detailed before any surface mining permit is ever issued. Creating land reclamation procedures for a future mine requires estimating how much acreage will be disturbed, and any future costs associated with reclaiming this disturbed acreage.

United States Public Law 95-87 also mandates that a land reclamation bond (this is a sum of money) be posted by any company intending to conduct surface

Figure 2-23, Mountain Top Mining, *Credit/Courtesy of Vivian Stockman/www.ohvec.org, Flyover courtesy SouthWings.org*

mining operations. The OSM issues bonding guidelines (United States Office of Surface Mining Reclamation and Enforcement, 2011). Bonds are payable to the United States, or to the state in which the mine is to be located, and covers the cost of land reclamation in the event that the mining company dissolves before reclamation is complete, or simply refuses to conduct the appropriate reclamation actions. These bonds are financial obligations, and on company balance sheets they are often called "asset retirement obligation (ARO) liabilities." Examining the annual reports of coal companies reveals the extent of their ARO obligations. For example, in 2010 Peabody Energy, the world's largest private-sector coal company, carried $704.4 million is surety bonds and bank guarantees to secure reclamation obligations or activities (Peabody Energy, 2011, p. 62).

However it is mined, coal must be transported to consumers. Most coal consumers are large-scale coal-fired electric power plants (about 90% of all coal in the United States is burned to generate electricity). These utilities need a constant supply of fuel all year long in order to generate the electricity upon which we all depend. Depending on location and market conditions, coal is shipped to power plants via rail, barge, truck, collier (this is a type of cargo ship), slurry pipeline, or a combination of these various modes. The cost of shipping is added to the cost of the coal itself, and both these costs have been steadily dropping (in inflation-adjusted dollars) over the past several decades (United States Energy Information Administration, 2004). Shipping has been getting cheaper because of competition amongst coal-carriers, and coal has been getting cheaper because of the vast reserves and efficient extraction currently underway in the western United States. Multimode (any combination of rail, truck, barge, and collier) and rail currently dominate the coal transportation sector, but barges are the cheapest means of moving this energy source (United States Energy Information Administration, 2011f).

Figure 2-24,
Grassland Reclamation
at the North Antelope
Rochelle Mine (NARM),
*Credit/Courtesy Peabody
Energy*

Figure 2-25, Grassland
and Pond Reclamation
at the North Antelope
Rochelle Mine (NARM),
*Credit/Courtesy Peabody
Energy*

Coal's Virtues and Vices

Like all energy resources, coal has important virtues as well as challenges. Perhaps most significant amongst its virtues is the fact that the United States has an abundant supply. **The United States Energy Information Administration estimates that there are 261 billion short tons (261,000,000,000) of estimated recoverable coal reserves left in the United States** (United States Energy Information Administration, 2011g). To give this gigantic number some perspective, consider

that the United States mined just over 1 billion of these 261 billion tons 2009. It is perhaps easier now to understand coal's R/P ratio (previously discussed in Chapter 1), where about 250 years of coal remain in the United States at the current consumption rate.

A second virtue of coal is that it is an inexpensive energy source when considering cost per short ton. Prices of course fluctuate, and coal often competes with natural gas, but for baseload electrical generation coal is often the preferred source of energy. **Baseload elec-**

Figure 2-26, Coal Trains Loading at the North Antelope Rochelle Mine (NARM), *Credit/Courtesy Peabody Energy*

trical generation plants supply the minimum amount of demanded electricity by continuously operating and producing electricity at a constant rate. In 2009, the nominal, average price for 1 short ton of coal was $32.92 (United States Energy Information Administration, 2010b) ("nominal" means the price paid in 2009, without any adjustments for inflation or purchasing power). This essentially means that in 2009, a single pound of "average" coal (containing roughly 10,000 Btu) cost less than 2 pennies.

A third important virtue of coal is that it is a domestic—and thus secure—energy resource. This is important when considering energy security. Energy-rich countries around the world may or may not have governments and/or social policies that are congruent with interests in the United States, and may or may not be politically stable. For example, Saudi Arabia, an important oil provider to the United States, is a non-democratic kingdom where women are not allowed to drive cars. While we depend upon Saudi oil, the kingdom's gender policies are abhorrent to most United States consumers. In another case, Nigeria's Niger River Delta holds vast petroleum resources, and yet the country is riddled with corruption and commits what some would argue are war crimes when it puts down revolutionary uprisings within the oil-rich delta region. These and other country-specific, internal issues make for challenging and risky long-term trade relationships and thus increase energy security concerns. United States policy-makers know, however, that no matter the situations that may exist abroad, the country has a safe and abundant domestic coal supply.

A final virtue of coal is that it is a solid, stable material. This makes it easy to mine and transport because companies can literally scoop it up and toss it onto a train or barge. There are no worries about leaking cargo, exploding pipelines, radioactivity, or terrorism when it comes to coal. It is a rock, after all, and it is treated accordingly, roughly, like a rock. Such rough treatment is impossible with petroleum, natural gas and uranium—all of which are valuable energy resources but require considerably more care in mining, transporting, and processing.

For all its virtues, coal is not without its challenges. Chief among its problems is greenhouse gas (GHG) emissions. When coal (or any fossil fuel, for that matter) is burned, an unavoidable byproduct of combustion is carbon dioxide (CO_2). This gas, along with a host of others, captures the energy re-radiated from the earth's land and water surfaces and induces long-term increases in global average temperatures (Intergovernmental Panel on Climate Change, 2007, pp. 30-33). Climatologists watch carbon dioxide emissions closely because such massive volumes are emitted into the atmosphere every year, and as a result, global sea and air temperatures are rising. **In 2009 coal-fired power plants in the United States emitted 1,876.8 million metric tons (1,876,800,000 or simply 1.9 billion metric tons) of carbon dioxide** (United States Energy In-

formation Administration, 2011h, p. 21) (Note that 1 "metric ton" is 2,204.62 pounds).

2.3 pounds of CO_2 are emitted into the atmosphere for every kilowatt-hour generated by coal (Carbon Dioxide Information Analysis Center, 2011). Thus, if a coal-fired power plant has a nameplate capacity of 500 MW, it emits about 4 million short tons of CO_2 into the atmosphere every year. Calculation 2-1 illustrates how this number is calculated. Consider this emission amount with the fact that, in 2009, the United States as a whole carried a nameplate capacity of 338,723 MW worth of coal-fired power plants (United States Energy Information Administration, 2011i). These fig-

Figure 2-27, A Coal Train Leaving the North Antelope Rochelle Mine (NARM), *Credit/Courtesy Peabody Energy*

Figure 2-28, A Coal Barge on the Monongahela River in Pennsylvania,
Credit/Courtesy of Jeffrey Giles - Pittsburgh, PA

ures and calculations make it a little easier to grasp the enormity of coal's CO_2 challenge, and to appreciate how the United States produced 1.9 billion metric tons (1,900,000,000) of CO_2 from coal alone in 2009.

The volume of CO_2 produced from burning coal and other fossil fuels is an increasingly worrisome issue that policy-makers, engineers and power producers are now tackling. Certainly one solution is to stop burning fossil fuels, but this idea does not gain any traction because fossil fuels are abundant, densely-packed with energy, and quickly halting their combustion would destroy any developed economy. Technological solutions are often a preferred route to solving the "carbon problem", and these techniques frequently rely on capturing and then injecting the CO_2 back

---

Calculating the Annual CO_2 Emissions from a 500 MW Coal-Fired Power Plant

A) 500 MW = 500,000,000 Watts
B) 500,000,000 Watts x 24 hours in a day = 1.2×10^{10} Watt-hours
C) 1.2×10^{10} Watt-hours / 1,000 = 12,000,000 kWh in a day
D) 12,000,000 x .80 capacity factor = 9,600,000 kWh in a day
E) 9,600,000 kWh x 2.3 pounds of CO_2 = 22,080,000 pounds of CO_2 in a day
F) 22,080,000 pounds / 2,000 = 11,040 short tons of CO_2 in a day
G) 11,040 tons x 365 = 4,029,600 short tons of CO_2 in a year

Calculation 2-1, Calculating the CO_2 Emissions from a 500 MW Coal-Fired Power Plant

into the ground in a process collectively referred to as carbon capture and storage technology (CCS-technology).

Commercial-scale CCS began in 1991 when Norway's Statoil began injecting CO_2 beneath the North Sea in order to avoid a \$55.00/ton carbon tax (Luby & Systa, 2007). Since then, seven other CCS commercial-scale projects have become operational, and 37 more are currently in the planning-phase (World Coal Association, 2011). After employing a variety of capture and transportation techniques, most of these projects focus on injecting CO_2 into: 1) deep saline formations, 2) depleted oil and gas fields, 3) unmineable coal seams. These are geologic sequestration techniques, and all of them depend on there being an impermeable caprock beneath the ground, a lack of faults, a lack of significant seismicity, a lack of any uncapped wellbores that lead to the surface, as well as having sufficient storage capacity. After meeting these criteria

an appropriately-selected site may expect to operate for several decades before closure, and then be monitored for several more decades, or even centuries, in order to ensure the site's integrity and safety (International Energy Agency, 2008, pp. 125-129).

One significant hurdle to CCS technology that is often overlooked is the simple scale of CCS need. Recall from our above calculations that a single 500 MW coal-fired power plant emits about 4 million short tons of CO_2 into the atmosphere every year. Recall also that United States' coal-fired power plant fleet emitted 1.9 *billion* metric tons of CO_2 in 2009. These emission numbers contrast sharply with the current storage capacity of the world's existing commercial-scale CCS projects. In 2011, the total storage capacity of *all* commercial-scale CCS projects was just 33 million metric tons per year (Global CCS Institute, 2011, pp. 10-11). Considerable work thus lies ahead as we continue enjoying the energy that fossil fuels provide, while simultaneously

Figure 2-29, Aerial Image of Kingston Ash Slide,
Credit/Courtesy of Tennessee Valley Authority

Figure 2-30, Cleanup Begins Following the Kingston Ash Slide, *Credit/Courtesy of Tennessee Valley Authority*

working out how to best mitigate the challenges associated with their use.

Another challenge associated with using coal as an energy source is that it produces ash when it burns. When this ash is carried up a smokestack, it is called "fly ash," and when it collects at the bottom of a combustion chamber it is called "bottom ash." Both fly and bottom ash must be collected and safely disposed of to prevent it from escaping into the atmosphere and becoming an airborne pollutant. Fortunately, electric power plants are well-positioned to meet this challenge and today capture most of the ash before it escapes into the air. About 75% of it is disposed of in landfills, with landfill size requirements varying depending on the size of the nearby power plant. The remaining 25% of coal ash is used in applications such as a cement substitute in the creation of concrete "cinder" blocks, wallboard, blasting grits and as a snow and ice control measure for wintertime roads (United States Environmental Protection Agency, 2011).

In late 2008 coal ash storage sites came under scrutiny because 5.4 million cubic yards of wet coal ash flooded out of a retaining pond near the Kingston Fossil Plant in Harriman, Tennessee. This was enough ash to bury 3,000 square acres 1 foot deep (Shaila, 2008). Homes were destroyed in this flood, and concerns about water and air quality arose as local residents learned that coal ash contains trace quantities of radioactive uranium, thorium and potassium, as well as toxic heavy metals such as arsenic, lead, mercury and selenium (PBS NewsHour, 2009). The Tennessee Valley Authority (TVA), owner of the plant, immediately began testing air and water for contaminants, cleaning up the ash, buying properties rendered unfit for habitation, and commenced a plan to convert all of its 24 coal ash ponds into dry storage sites by 2018 (Tennessee Valley Authority, 2011).

Coal-fired power plants meet additional challenges by employing electrostatic precipitators (ESP), flue-gas desulfurization (FGD) and selective-catalytic reduction (SCR) techniques. Electrostatic precipitators add a charge to fly ash particles, thereby making them "stick" to rods carrying an opposite charge. Ash is relatively easily collected in this manner. Flue-gas desulfurization "scrubbers" spray limestone slurry over flue gas. The limestone slurry chemically combines with the gaseous sulfur oxide (SO_2) and then precipitates out as solid calcium sulfite ($CaSO_3$). This process removes sulfur emissions from coal-fired power plants, and has an additional benefit of being a valuable additive in the wallboard manufacturing industry. Selective-catalytic reduction reduces nitrous oxide emissions by spraying anhydrous ammonia over flue gas. The ammonia chemically combines with the GHG nitrous oxide compounds and creates benign nitrogen gas and liquid water. These and many other methods developed since the launch of the 1986 Clean Coal Technology Demonstration Program have significantly contributed to today's coal-fired power plants being "dramatically cleaner and more efficient than plants based on older technology" (National Energy Technology Laboratory, 2011).

From the above material it is apparent that burning coal produces significant gaseous as well as solid pollutants. Many of these are effectively controlled using technologies with proven track records. What if, however, a fuel could be burned in power plants that was hotter than coal, burned cleaner than coal, but still had all the associated benefits of coal? Such a fuel does exist—it is called synthetic gas, town gas, or simply "syngas"—and it is already established as a fuel for today's electrical power generation plants. Syngas can be created from any carbonaceous matter (materials rich in carbon that are also combustible). When treating crushed coal with oxygen (O_2) and steam (H_2O), under high pressure and temperature, some of the coal's carbon is liberated, only to immediately combine with the oxygen to create carbon monoxide (CO). Hydrogen (H_2) is also liberated in the process. The end result is syngas, composed of carbon monoxide and hydrogen. The process can also be described as follows: Coal + O_2 + H_2O → H_2 + CO.

Notice that there is no sulfur or nitrogen in syngas, and since it is indeed a gas, there is no associated ash when it is burned. These facts alone make syngas an attractive fuel option, even though carbon dioxide (CO_2) remains as a byproduct of its combustion. Rather than using the syngas to heat water, to generate steam, to turn a turbine, syngas is burned directly inside a gas turbine, thereby spinning the turbine's axis and wholly skipping the water-to-steam state change process. This of course increases a power plant's efficiency because energy is not spent repeatedly changing the state of water. Gas turbines are also efficient because today's models employ very high gas pressures and firing temperatures, which means that less fuel needs to be burned in order to induce a high spin.

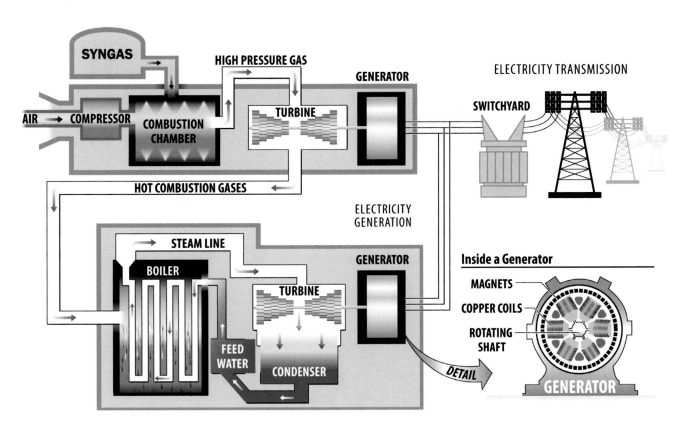

Figure 2-31, Integrated Gasification Combined Cycle (IGCC) Power Plant Model Using Syngas, *Credit/Courtesy of National Energy Education Development*

For example, Mitsubishi Heavy Industries makes the M701G2 heavy frame "G-Class" turbine, with a firing temperature of 1,500 °C, an exhaust temperature of 587 °C, and a thermal efficiency of 39.5% (Langston, 2010). The company also creates the latest "J-Class" gas turbine with a 100 °C hotter firing temperature and significantly higher pressure than the Mitsubishi G-Class (Turbomachinery: the Global Journal of Energy Equipment, 2009, p. 14). These efforts are mirrored by other gas turbine manufacturers such as General Electric and Siemens. All these modern efforts greatly eclipses the first gas turbine (built in 1903 by Jens William Aegidus), which had a firing temperature no greater than 400 °C and a thermal efficiency less than 20%.

There is yet another benefit of using syngas instead of coal to generate electricity. Today's syngas turbines can work in the same power plant right next to a traditional steam turbine. Why would a single power plant want to have two different turbine types? **In an Integrated Gasification Combined Cycle (IGCC) Power Plant, exhaust from a gas turbine (>500 °C) is used to create the steam that spins a steam turbine.** It is both foolish and wasteful to simply vent the hot exhaust from a gas turbine when that energy can be put to good use. By capturing this exhaust heat, IGCC plants radically boost their energy efficiency up to 60% (recall that the ageing coal-fired steam-driven power plant fleet in the United States is about 33% efficient). For example, the General Electric 50-Hz Frame 9FB turbine approaches 58% efficiency when in combined-cycle configuration, while the Siemens SGT5-8000H 340 MW turbine exceeds 60% efficiency when in combined-cycle configuration (Zachary, 2008).

With all the benefits of IGCC technology, one might reasonably expect these kinds of power plants to be the industry standard. Unfortunately, this is not yet the case. In 2009 only five IGCC plants existed in the world. Two of the plants are based in the United States, one in West Terra Haute, Indiana (the Wabash River Facility), and the other in Mulberry, Florida (the Polk Power Station), near Tampa. These two power plants came online in the mid-1990s with extensive financial backing from the United States Department of Energy (DOE). For example, the Wabash River Facility received $219,100,000 (50% of its total cost) (National Energy Technology Laboratory, 2003, p. 4), while the Polk Power Station received $ 150,894,223 (49% of its total cost) (National Energy Technology Laboratory, 2003b, p. 1).

Figure 2-32, IGCC Power Plant in Puertollano, Spain, Using Both SGT5-3000E Gas Turbines and SST5-5000 Steam Turbines, *Credit/ Courtesy of Siemens*

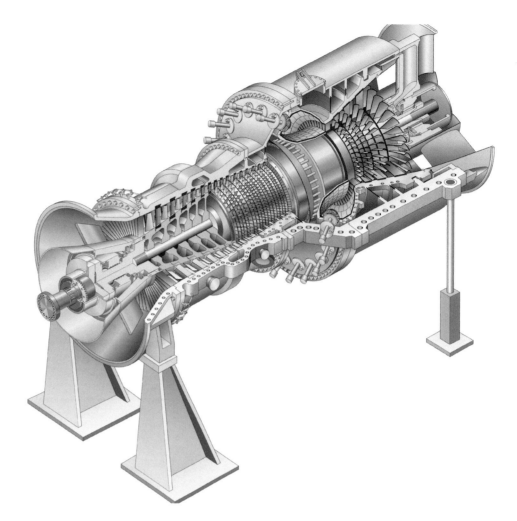

Figure 2-33, SGT5-3000E Gas Turbine CAD Drawing, *Credit/Courtesy of Siemens*

The dearth of IGCC power plants seems counterintuitive given the potential of IGCC. Reasons for the slow adoption and high prices are numerous, including technological hurdles (IGCC remains an evolving technology), high plant-specific price tags (no industry standard IGCC power-plant model exists), and permitting challenges (Neville, 2009, p. 2). Significant growth is occurring, however, with 11 IGCC projects under construction worldwide as of 2010, and global capacity projected to grow by over 51,000 MW (a 72% increase) from 2011 through 2016 (National Energy Technology Laboratory, 2010).

Important Ideas:

- Coal is a combustible sedimentary rock of biological origin.

- Fossil fuels are energy resources that come from the transformed remains of plants and animals that lived millions of years ago.

- The Carboniferous period lasted 60 million years.

- The United States produced about 69% more coal in 1990 than in 1974.

- The bulk of United States coal production currently occurs west of the Mississippi River, specifically in the Powder River and Unita Basins.

- The majority of coal produced in the United States comes from surface mines.

- Conventional mining creates a "room and pillar" layout in the midst of a coal bed.

- Longwall mining extracts entire panels of coal, leaving no space or support pillars.

- Contour mining, as the name suggests, occurs along "contours" in sloping terrain.

- Highwall mining often occurs as a final extraction method when overburden becomes too thick and thus too costly for removal, especially in contour mining operations.

Figure 2-34, SGT5-8000H 340 MW Gas Turbine, *Credit/ Courtesy of Siemens*

Figure 2-35, Turbine Components for a 50% Efficient Coal-Fired Power Plant, *Credit/Courtesy of Siemens*

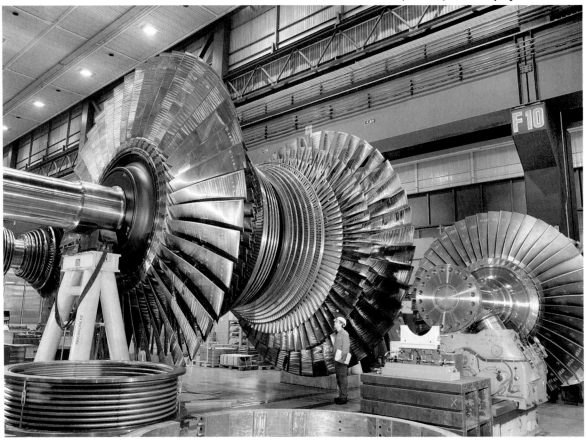

- Area mining occurs in places with relatively flat terrain and proceeds by removing vast, sectional strips of overburden.

- Mountaintop mining (MTM) is a form of surface mining that involves removing approximately 400 vertical feet of mountain peak to access numerous beds of coal.

- The Surface Mining Control and Reclamation Act of 1977 (United States Public Law 95-87) states that society and the environment be protected from the adverse effects of surface mining.

- The United States Energy Information Administration estimated that 261 billion short tons (261,000,000,000 short tons) of proven coal reserves remained in the United States in 2009.

- Baseload electrical generation plants supply the minimum amount of demanded electricity by continuously operating and producing electricity at a constant rate.

- In 2009 coal-fired power plants in the United States emitted 1,876.8 million metric tons (1,876,800,000 or simply 1.9 billion metric tons) of carbon dioxide.

- 2.3 pounds of CO_2 are emitted into the atmosphere for every kilowatt-hour generated by coal.

- In an Integrated Gasification Combined Cycle (IGCC) Power Plant, exhaust from a gas turbine (>500 °C) is used to create the steam that spins a steam turbine.

Works Cited

Carbon Dioxide Information Analysis Center. (2011). *Frequently Asked Global Change Questions*. Retrieved August 30, 2011, from How much CO2 is emitted as a result of my using specific electrical appliances?: http://cdiac.ornl.gov/pns/faq.html

Copeland, C. (2008, April 9). *Congressional Research Service, Mountaintop Mining: Background on Current Controversies*. Retrieved August 29, 2011, from National Biological Information Infrastructure: http://www.nbii.gov/images/uploaded/156209_1213647748387_Mountaintop_Mining.pdf

Gallagher, M. (2011, November 18). Peabody Energy, Director Corporate Communications. (T. Erski, Interviewer)

Global CCS Institute. (2011). *The Global Status of CCS: 2011*. Retrieved November 29, 2011, from Chapter 2: Projects: http://cdn.globalccsinstitute.com/sites/default/files/publications/22562/global-status-ccs-2011.pdf

Intergovernmental Panel on Climate Change. (2007). *Climate Change 2007: Synthesis Report*. Retrieved December 1, 2011, from Observed changes in climate and their effects, 1.1 Observations of climate change: http://www.ipcc.ch/pdf/assessment-report/ar4/syr/ar4_syr.pdf

International Energy Agency. (2007). *Fossil-Fuel Fired Power Generation: Case Studies of Recently Constructed Coal and Gas Fired Power Plants*. Retrieved January 25, 2012, from Pulverised Coal Combustion: http://www.iea.org/textbase/nppdf/free/2007/fossil_fuel_fired.pdf

International Energy Agency. (2008). *CO2 Capture and Storage: A Key Carbon Abatement Option*. Retrieved November 29, 2011, from Site Selection, Monitoring and Verification: http://www.iea.org/textbase/nppdf/free/2008/CCS_2008.pdf

Labbe, D. (2009, November). Nox, Sox, & CO2 Mitigation Using Blended Coals. *Power Engineering, 113*(11), 200-210.

Langston, L. S. (2010). *Efficiency by the Numbers*. Retrieved August 30, 2011, from Mechanical Engineering: The Magazine of ASME, Web Exclusive: http://memagazine.asme.org/Web/Efficiency_by_Numbers.cfm

Lind, G. H. (2005, September). Risk Management of Underground Coal Pillar Extraction in South Africa. *International Journal of Surface Mining, Reclamation & Environment, 19*(3), 218-233.

Luby, P., & Systa, M. (2007, April 15). *Power Magazine*. Retrieved November 29, 2011, from Exploring the many carbon capture options: http://www.powermag.com/coal/Exploring-the-many-carbon-capture-options_189.html

McQuaid, J. (2009, January). Mining the Mountains. *Smithsonian, 39*(10), pp. 74-85.

National Energy Technology Laboratory. (2003). *Clean Coal Technology Demonstration Program (CCTDP) - Round 4, Advanced Electric Power Generation - Integrated Gasification/Combined Cycle*. Retrieved August 30, 2011, from Wabash River Coal Gasification Repowering Project, Project Brief: http://www.netl.doe.gov/technologies/coalpower/cctc/cctdp/project_briefs/wabsh/documents/wabsh.pdf

National Energy Technology Laboratory. (2003b). *Clean Coal Technology Demonstration Program (CCTDP) - Round 3, Advanced Electric Power Generation - Integrated Gasification/Combined Cycle*. Retrieved August 30, 2011, from Tampa Electric Integrated Gasification Combined-Cycle Project, Project Brief: http://www.netl.doe.gov/technologies/coalpower/cctc/cctdp/project_briefs/tampa/documents/tampa.pdf

National Energy Technology Laboratory. (2010). *2010 Worldwide Gasification Database* . Retrieved August 30, 2011, from Worldwide Growth Planned Through 2016: http://www.netl.doe.gov/technologies/coalpower/gasification/worlddatabase/growthworld.html

National Energy Technology Laboratory. (2011). *Technologies: Coal and Power Systems*. Retrieved August 30, 2011, from http://www.netl.doe.gov/technologies/coalpower/index.html

Neville, A. J. (2009, August 1). *IGCC Update: Are We There Yet?* Retrieved August 30, 2011, from Power: Business and Technology for the Global Generation Industry: http://powermag.com/coal/IGCC-Update-Are-We-There-Yet_2063.html

PBS NewsHour. (2009, February 2). Retrieved August 30, 2011, from Tennessee Coal Ash Disaster Raises Concerns about Similar Sites Nationwide: http://www.pbs.org/newshour/bb/environment/jan-june09/coalash_02-02.html

Peabody Energy. (2011). *2010 Annual Report*. Retrieved November 22, 2011, from http://www.peabodyenergy.com/mm/files/Investors/Annual-Reports/2010BTUAnnualReport.pdf

Shaila, D. (2008, December 28). *Tennessee Ash Flood Larger Than Initial Estimate*. Retrieved February 19, 2010, from The New York Times: http://www.nytimes.com/2008/12/27/us/27sludge.html

Tennessee Valley Authority. (2011, June 6). *Fact Sheet: Kingston Ash Recovery Project*. Retrieved August 30, 2011, from http://www.tva.gov/kingston/pdf/Kingston%20Ash%20Recovery%20Project%20Fact%20Sheet%20Final%2006-06-2011.pdf

Turbomachinery: the Global Journal of Energy Equipment. (2009, March/April). *Breaking the Efficiency Barrier*. Retrieved October 25, 2010, from Turbomachinerymag.com: http://www.turbomachinerymag.com/sub/2009/March-.pdf

United States Department of Labor. (2011). *Mine Safety and Health Administration*. Retrieved August 29, 2011, from Coal Mine Safety and Health: http://www.msha.gov/PROGRAMS/COAL.HTM

United States Energy Information Administration. (2004, September 17). *Independent Statistics and Analysis*. Retrieved August 29, 2011, from Coal Transportation: Rates and Trends: http://www.eia.gov/cneaf/coal/page/trans/ratesntrends.html

United States Energy Information Administration. (2006, October). *Coal Production in the United States, An Historical Overview*. Retrieved August 29, 2010, from http://www.eia.doe.gov/cneaf/coal/page/coal_production_review.pdf.

United States Energy Information Administration. (2010, August 19). *Sources & Uses, Coal, Data*. Retrieved August 29, 2011, from U.S. Coal Production, Annual: http://www.eia.gov/totalenergy/data/annual/txt/ptb0702.html

United States Energy Information Administration. (2010b, August 19). *Independent Statistcis and Analysis*. Retrieved August 29, 2011, from Coal Prices, 1949-2009: http://www.eia.gov/totalenergy/data/annual/txt/ptb0708.html

United States Energy Information Administration. (2011). *Independent Statistcis and Analysis*. Retrieved August 29, 2011, from Glossary: http://www.eia.doe.gov/glossary/index.html

United States Energy Information Administration. (2011b, August 22). *Independent Statistics and Analysis, Coal News and Markets*. Retrieved August 26, 2011, from http://www.eia.gov/coal/news_markets/

United States Energy Information Administration. (2011c, February 3). *Independent Statistics and Analysis*. Retrieved August 29, 2011, from Coal Production and Number of Mines by State and Mine Type: http://www.eia.gov/cneaf/coal/page/acr/table1.html

United States Energy Information Administration. (2011d, February 3). *Independent Statistics and Analysis.* Retrieved August 29, 2011, from Coal Mining Productivity & Employees by State and Mine Type: http://www.eia.gov/cneaf/coal/page/acr/table21.html

United States Energy Information Administration. (2011e). *Independent Statistics and Analysis*. Retrieved August 29, 2011, from Glossary: http://www.eia.gov/tools/glossary/index.cfm?id=L

United States Energy Information Administration. (2011f). *Independent Statistics and Analysis*. Retrieved August 29, 2011, from Average Utility Contract Coal Transportation, Rate per Ton by Transportation Mode, 1979-1997: http://www.eia.gov/cneaf/coal/ctrdb/tab35.html

United States Energy Information Administration. (2011g, May 18). *Independent Statistics and Analysis*. Retrieved August 29, 2011, from Coal Explained: How Much Coal is Left?: http://www.eia.gov/energyexplained/index.cfm?page=coal_reserves

United States Energy Information Administration. (2011h, March). *Independent Statistics and Analysis*. Retrieved August 30, 2011, from Emissions of Greenhouse Gases in the United States 2009: http://www.eia.gov/environment/emissions/ghg_report/pdf/0573(2009).pdf

United States Energy Information Administration. (2011i, April). *Independent Statistics and Analysis*. Retrieved August 30, 2011, from Existing Capacity by Energy Source: http://www.eia.gov/cneaf/electricity/epa/epat1p2.html

United States Environmental Protection Agency. (2011, July 8). *Coal Ash*. Retrieved August 30, 2011, from http://www.epa.gov/radiation/tenorm/coalandcoalash.html

United States Geological Survey. (2011, July 5). *Coal Assessment: Overview*. Retrieved August 29, 2011, from http://energy.usgs.gov/Coal/AssessmentsandData/CoalAssessments.aspx

United States Government. (1993, December 31). *Public Law 95-87, Surface Mining Control and Reclamation Act of 1977*. Retrieved August 29, 2011, from Office of Surface Mining Reclamation and Enforcement (OSM): http://www.osmre.gov/topic/SMCRA/SMCRA.pdf

United States Office of Surface Mining Reclamation and Enforcement. (2011). *Handbook for Calculation of Reclamation Bond Amounts*. Retrieved August 29, 2011, from http://www.techtransfer.osmre.gov/NTTMainSite/Library/hbmanual/bondcal/bondcal.pdf

World Coal Association. (2011). *Carbon Capture & Storage Projects*. Retrieved November 29, 2011, from CCS Map: http://www.worldcoal.org/carbon-capture-storage/ccs-map/

World Coal Association. (2012). *Coal & Electricity*. Retrieved January 25, 2012, from How is Coal Converted to Electricity?: http://www.worldcoal.org/coal/uses-of-coal/coal-electricity/

Zachary, J. (2008, April 15). *Turbine Technology Maturity: A Shifting Paradigm*. Retrieved August 30, 2011, from Power: Business and Technology for the Global Generation Industry: http://powermag.com/issues/features/Turbine-technology-maturity-A-shifting-paradigm_68_p2.html

The West Azeri Platform, *Credit/Courtesy of BP,*
© BP p.l.c.

Chapter 3
Petroleum

Petroleum Geology

Like coal, petroleum (usually simply called oil) is a non-renewable fossil fuel energy source. Unlike coal, it is a liquid. This "rock-oil" is literally found inside rocks, the most important type being sandstone. Petroleum is created from the transformed remains of microscopic sea plants and animals (collectively referred to as plankton) that died between 299 and 416 Ma during both the Carboniferous and Devonian geologic time periods. You'll undoubtedly recognize the Carboniferous geologic time period from the discussion about coal in Chapter 2. In the case of petroleum's creation, however, the Devonian's 57 million years are added to the 60 million years of the Carboniferous period, for a total plankton collection time of about 117 million years. During this vast passage of time, enormous quantities of plankton accumulated at the bottom of shallow oceans. Burial occurred due to sediment settling, especially at the mouths of rivers where continental detritus washed into the seas. Eventually, the heat and pressure from the overlying material (this being several miles thick) transformed the plankton into a viscous liquid rich in hydrocarbons.

Hydrocarbons are molecular combinations of hydrogen and carbon that can exist as gases, liquids or solids. The reason hydrocarbons come in a variety of states is because some are composed of relatively few atoms, while others contain many. When these various combinations are brought up to the earth's surface, the ambient pressure and temperature dictate their physical states. For example, propane (C_3H_8) is a gas at the earth's standard ambient temperature and pressure. Propane is commonly used as a fuel for heating and cooking, and is often purchased in convenient 20 pound steel cylinders which, if carried, one might hear a "liquid" sloshing about. Why is the propane a liquid in the cylinder, but a gas when we burn it? The pressure inside the steel cylinder is much higher than the air pressure outside the cylinder. This high pressure pushes the C_3H_8 molecules together until they condense into a liquid. Upon pressure release, a physical state change occurs in the propane and it quickly becomes a gas. Paraffin waxes, on the other hand, have many more atoms of hydrogen and carbon than propane. Their chemical formulas range from is $C_{20}H_{42}$ to $C_{40}H_{82}$, and they exists as a solid at the earth's ambient pressure and temperature. It too is burned, as we all know when lighting a candle. Both propane and paraffins, and many other kinds of hydrocarbons, are extracted from the ground when companies tap into an oil reservoir to extract the transformed remains of plankton.

Heat, pressure and time transform the microscopic sea plants and animals into petroleum. Petroleum geologists frequently refer the place where this happens as the "oil window". **The oil window is between 1.5 to 3 miles deep where sufficient temperatures and pressures exist to transform organic-rich sediment into oil**. Temperature range at these depths is between 150 °F and 300 °F (United States Congress, Office of Technology Assessment, 1989, p. 104) (recall that a common temperature at which a kitchen stove bakes cookies is 350 °F). To conceptualize the

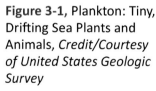

Figure 3-1, Plankton: Tiny, Drifting Sea Plants and Animals, *Credit/Courtesy of United States Geologic Survey*

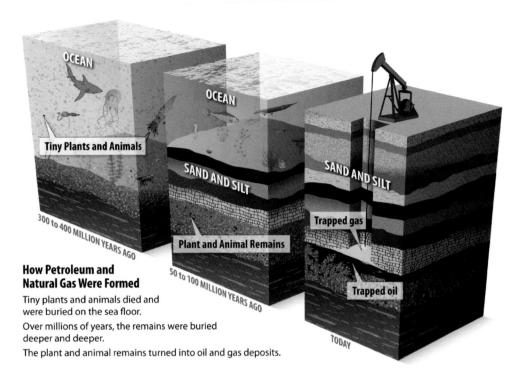

How Petroleum and Natural Gas Were Formed

Tiny plants and animals died and were buried on the sea floor.

Over millions of years, the remains were buried deeper and deeper.

The plant and animal remains turned into oil and gas deposits.

Figure 3-2, Model of Petroleum & Gas Formation, *Credit/ Courtesy of National Energy Education Development Project*

pressures within the oil window, consider that the average pressure of our atmosphere at sea level is 14.7 pounds per square inch (psi), or simply "1 standard atmosphere" (1 atm). Just 1 mile inside the earth's crust pressure increases to about 6,000 psi. At 3 miles down, the average pressure is around 20,000 psi.

A common misconception about petroleum is that it exists in vast underground, cave-like cisterns, and it is simply a matter of puncturing these cisterns with an oil well in order to suck out the valuable fluid. **While pockets or cavities of petroleum do exist, the vast majority of oil resources are locked inside solid rock.** These petroleum-saturated rocks are called "oil reservoirs." It is admittedly odd to consider a solid rock being saturated with a fluid. We tend to think of rocks as both perfectly rigid and perfectly solid. All rocks are not the same, however. Consider the igneous rock basalt. Basalt is formed when specific molten minerals solidify. One of the reasons basalt is dense (in addition

to its mineral content) is because very little space exists between its crystallized mineral grains. Sandstone, on the other hand, is a sedimentary rock. Sandstone is formed when solid grains of quartz are compressed and cemented together. Sandstone, compared to basalt, is not very dense (and also contains less-dense minerals) and thus literally feels lighter than basalt. One of the reasons sandstone is less dense is because a comparatively greater amount of space exists between its component mineral grains.

The spaces between and among the cemented mineral grains of "solid" rock, then, are what 1) hold oil and also 2) allow oil to flow into an well's bore hole. Sandstone, then, is appropriately compared to your everyday kitchen sponge. A sponge is a "solid" object that we can pick up. At the same time, if placed in a sink full of water, it becomes saturated. If lifted out of the water, much of the fluid drips out as the water moves through the sponge and back into the sink.

Depth (km)	Depth (m)	Pressure (psi)	Temperature (°F)	Temperature (°C)
1	0.62	4,134	104	40
2	1.24	8,267	149	65
3	1.86	12,401	194	90
4	2.49	16,534	239	115
5	3.11	20,668	284	140
6	3.73	24,801	329	165

Table 3-1, Temperatures and Pressures in the Oil Window

The sponge has the ability to hold water, as well as to move water, under the right circumstances. Geologists refer to this ability in rocks as porosity and permeability, both of which are carefully quantified before any well comes on-line. **Porosity is the percentage of void spaces inside a rock. Permeability is a measure of a rock's ability to transmit a fluid.** Oil reservoirs must be both porous and permeable in order to be considered economically viable for petroleum extraction.

Figure 3-3, A Close-up View of Sandstone's Many Voids, *Credit/Courtesy of CO2CRC*

Porosity and permeability are not the only geologic requirements of an oil reservoir. Since gases and liquids are less dense than rock, they naturally migrate through younger rock layers as they migrate up to the earth's surface. In the absence of an impermeable barrier, these resources literally seep up to the surface and (in the case of gases) migrate into the atmosphere or (in the case of oil) pool in shallow depressions on the surface. **An impermeable "caprock" stops the migration of gas and oil to the earth's surface.** This caprock is often shaped like an upside-down bowl (geologists call this an "anticline") under which the gas and oil collect. Sometimes faulting—the cracking and rearranging of the earth's surface—also creates geologically appropriate oil traps. Whatever the case, when boring into these reservoirs it is common to first encounter a gas-rich area (called the "gas cap") and then the petroleum-rich area. This predicable arrangement occurs due to the respective densities of gas and liquid. Gas, being less dense, most readily migrates upwards, and passes through the liquid petroleum en route to the caprock barrier.

The Early Market & Standard Oil

Until the middle of the nineteenth century, petroleum was scooped up from oil seeps whose reservoirs lay close to the surface and were overlain with weak and often fractured caprocks. Seeps are depressions in the ground where oil naturally pools after rising to the earth's surface. People commonly used petroleum to waterproof boats, light lamps, and treat skin irritations and digestion problems. The rudimentary collection methods prior to drilling, however, precluded acquiring large-enough quantities to create any significant marketplace for the product. This radically changed in 1859 when Edwin Drake struck oil in Pennsylvania with the world's first oil drilling rig. An oil-rush began within the year.

Refineries bloomed in tandem with the world's first oil rush as entrepreneurs hurried to fill kerosene orders. Until this time, home lighting primarily occurred from the combustion of whale oil, an increasingly scarce—and thus expensive—commodity. At pennies on the dollar, kerosene became the fuel of choice. Not only did this resource-switch result in the demise of the whaling industry, it also birthed a global oil addiction of which we are all still bound to as the 21st century matures. The petroleum marketplace roared into existence in the late 1800s, quickly followed by boom and bust cycles as overproduction, scarcity, and less-than-transparent market maneuvering took hold.

Figure 3-4, An Idealized Oil Trap

Non-permeable cap rock
Natural gas
Oil (migrated from source rock)
Brine (salt water)
Porous & permeable rock
Organic-rich source rock exposed to heat & pressure

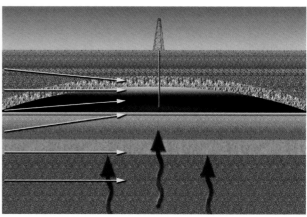

During this time John D. Rockefeller rose above all other refinery owners as he consolidated and thereby monopolized the industry. In 1870 he and six other entrepreneurs founded Standard Oil Company. It began with a single refinery in Cleveland, Ohio, and by destroying or absorbing its competitors, Standard Oil quickly controlled about 90% of the refining capacity in the United States. By the early twentieth century the company had expanded to control many of the ships, railroads and pipelines that moved crude oil.

While market consolidation did have the effect of driving down crude oil and kerosene prices, it also threatened the very principle upon which capitalism stands—healthy competition within the marketplace. Answering the rising clamor of resentment against Standard Oil's monopolistic business practices in general, and Rockefeller's colossal wealth in particular, the United States began litigation proceedings in 1909 under the Sherman Antitrust Act. **By 1911 the United States Supreme Court ruled that Standard Oil would disassemble into 34 separate companies** (Standard Oil Co. of New Jersey v. United States, 1910).

Figure 3-5, Standard Oil Refinery #1, Cleveland, Ohio, *Credit/Courtesy of Case Western Reserve University*

Petroleum Grading and Refining

All crude oil is not the same. There are over 150 different types of petroleum traded in the world today, and each type contains varying amounts of carbon, hydrogen, nitrogen, oxygen, sulfur and metals. These elements are chemically combined to create an extensive assortment of hydrocarbons with a wide range of densities. **Crude oils are labeled "light", "intermediate" or "heavy" depending on their density, which is mathematically calculated using a simple formula established by the American Petroleum Institute (API)** (United States Energy Information Administration, 2011). The API is an oil and gas trade association and lobbying group based in Washington, D.C. that, among

other agendas, creates standards and recommended practices ranging from the quality of drill bits to environmental protection. API standards are frequently adopted by governments and industries worldwide. The formula used to calculate the density of crude oil yields a value called an "API Degree".

API Degree = (141.5 / specific gravity at 60 °F) – 131.5

To calculate a crude oil's API Degree, its specific gravity is first determined. Specific gravity is the ratio of oil's density compared to water's density. For example, if the oil is the same density as water, its specific gravity is 1; if its density is less than water, its specific gravity might be .934; and if its density is more than water, its specific gravity might be 1.014. Once the specific gravity is calculated, the value is simply inserted into the above equation where just two more steps are required (first divide, then subtract). API Degrees can range from 8 to 58, with 8 being the most dense (thus "heavy"), and 58 being the least dense (thus "light"). Intermediate density grades generally score 22 to 38. If a crude oil scores an API Degree of 10, it is the same density as water.

Obviously, then, some of the crude oils actually sink in water, while most float. API degree is important because it takes comparatively less energy to refine lighter crudes, and thus lighter crudes generally—but not always—price higher compared to heavier crudes. You may have heard oil market commentators using the term "West Texas Intermediate" (WTI) when discussing the price of oil. WTI is a benchmark crude oil, meaning its price is a standard against which other crude oils are priced on exchanges around the world. WTI has an API degree of about 40. Brent, another benchmark crude oil, is a blend of oils pumped from many different fields in the North Sea. Brent blend has an API Degree of about 38, which makes it light, but not as light as WTI. Contrasting both WTI and Brent is yet another benchmark crude oil called Dubai Fateh, with an API degree of about 32. (Hyne, 2001, p. 4) Today, the prices of nearly 40 different benchmark crude oils are continually monitored by global markets (United States Energy Information Administration, 2011b).

In addition to density, crude oil is also graded on its sulfur content. WTI and Brent Blend are both "sweet" crude oils, while Dubai's Fateh is "sour". **Sweet crude oil contains less than 0.5% sulfur, while sour crude oil contains greater than 0.5% sulfur.** Sulfur is an impurity that must be removed during the refining process, especially if making gasoline. Since time and money must be spent to refine out the sulfur, sour crudes are generally cheaper than sweet crudes.

Table 3-2,
Selected Crude Oils,
API Degree & Price,
August 26, 2011

Name	API Degree	Price ($/Bbl)
West Texas Intermediate	40	98.04
Brent Blend	38	110.58
Dubai	32	103.88
Canada, Heavy Hardisty	22	74.45
Saudi Arabia, Arabian Heavy	27	101.46
Saudi Arabia, Arabian Light	34	105.61

Heavy or light, sour or sweet, petroleum must be refined in order to separate-out its many useful molecules. Hydrocarbon separation takes place inside "fractional distillation columns", which are tower-like structures measuring about 60 to 80 meters high. A single refinery usually contains numerous distillation columns where hydrocarbons are refined many times over in pursuit of particular products. A boiler is located near each column, and it is here that newly arrived crude oil is literally boiled using coal or natural gas. **Boilers heat the crude to over 700 °F. As the oil's temperature rises, various liquid hydrocarbons change state and become gas** (this is similar to boiling water on your stove, where liquid water steadily changes state into water vapor).

Gases are then routed from the boiler to the bottom of the distillation column where they immediately begin rising due to their relative density. As the gases rise, they cool and eventually reach their unique condensation temperatures at specific heights within the distillation column. Here they again change state, this time back into a liquid, but now nicely separated from the crude oil feedstock. The liquids are collected on distillation trays situated at different heights, and are then routed through pipes for further refining, or stored to be sold on the petrochemical marketplace. Each refined liquid product is considered a "fraction" of the crude oil feedstock—hence a "fractional" distillation column (National Petrochemical & Refiners Association, 2011).

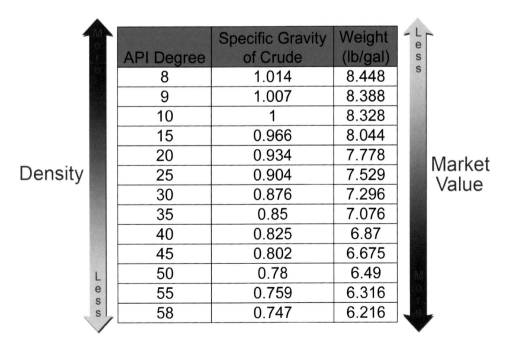

API Degree	Specific Gravity of Crude	Weight (lb/gal)
8	1.014	8.448
9	1.007	8.388
10	1	8.328
15	0.966	8.044
20	0.934	7.778
25	0.904	7.529
30	0.876	7.296
35	0.85	7.076
40	0.825	6.87
45	0.802	6.675
50	0.78	6.49
55	0.759	6.316
58	0.747	6.216

Figure 3-6, API Degrees, Density & Comparative Market Values

Figure 3-7, East Plant of the Texas City Refinery,
Credit/Courtesy of BP, © BP p.l.c.

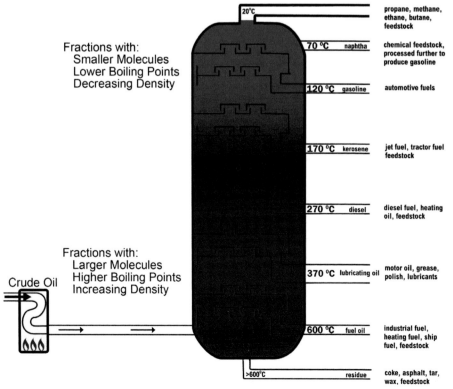

Figure 3-8,
Fractional Distillation Column,
*Credit/Courtesy of National
Energy Education Development
Project (adapted)*

The lighter hydrocarbons, such as propane (C_3H_8) and butane (C_4H_{10}), boil off first and condense high up within the fractional distillation column. Hydrocarbons such as octane (C_8H_{18}), and decane ($C_{10}H_{22}$) boil off later, only to rise and then condense to about the middle of the column. Heavier and more complex hydrocarbons, containing more carbon atoms, condense in the lower portions of the distillation column. These are used as lubricating oils or heavy fuel oils. Finally, petroleum coke, asphalt and tar, which have very high boiling points and the greatest number of carbon atoms, are drawn off as "residuals" or "bottom of the barrel" petroleum products because they literally collect at the bottom of the fractional distillation column (Chevron U.S.A., 2010).

Recall that one barrel (Bbl) contains 42 gallons of crude oil. When refined, this feedstock creates a great variety of useful products. **About 45% (19 gallons) of a barrel of crude oil becomes gasoline** (United States Energy Information Administration, 2011c). Diesel fuel, chemicals for the plastics industry, jet fuel, solvents, asphalt and many types of lubricants are also created. Chevron, one of the thirty companies comprising the Dow Jones Industrial Average, is a global energy company involved in exploration, extraction, refining, and marketing of crude oil. A refining statement from the company illustrates the usefulness of just a single barrel of petroleum:

Here's what just one barrel of crude oil can produce:

- Enough liquefied gases (such as propane) to fill 12 small (14.1 ounce) cylinders for home, camping or workshop use.

- Enough gasoline to drive a medium-sized car (17 miles per gallon) over 280 miles.

- Asphalt to make about one gallon of tar for patching roofs or streets.

- Lubricants to make about a quart of motor oil.

- Enough distillate fuel to drive a large truck (five miles per gallon) for almost 40 miles. If jet fuel fraction

is included, that same truck can run nearly 50 miles.

- Nearly 70 kilowatt hours of electricity at a power plant generated by residual fuel.

- About four pounds of charcoal briquettes.

- Wax for 170 birthday candles or 27 wax crayons.

Chevron U.S.A., 2011

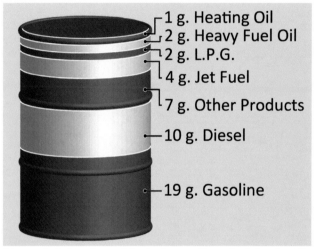

Figure 3-9, Products Made from a Barrel of Crude Oil

Oil Consumption, Production & Reserves

Oil consumption and production is often measured in "barrels per day" and is abbreviated "bpd". Since a barrel contains only 42 gallons, and there are billions of petroleum users on the earth, oil figures are often large and thus a bit cumbersome. **The United States consumed about 19 million bpd in 2010** (United States Energy Information Administration, 2011d). This number is reported by the United States Energy Administration as "19,180.126 thousand barrels per day"—a somewhat awkward way of reporting because it literally reads as: nineteen thousand, one hundred eighty…thousand. The "thousand" in this statement, however, directs us to move the decimal point three places to the right, or if we prefer, multiply the reported number by 1,000. Either step yields about 19 million bpd. Note that the same figure might also be stated as 19.1 million bpd. The "million" in this statement directs us to move the decimal six places to the right, or if we prefer, multiply the reported number by 1,000,000. This is undoubtedly obvious to the experienced statistics-hound, but it is worth reminding us all to pay careful attention to the reporting units. For example, if we simply read the data as 19,180.126 bpd, we might mistakenly report that in 2010 the United States consumed less than 20,000 bpd, a woefully incorrect number.

Of the 19 million bpd consumed in the United States in 2010, 9.2 million bpd was imported (United States Energy Information Administration, 2011e). **The United States had to import about 48% of its crude oil in 2010**. What often surprises people is the fact that Canada, not Saudi Arabia, is the country from which we import most of our petroleum. **About 21% of the crude oil imported into the United States in 2010 came from Canada, compared to 11% from Saudi Arabia.** Other significant countries from which the United States imported oil in 2010 are Mexico (13%), Nigeria (11%) and Venezuela (10%). Another often revealing fact is that the United States, with about 312 million people on a planet with over 6.9 billion (United States Census Bureau, 2011), consumed about 22% of the world's total petroleum consumption in 2010 (United States Energy Information Administration, 2011f). **With only 4.5% of the world's population, the United States consumed 22% of the world's petroleum in 2010.**

The United States has an oil consumption habit that is not satisfied with the country's domestic proven reserves. Crude oil production in the United States peaked in 1970 at about 9.6 million bpd. 41 years later, in 2010, only 5.4 million bpd was produced, a 56% decrease (United States Energy Information Administration, 2011g). This declining production is cause for concern because petroleum and its associated products, from gasoline to plastics to fertilizers, are such integral parts of the United States economy.

Calls to increase domestic oil production are constant and become especially loud when oil prices rise and consumers experience a price increase at the gasoline

pump. Advocates of increasing domestic oil extraction point to the fact that vast regions in the United States remain untapped not because there is a lack of oil, but rather because there is a lack of political will coupled with onerous permitting regulations (Murkowski, 2011). Advocates also point to a recent report by the United States Congressional Research Service that states that the country is endowed with about 163 billion barrels of oil, a number compiled from proven reserves as well as undiscovered but technically recoverable oil resources from both on and offshore basins (Whitney, Behrens, & Glover, 2010, p. 10).

In contrast, critics of increased domestic oil drilling argue that even if all restrictions and permitting regulations were lifted, any resultant oil gusher would in reality be only a trickle. The reported 163 billion barrels of recoverable oil, for example, would provide only about 23 years of oil to the United States at our current consumption volume if no foreign oil was imported. The 23 year value is obtained by dividing 163 billion barrels by the country's current annual consumption, about 7 billion barrels.

So where does the United States stand in terms of oil production going into the future? In short, both advocates and critics of increasing domestic U.S oil production have valid, factually-based arguments. Before taking a side, however, and arguing for or against increased domestic drilling, it is important that we acquire some important facts and thereby argue from a position of sound reasoning rather than passionate hyperbole.

As of 2012, for example, advances in drilling and extraction technology have "been instrumental in reversing the 2-decade-long decline in U.S. oil production" (Kerr, 2012, p. 522). The decline reversed in 2009 when about 8% more domestic oil was produced compared to 2008. Increased production continued in 2010, albeit modestly, with nearly 2% more oil produced compared to 2009 (United States Energy Information Administration, 2012). These production values should be tempered with the understanding

Country	Thousands of Bbls/day
Canada	1,970
Mexico	1,152
Saudi Arabia (OPEC member)	1,082
Nigeria (OPEC member)	983
Venezuela (OPEC member)	912

Table 3-3, Top Sources of Imported Petroleum to the United States in 2010

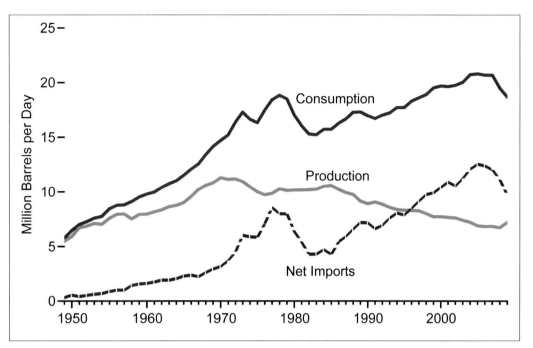

Figure 3-10, United States Oil Consumption, Production & Imports, 1949-2009, *Credit/Courtesy of United States Energy Information Administration*

that drilling depths and costs are both increasing. For example, the average depth of crude oil exploratory and developmental wells in 2000 was 4,593 feet, compared to 5,094 feet in 2007 (United States Energy Information Administration, 2012b). Also, the nominal cost per crude oil well in 2000 was $593,000, compared to $4,000,000 in 2007 (United States Energy Information Administration, 2012c). So, the United States has significant domestic oil supplies and these supplies are taking longer, and becoming more expensive, to extract.

While 31 states produce crude oil, Texas, Alaska and California respectively produced the most petroleum in 2010 (United States Energy Information Administration, 2011h). This makes sense given that these three states also contain significant proven reserves. For example, in 2009 proven reserves in Texas totaled 5 billion barrels, Alaska totaled 3.5 billion barrels and California totaled 2.7 billion barrels (United States Energy Information Administration, 2010).

While the above three states annually experience

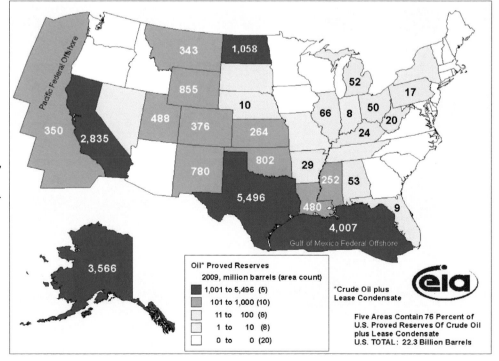

Figure 3-11, United States Crude Oil Proven Reserves, 2009, *Credit/Courtesy of United States Energy Information Administration*

steady or diminishing reserves, North Dakota is proceeding in the opposite direction. In 2004 the state had only 389 million barrels of proved reserves. In 2009, by contrast, it contained over 1 billion. This surprising leap is due to the recently assessed and tapped Bakken Formation located in North Dakota and Montana. In 2008, the United States Geologic Survey (USGS) announced that this formation contained up to 4.3 billion barrels of recoverable oil (United States Geological Survey, 2008). There is likely more because new drilling techniques are opening ever-greater portions of the basin to extraction efforts, and new scientific and technical information suggest that an increase in proven reserves is warranted. The USGS begins its new assessment of the Bakken Formation in October, 2011 and expects to complete it within two years (United States Geolical Survey, 2011).

Domestic production aside, when we think of oil the Organization of Petroleum Exporting Countries (OPEC) comes to mind. OPEC is a cartel—an agreement among otherwise competing countries to exert control over a marketplace. **OPEC's 12 member countries agree to coordinate their oil production in order to limit supply and thereby manipulate the price of crude oil in the global marketplace.** OPEC member countries are Algeria, Angola, Ecuador, Iran, Iraq, Kuwait, Libya, Nigeria, Qatar, Saudi Arabia, United Arab Emirates and Venezuela (Organization of Petroleum Exporting Countries, 2011). In 2010, OPEC member countries supplied the United States with 4.5 million bpd of crude oil, or about 49% of all crude oil imports that year (United States Energy Information Administration, 2011i).

OPEC's oil production is established by monthly production quotas (sometimes called "targets"), measured in bpd. When gasoline prices are high, consumers sometimes see these quotas as OPEC's attempt to grab every last penny from western world consumers. Not only would this strategy fail, however, it is counterproductive to the long-term stability of OPEC member countries. Rather, production quotas are intended to provide enough petroleum for consumers and enough revenue for producers so that each side avoids wild price swings. This is arguably preferable to the boom-and-bust cycle that characterizes other commodities, where planning future expenditures or revenues is next to impossible because of market volatility. On the other hand, the cartel is undoubtedly engaged in anti-competitive market practices which is anathema to the economic philosophy of much of the western world (consider, for example, the forced breakup of Standard Oil in 1911).

In 2010, total world oil production was 86.7 million bpd (United States Energy Information Administration, 2011j), and of this total, OPEC supplied 29.7 million bpd (United States Energy Information Administration, 2011k). **In 2010, OPEC member countries supplied about 34% of the world's crude oil.** Due to geology, each OPEC member country produces different amount of crude oil. For example, in 2010 Ecuador and Qatar produced 490,000 and 850,000 bpd respectively (relatively low numbers), while Kuwait and Iran produced 2.28 and 3.75 million bpd respectively (relatively high numbers). Saudi Arabia, on the other hand, is the undisputed king of production, averaging an annual production of about 8.7 million bpd over the last eleven years. Saudi Arabia's massive production ability and surplus capacity allow it to be the "swing producer" of OPEC, releasing more oil into the marketplace when supplies are tight, and less when supplies are abundant. In 2010, for example, it could have produced another 3.41 million bpd (United States Energy Information Administration, 2011k). If it had gone ahead and produced this, it would have undoubtedly lowered prices as more crude oil supply entered the world marketplace.

Year	Million Bbls/day
2000	28.19
2001	27.35
2002	25.57
2003	27.20
2004	29.50
2005	30.83
2006	30.45
2007	30.06
2008	31.27
2009	29.10
2010	29.77

Table 3-4, OPEC Crude Oil Production, 2000-2010

Saudi Arabia out-produces all other OPEC members because it has 260 billion barrels of proven oil reserves, of which an estimated 70 billion are contained in the world's largest oil field, Ghawar, measuring 1,260 square miles (United States Energy Information Administration, 2011l). There is some discussion today, however, that after more than 50 years of production Ghawar's oil abundance may be diminishing. This is difficult to assess because the Saudi government is secretive about production numbers and swears that the field is as productive as ever. Various assessment techniques have, however, convinced some analysts that all is not well with the field. In 2007, for example, it was discovered that the number of oil rigs had tripled from 2005-2007, and that this might indicate a rush to replace areas with diminishing production (Hamilton, 2007, p. 42). This finding is coupled with satellite imagery verifying a recent 10% increase in drilling sites in Ghawar, again raising the possibility of diminishing petroleum reserves in this field (King, 2008). It should be noted, however, that until verifiably transparent reporting occurs, these findings remain conjecture.

Figure 3-12, Ghawar Field, Saudi Arabia

Oil is a wasting asset—it is consumed faster than it is replenished. This does not mean, however, that "the world is going to run out of oil." This statement makes for juicy media headlines, but is erroneous and counterproductive hyperbole that is potentially dangerous. Present-day "abandoned" wells, for example, retain as much as 78% of their oil (Sandrea & Sandrea, 2007). Decades ago, this percentage was higher. As each year passes, increasingly sophisticated extraction technology matures and thereby brings previously inaccessible oil to market. Also, as oil inevitably becomes scarce, prices accordingly rise. Rising prices spur new exploration, conservation, and use of alternative fuels. The public grows into these changes until they become part of the mainstream. Witness the rise of hybrid vehicles. In the year 2000, these dual gas-electric vehicles existed only as showpieces. Today they are readily available on the mass market and easily spotted on roads all across the United States. **Rather than the world "running out of oil," there will gradually come a time when oil extraction is no longer economically feasible.** People argue about the date, duration and effect of this future time, and with good reason given the many unknown variables that must be wrestled into any proven reserve projection.

The term "peak oil" refers to the year and volume of maximum petroleum extraction. It is used in reference to the now famous paper presented in 1956 by M. King Hubbert, who accurately predicted that the United States would reach peak oil in 1970 (Hubbert, 1956, pp. 22-24). Since that year petroleum production has declined from 3.5 billion Bbls in 1970 to 1.9 billion Bbls in 2010 (United States Energy Information Administration, 2011m). This decline was partially slowed in the early 1980s as Alaskan oil first came to market.

Since confirmation of Hubbert's prediction for the United States, industry analysts and policy makers continually question when peak oil will occur for the entire planet. In 2007, the United States Government

Figure 3-13, Peak Oil in the United States, *Credit/Courtesy of United States Energy Information Administration*

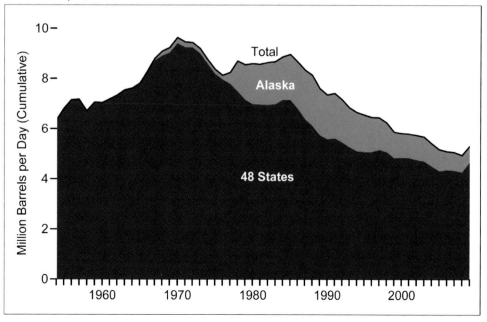

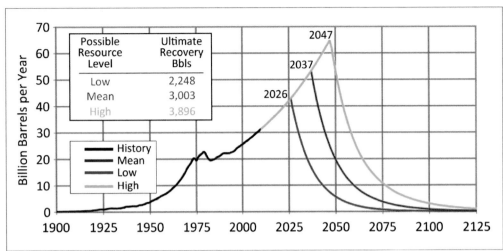

Figure 3-14, Peak Oil Scenarios for the Earth, *Credit /Courtesy of United States Energy Information Administration (adapted)*

Accountability Office (GAO) issued a report examining when global oil production will peak, how improving transportation technology might mitigate the consequences of the peak, and how to reduce uncertainties about the timing of the peak (United States Government Accountability Office, 2007). Other public agencies have also contributed their expertise to the study of global peak oil, including the United States Geologic Survey (USGS) as well as the Energy Information Administration (EIA). While scenarios differ depending on variables such as improving extraction technologies, new oil field discoveries, investment decisions, political issues and environmental concerns, average findings are that global crude oil production is "expected to peak in 2037 at a volume of 53.2 billion barrels per year" (Wood, Long, & Morehouse, 2004). As this date nears, prices are projected to rise to $125.00 per barrel by 2035 (United States Energy Information Administration, 2011n, p. 25).

Oil Wells

On land or offshore, oil wells are created and brought into production with an assortment of equipment unique to the industry. An oil derrick (the tallest and most visible structure) is a tall steel frame that provides the necessary support to lift and control the variety of steel piping used throughout the drilling and production phases. A drill string is a long column of hollow steel pipe made up of many hundreds of individual, 30 foot-long segments, that are threaded together by roughnecks—men working on the drilling rig. **When drilling a well, the entire length of the drill string spins clockwise.** The drill string spins due to the torque provided by a rotary table, which is a circular table through which the pipe runs.

At the end of the drill string is the drill bit, a very com-

mon type being the "rotary tricone". This hardened steel bit has three cone structures at its head—hence "tricone"—and is simply threaded onto the end of the drill string. The cones spin because the bit spins, and the bit spins because the drill string spins. The three cones are not automated and do not spin on their own. As the bit spins, each cone rotates and thereby chips pieces of rock, called cuttings, out from the bottom of the wellbore (Hyne, 2001, pp. 258-260). Bits range in

Figure 3-15, The Mad Dog Platform, Offshore Oil Rig, *Credit/Courtesy of BP, © BP p.l.c.*

Figure 3-16, Rotary Table, *Credit/Courtesy of Frank Wallace*

diameter from about 5 to 36 inches, and different diameters are used at different stages of drilling. Larger diameter bits are used first, and as the wellbore deepens, increasingly smaller bits are employed. Thus a well's diameter shrinks with increasing depth.

The spinning bit creates rock cuttings which must be cleared from the bore hole. This is accomplished using drilling mud. Mud is pumped down the length of the drill pipe and out through small jets in the bit. It then flows back up the bore hole on the outside of the drill pipe—this space is called the annulus—thereby carrying the rock cuttings up for analysis and disposal. Geologists and chemists not only examine the rock cuttings, but also the mud for any signs of hydrocarbons or unexpected rock changes within the wellbore.

When imagining the drilling process, two things readily come to mind. First, the sheer length of the drill string is impressive. The average depth of exploratory and development crude oil wells in 2008 was 5,094 feet (United States Energy Information Administration, 2011o) (for comparison, realize that 1 miles is 5,280 feet). Second, the borehole diameter must be wider than the drill pipe. This is essential to clear the borehole of rock cuttings and later to "case" the well. **Casing a well means lining it with steel "casing" pipe and then permanently cementing that pipe to the borehole wall.** This is necessary to prevent the well walls from collapsing and to control the rate of resource extraction.

Oil production begins after the casing is perforated by firing charges at specific depths where mud and cutting analysis revealed petroleum concentrations. The firing charges not only open holes in the steel and concrete lining, they also penetrate and fracture the rock, thereby creating low-pressure pathways into which fluids move. Oil then flows in a controlled manner from the reservoir rock, through the cement casing, through the steel casing pipe, and finally into the well.

If there is enough pressure in the reservoir, the oil will

Figure 3-17, Drill Pipe, *Credit/Courtesy of BP, © BP p.l.c.*

flow up to the surface without pumping. This is the likely case in newly tapped wells where the reservoir contains enough natural pressure to push the fluids up to the surface. Most mature wells, however, do not have enough pressure and thus the oil must be artificially lifted from the well. This is done with a variety of techniques, including pressurizing a well with gas, using pumping rods, and installing down-hole pumps directly in the oil at the bottom of a production zone.

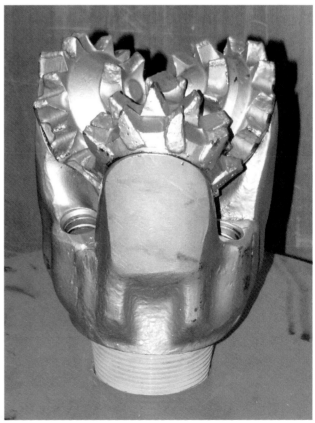

Figure 3-18, Tricone Drill Bit, Credit/Courtesy of Wikipedia

Ensuring that a high-pressure well does not get out of control and create a "gusher"—an out-of-control well literally gushing oil—requires that every well have a blowout preventer bolted to the top of the wellhead. **A blowout preventer is a massive, high-pressure steel pipe designed to quickly close a well. Blowout preventers are situated within a blowout preventer stack that contains hydraulic rams that can close around, or even cut across, a well pipe.** If an oil well gets out of control, blind shear rams—usually three separate rams per blowout preventer—can literally crush the well's steel pipe together and seal it at the wellhead, thereby stopping the flow of oil. Unfortunately, this is not what happened in 2010 when a blowout preventer failed to seal BP's Macondo well, resulting in many millions of gallons of oil spilled into the Gulf of Mexico (this event is explored in more detail later).

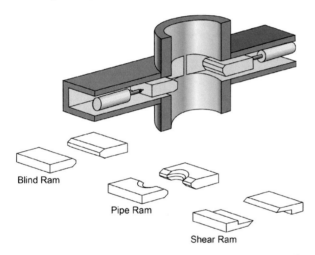

Figure 3-19, Model of a Blowout Preventer, *Credit/ Courtesy of Wikipedia (adapted)*

Oil wells are generally classified as "exploration," "appraisal" or "production" wells. Exploration wells are also called "wildcat" wells and occur in untapped and often remote areas where geologists suspect oil exists. These endeavors carry high financial risk, but also potentially high financial rewards. Appraisal wells occur after oil is discovered and are used to evaluate a reservoir's extent with regular step-out wells. These endeavors examine a reservoir's oil type, water and gas content, and porosity and permeability. Finally, production wells actually bring the oil to the surface in order to sell it on the oil market.

Pipelines and Tankers

About 170,000 miles of petroleum transmission pipelines exists in the United States (American Petroleum Institute, 2011). This network connects petroleum producers to refiners, and refiners to purchasers and retailers of petroleum products. Pipelines are the cheapest way to move petroleum on land, the other methods being rail lines and trucks. This cost effectiveness is frequently publicized by the Association of Oil Pipe Lines (AOPL), a lobbying group located in Washington, D.C. The AOPL states that it costs about $1.00 to transport a barrel of petroleum products from Houston to New York harbor, which only adds an additional 2.5 cents to the price of a gallon of gasoline at a local gas station (Association of Oil Pipelines, 2011).

Modern pipelines are 48 inches in diameter and carry much more than just crude oil. A single pipeline, for instance, might carry jet fuel, regular gasoline, premium gasoline, diesel and dozens of other refined petroleum products. This technique is called "batching" and it is an efficient way to use a pipeline even though some undesirable mixing (called "transmix") occurs during transport. Transmix is ultimately separated out into its appropriate refined products at numerous transmix processing plants before final delivery (Kinder Morgan, 2011).

Pipelines in the United States are privately owned, for-profit enterprises that are regulated as common carriers by the Federal Energy Regulatory Commission (FERC). A "common carrier" is a company that transports products without favoring one firm over another, and charges all customers the same, transparent rates. Petroleum pipelines in the United States have been regulated as common carriers since 1906 when the United States government passed the Hepburn Act in an effort to limit Standard Oil's market dominance over independent oil and refining companies. FERC regulates pipelines as part of its mission to ensure that United States consumers get reliable energy supplies delivered at reasonable costs. In this capacity FERC reviews pipeline tariffs, monitors the performance of pipelines and the compliance of pipeline companies, and conducts pipeline depreciation studies (Federal Energy Regulatory Commission, 2010).

For an example of a pipeline company and its common carrier rates, consider Plains All American Pipeline (PAA). PAA is a publically traded pipeline company that owns 17,000 miles of crude oil and refined product pipelines. The tariffs PAA charges to move crude oil through its pipelines are transparent— they are posted online and are the same for every customer. For instance, to move crude petroleum from St. James, Louisiana, to Patoka, Illinois, PAA charges 54.53 cents per barrel (Plains All American Pipeline,

Plains Pipeline, L.P. F.E.R.C. No. 87.1.0

Figure 3-20, Pipeline Tariff from St. James, Louisiana, to Patoka, Illinois, *Credit/ Courtesy of Plains All American Pipeline*

LIST OF POINTS FROM AND TO WHICH RATES APPLY AND RATES ON CRUDE PETROLEUM IN CENTS PER BARREL OF 42 UNITED STATES GALLONS

ROUTE NO.	ORIGIN	DESTINATION	THROUGH RATE TO ESTABLISHED DESTINATION
02	St. James, St. James Parish, Louisiana	Patoka, Marion County, Illinois	[I] 58.28
03	Liberty, Amite County, Mississippi		[I] 58.28

2010). To put this into another perspective, consider that it costs any customer about 55 cents to move 42 gallons of crude oil 688 miles, or about 1.3 cents to move one gallon 688 miles. Stated in this manner, it is perhaps easier to understand the cost-effectiveness of pipelines.

Perhaps the best known pipeline in the United States is the Trans-Alaska Pipeline System (TAPS). This pipeline was designed and constructed by the Alyeska Pipeline Service Company which continues to operate and maintain the line today (Alyeska is an Aleut word meaning "mainland"). The company is a consortium, of which BP Pipelines (Alaska) Inc. is the largest stakeholder, owning 46.93% (Alyeska Pipeline Service Company, 2011). Pipeline construction took about three years, beginning on the heels of the 1970s oil crisis, and finishing in 1977. Crude oil moves south 800 miles from Prudhoe Bay, located on the Arctic Ocean, through 12 pumping stations (abbreviated PS 1, PS 2, etc.), to Valdez, Alaska. The 48 inch pipeline carries about 17% of the country's domestic petroleum supply and runs both above and below ground while navigating across Alaska's extreme topography and climate. Transit time for a barrel of oil takes 9 days from PS1 to the Valdez Marine Terminal (VMT) (Alyeska Pipeline Service Company, 2004). Since 1977 the pipeline has moved over 16 billion barrels of oil (16,274,848,255 as of 2010), with peak throughput occurring in 1988 with 744,107,855 barrels (Alyeska Pipeline Service Company, 2011b).

When pipelines are unavailable, tankers transport oil from producers to consumers. Oil tankers are generally classified by their deadweight tonnage (DWT). DWT is the weight of the cargo (the crude oil) and the fuel (normally diesel) needed to transport that cargo. The industry workhorse is the Very Large Crude Carrier (VLCC), with a DWT of between 160,000 – 319,999 long tons DWT (a "long ton" is 2,240 pounds) (London Tanker Brokers' Panel, 2011). **The average 300,000 ton VLCC carries about 2,000,000 barrels of oil** (Center for Tankship Excellence, 2011). Bigger vessels maximize economies of scale—it is cheaper per unit when moving greater quantities of oil. Thus the industry generally prefers larger ships. There are limits,

Figure 3-21, Trans-Alaska Pipeline & Pump Stations Map

Figure 3-22, Building the Trans-Alaska Oil Pipeline, *Credit/Courtesy of Bob Heims, United States Army Corps of Engineers, Alaska District*

Figure 3-23, Trans-Alaska Oil Pipeline, *Credit/Courtesy of BP, © BP p.l.c*

Figure 3-24, Trans-Alaska Oil Pipeline, *Credit/Courtesy of United States Bureau of Land Management*

Figure 3-25, Tanker Loading at the Valdez Oil Terminal, *Credit/Courtesy of Joint Pipeline Office*

however, since tanker size can easily exceed port facility limits, and larger vessels may draw too deeply for some harbors. Ultra Large Crude Carriers (ULCC), for example, have a DWT of up to 549,999 long tons. The largest ever built was the *Seawise Giant*, measuring over 1,500 feet long and drawing 81 feet of water. This ship was built in 1979 and after 30 years of service was

scrapped in India in 2010 under the name *Mont* (Maritime Connector, 2011). Some VLCCs are also too large for a country's port facilities and so require offshore mooring systems to transfer petroleum. In the United States, the Louisiana Offshore Oil Port (LOOP) is an example, where the largest takers tie up to a "single point moor" (SPM) and offload petroleum via flexible hoses (Louisiana Department of Transportation and Development, 2011).

Petroleum's Virtues and Vices

It is difficult to overstate the importance of petroleum—the world's economy is absolutely dependent upon this fossil fuel. A completely integrated infrastructure exists in the developed world to transport, refine and distribute petroleum products. Finished goods, many of which are created from petroleum, move from manufacturers to consumers via the same basic resource. When oil prices rise, national econo-

Figure 3-26, VLCC British Pioneer, *Credit/Courtesy of BP, © BP p.l.c*

Figure 3-27, VLCC British Progress, *Credit/Courtesy of BP, © BP p.l.c*

Figure 3-28, VLCC AbQaiq, *Credit/Courtesy of United States Navy News Service*

Figure 3-29, Single Point Mooring Facility, *Credit/ Courtesy of Wikimedia*

mies often suffer as higher energy prices make nearly everything else more expensive.

This utter dependency on petroleum often provokes considerable anxiety in energy security analysts as they conduct long-term supply scenarios. Consider that state-owned, national oil companies (NOCs) own the largest proven reserves of petroleum on the planet. Companies such as the National Iranian Oil Company (NIOC), Saudi Aramco, the China National Petroleum Corporation (CNP), Petroleum of Venezuela (PDVSA) and the Nigerian National Petroleum Corporation (NNPC) are just a few examples. In terms of reserves, these NOCs hold 77% of the world's oil and thus dwarf publically traded and better recognized oil companies such as Royal Dutch Shell, Chevron, Exxon Mobil and BP (Helman, 2010). Many NOCs are OPEC members, and many are located in the politically-volatile Middle East. These facts should at least give us pause when considering future access to petroleum because NOCs are beholden to national interests and not necessarily to markets.

Energy security analysis also watch the rapid economic growth in China and India and the consequential rising demand for vehicles, and thus for petroleum. They note that most of the growth in petroleum consumption since the turn of this century has occurred in countries outside the Organization for Economic Co-operation and Development (OECD) (Ruhl, 2010). Envisioning the next 30 to 50 years, therefore, it is reasonable to see tightening supplies, rising prices, and states—not the free market—playing an increasingly active role, or even outright determining, the extraction and allocation of this finite resource. This is indeed already happening. Take, for an example, one of China's state-run energy firms, China National Offshore Oil Corporation (CNOOC) recently locking in oil extraction rights in Sudan, Ethiopia and Nigeria (Economist, 2006).

Immediate panic is unwarranted, however. There is still a lot of petroleum out there, and this, along with

synthetics and ever-increasing efficiencies, will dampen the inevitable tightening of supplies. For example, Canada's Athabasca oil sands hold 170 billion Bbls of proven reserves (Government of Alberta, 2011). This is second only to Saudi Arabia's proven oil reserves of about 260 billion Bbls. Thus Canada, already an important source of imported oil for the United States, will undoubtedly remain so and even grow in significance as the decades pass. There is also speculation the recently tapped Bakken Formation in North Dakota and Montana may hold over 20 billion barrels—this is something that the USGS is presently investigating.

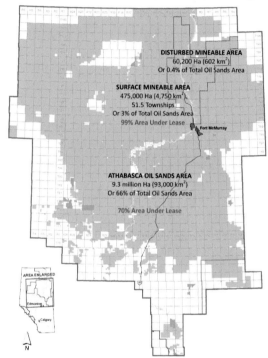

Figure 3-30, Athabasca Oil Sands, Canada, *Credit/ Courtesy of Government of Alberta*

The United States also contains massive oil shale deposits, which are unconventional sources of petroleum but may contain approximately 1.8 trillion barrels of economically recoverable oil (United States Department of Energy, 2007). Since the United States consumes about 20 million bpd, in one year approximately 7,300,000,000 Bbls are consumed. A simplistic future scenario might find oil supplies tightening and the country tapping its oil shale reserves to satisfy 50% of its demand (3,650,000,000 Bbls per year). In this case, the oil shale reserves R/P ratio equation yields this resource lasting 500 years ($1.8 \times 10^{12} / 3.6 \times 10^9 = .5 \times 10^3 = 500$ years). Adding comfort to nervous energy analysts is the fact that the United States government controls access to 1.9 million of acres of public land that is commercially attractive for oil shale development (United States Department of the Interior, 2011).

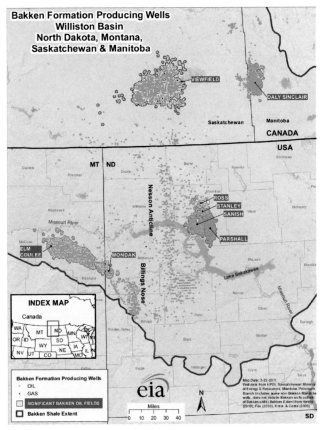

Bakken Formation Producing Wells
Williston Basin
North Dakota, Montana,
Saskatchewan & Manitoba

INDEX MAP
Canada

Bakken Formation Producing Wells
· OIL
· GAS
☐ SIGNIFICANT BAKKEN OIL FIELDS
☐ Bakken Shale Extent

eia

Miles
0 10 20 30 40

Figure 3-31, The Bakken Formation, *Credit/Courtesy of United States Energy Information Administration*

Like coal, burning petroleum products releases carbon dioxide (CO_2) into the atmosphere. Climate scientists are especially concerned about gasoline because millions of vehicles are on the road and the demand for gasoline keeps rising. **Every gallon of motor gasoline that is burned emits 8.91 kilograms (19.64 pounds)**

of CO_2 into the atmosphere (United States Energy Information Administration, 2011p). The United States is by far the largest contributor of CO_2 emissions from the consumption of petroleum because of its heavy reliance on gasoline-fueled passenger vehicles— about 137 million as of 2008 (United States Department of Transportation, 2010). In 2009, 2,319 million metric tons of CO_2 was emitted from the consumption of petroleum in the United States. China followed with 1,060 million metric tons. A distant third rank was Japan, with 511 million metric tons (United States Energy Information Administration, 2011q).

Mitigating CO_2 emission volume requires both policy and technological initiatives. On the policy side, federal lawmakers can mandate higher fuel efficiency standards to reduce the number of gallons of gasoline consumed. In 1975, for example, the United States government established the Corporate Average Fuel Economy (CAFE) standard, mandating a fleet average of 27.5 miles per gallon (mpg) by 1985 (National Highway Trafic Safety Administration, 2011). Another policy effort was the Energy Independence and Security Act of 2007. This legislation, signed into law December of 2007 by President George W. Bush, requires a fleet average of 35 mpg for the 2020 model year (110th Congress, 2007). More recently, President Barak Obama's administration increased fleet average fuel economy to 35.5 mpg by 2016. This latest, more aggressive timeline to increase fuel economy came about only after unprecedented collaboration amongst the following stakeholders: the Department of Transportation (DOT), the Environmental Protection Agency (EPA), auto manufacturers, the United Auto Workers, leaders

Figure 3-32, The 2012 Toyota Prius, An Exemplary Hybrid Electric Vehicle (HEV), *Credit/ Courtesy of Toyota Motor Sales, U.S.A., Inc.*

of the environmental community, the State of California and other state governments (The White House-Office of the Press Secretary, 2009).

Progressive policies such as these motivate improvements in vehicle technology which improve efficiency, reduce gasoline consumption and therefore decrease CO_2 emissions. The latest generation of hybrid electric vehicles (HEVs) illustrates this fact, with many models already exceeding the 2016 fuel economy targets. These vehicles combine the best qualities of internal combustion and electric engines because there are literally two engines in a single hybrid vehicle. HEV drivers, therefore, enjoy the extended range available from the internal combustion engine that runs on energy-dense gasoline, as well as the fuel economy and low carbon emissions from the electric engine. Both engines operate the vehicle, with an onboard computer determining their most efficient use (Union of Concerned Scientists, 2010). While consumer adoption of HEVs is currently still quite modest, a growing demand clearly exists that auto-makers are satisfying with increased models and availability. As the HEV fleet on United States roads increases over time, significant reductions in gasoline use, and thus CO_2 emissions per mile, are expected, depending on market penetration rates (Bandivadekar, et al., 2008, pp. 114-125).

Aside from energy security and carbon dioxide emissions, perhaps the most noticeable vice of petroleum is the fact that it can spill. Few energy-related incidents disturb people more than seeing crude oil spilling into the environment—especially into the ocean. Oil spills kill wildlife, ruin coastlines and destroy the livelihood of thousands of people in coastal communities. **As a society dependent upon oil, we have accepted the environmental and economic risks associated with offshore oil drilling, and transporting oil across oceans.** Most of the time there are no spills, but Mother Nature—and more often human error—make spills inevitable. The latest oil spills that have occurred in the United States are the 1989 Exxon Valdez spill in Prince William Sound, Alaska, and the 2010 Deepwater Horizon spill in the Gulf of Mexico.

Recall from the pipeline discussion that the Trans-Alaska Pipeline System (TAPS) moves oil from Prudhoe Bay to Valdez. Locating the terminal for TAPS at Valdez seems logical since Valdez is the northern-most ice-free port in the United States. Valdez is situated on the Bay of Valdez, a wide expanse of restricted water tucked into the northern extremity of Prince William Sound. Accessing the Bay, however, requires tankers navigate a single fiord—the "Valdez Narrows"—that is less than 1 mile across. Fully laden tankers must tran-

sit this same fiord, and then, about 16 miles further south, navigate past Bligh Reef, a treacherously shallow area about 2 miles west of Bligh Island. It was here on March 24, 1989 that the Exxon Valdez tanker ran aground and spilled 11 million gallons—about 261,000 barrels—of crude oil into the Sound (National Oceanic and Atmospheric Administration, 2011). Cleaning up the oil took years, and litigation lasted over twenty. The spill prompted state and federal legislation requiring greater safety oversight of tankers transporting crude oil within United States waters. Part of Alyeska Pipeline Service Company's response was to establish the Service Company's Ship Escort/Response Vessel System (SERVS), which, among other responsibilities, escorts laden tankers from the VMT through Prince William Sound (Alyeska Pipeline Service Company, 2009).

Figure 3-33, The Exxon Valdez Leaking Oil into Prince William Sound, *Credit/Courtesy of National Oceanic & Atmospheric Administration*

The Exxon Valdez spill also prompted federal legislation—the Oil Pollution Act of 1990—mandating double hulls for oil tankers trading within United States waters (the Exxon Valdez was a single hulled vessel) (United States Senate Committee on Environment and Public Works, 2000, pp. 266-267). Double hulls are considered safer in the event of some collisions because any petroleum will hopefully remain aboard the vessel, securely contained behind the second hull. The mandated 2 meter spacing between hulls is also used for ballast space, thus eliminating pollution that

Figure 3-34, Cleaning Up After the Exxon Valdez Oil Spill, *Credit/Courtesy of National Oceanic & Atmospheric Administration*

once came from filling empty oil tanks with seawater. The International Convention for the Prevention of Pollution from Ships, 1973, as modified by the Protocol of 1978 relating thereto (MARPOL), also adopted the double hulled standard for oil tankers. Thus, as of 2010, double hulls are the industry standard (International Maritime Organization, 2011).

The 261,000 barrels of oil spilled by the Exxon Valdez in 1989 is dwarfed by the 4,000,000 barrels of oil spilled from the Macondo well into the Gulf of Mexico in April, 2010. This incident occurred after the Deepwater Horizon—a floating offshore oil rig—caught fire

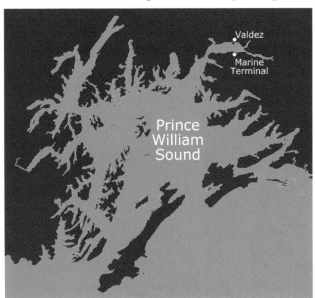

Figure 3-35, Prince William Sound

and exploded after the well's blowout preventer failed to control a sudden rush of escaping methane. Eleven people lost their lives during the event, and after about 36 hours the $560 million dollar rig sank to the ocean's bottom (National Commission on the BP Deepwater Horizon Oil Spill and Offshore Drilling, 2011, p. 1). The scope of the disaster widened after BP, the rig's operator, realized that the wellhead—about one mile beneath the ocean's surface—was uncontrollably gushing oil into the sea. 87 frantic days passed, with engineers working around the clock, before the well was finally sealed (BP, 2011).

Oil slicks during the late spring and summer of 2010 forced the closure of 88,522 square miles of the Gulf to any commercial or recreational fishing, and coastlines in Louisiana, Mississippi, Alabama and Florida witnessed surface oil and tarballs washing ashore for weeks (National Commission on the BP Deepwater Horizon Oil Spill and Offshore Drilling, 2011, pp. 187-198). Income from seafood and tourism dollars dried up, and executives of BP, Transocean (the rig's owner), and Halliburton (the firm in charge of casing the well), were all brought before congress and the courts to determine the extent of culpability. While these hearings proceeded, scientists found that the bulk of the Macondo well oil rose less than 2,000 feet off the ocean floor where it then coalesced into a massive plume 22-miles long by 1.2 miles wide and 650 feet high, 3,000 feet below the ocean's surface (National Oceanic and Atmospheric Administration,

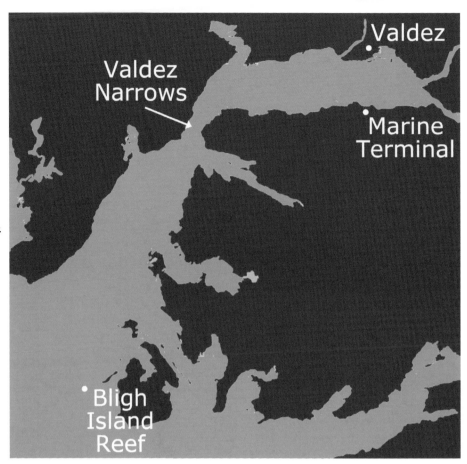

Figure 3-36, Valdez Narrows & Bligh Island Reef

2010). Surprisingly effective microbial digestion of the oil ensued over the course of a year, and on April 19th, 2011, the National Oceanic and Atmospheric Administration (NOAA) lifted the last restriction on fishing and recreation, thereby opening the entire Gulf to normal activity. Extensive testing by the NOAA and the United States Food and Drug Administration (FDA) determined that safe, near zero levels of oil and/or chemical dispersant where present in finfish and shrimp throughout the previously restricted Gulf waters (Gulf Coast Ecosystem Restoration Task Force, 2011).

Important Ideas:

- While pockets or cavities of petroleum do exist, the vast majority of oil resources are locked inside solid rock.

Figure 3-37, The Deepwater Horizon Ablaze, *Credit/Courtesy of United States Chemical Safety and Hazard Investigation Board*

Figure 3-38, The Failed Blowout Preventer from BP's Mocando Well, *Credit/Courtesy of Lawrence Livermore National Laboratory*

- Porosity is the percentage of void spaces inside a rock.

- Permeability is a measure of a rock's ability to transmit a fluid.

- An impermeable "caprock" stops the migration of gas and oil to the earth's surface.

- The oil window is between 1.5 to 3 miles deep where sufficient temperatures and pressures exist to transform organic-rich sediment into oil.

- By 1911 the United States Supreme Court ruled that Standard Oil would disassemble into 34 separate companies.

- Crude oils are labeled "light", "intermediate" or "heavy" depending on their density, which is mathematically calculated using a simple

Figure 3-39, Deepwater Horizon Oil Slick in Gulf of Mexico, *Credit/Courtesy of NASA Earth Observatory*

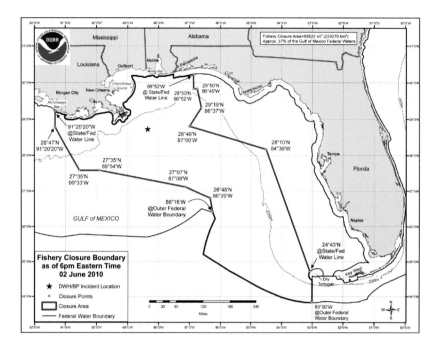

Figure 3-40, Gulf of Mexico Fishery Closure Map, June 2, 2010, *Credit/ Courtesy of National Oceanic and Atmospheric Administration*

formula established by the American Petroleum Institute (API).

- Sweet crude oil contains less than 0.5% sulfur, while sour crude oil contains greater than 0.5% sulfur.

- Boilers heat the crude to over 700 °F. As the oil's temperature rises, various liquid hydrocarbons change state and become gas.

- About 45% (19 gallons) of a barrel of crude oil becomes gasoline.

- Oil consumption and production is often measured in "barrels per day" and is abbreviated "bpd".

- The United States consumed about 19 million bpd in 2010.

- The United States had to import about 48% of its crude oil in 2010.

- About 21% of the crude oil imported into the United States in 2010 came from Canada, compared to 11% from Saudi Arabia.

- With only 4.5% of the world's population, the United States consumed 22% of the world's petroleum in 2010.

- While 31 states produce crude oil, Texas, Alaska and California respectively produced the most petroleum in 2010.

- OPEC's 12 member countries agree to coordinate their oil production in order to limit supply and thereby manipulate the price of crude oil in the global marketplace.

- OPEC member countries supply about 34% of the world's crude oil.

- Saudi Arabia out-produces all other OPEC members because it has 260 billion barrels of proven oil reserves, of which an estimated 70 billion are contained in the world's largest oil field, Ghawar, measuring 1,260 square miles.

- Rather than the world "running out of oil," there will gradually come a time when oil extraction is no longer economically feasible.

- The term "peak oil" refers to the year and volume of maximum petroleum extraction.

- When drilling a well, the entire length of the drill string spins clockwise.

- Casing a well means lining it with steel "casing" pipe and then permanently cementing that pipe to the bore hole wall.

- A blowout preventer is a massive, high-pressure steel pipe designed to quickly close a well. Blowout preventers are situated within a blowout preventer stack that contains hydraulic rams that can close around, or even cut across, a well pipe.

- About 170,000 miles of petroleum transmission pipelines exists in the United States.

- Pipelines in the United States are privately owned, for-profit enterprises that are regulated as common carriers by the Federal Energy Regulatory Commission (FERC).

- The average 300,000 ton VLCC carries about 2,000,000 barrels of oil.

- Every gallon of motor gasoline that is burned emits 8.91 kilograms (19.64 pounds) of CO_2 into the atmosphere.

- As a society dependent upon oil, we have accepted the environmental and economic risks associated with offshore oil drilling, and transporting oil across oceans.

Works Cited

110th Congress. (2007, December 19). *Energy Independence and Security Act.* Retrieved September 28, 2011, from Public Law 110–140—DEC. 19, 2007: http://frwebgate.access.gpo.gov/cgi-bin/getdoc.cgi?dbname=110_cong_public_laws&docid=f:publ140.110.pdf

Standard Oil Co. of New Jersey v. United States, 221 (United States Supreme Court May 15, 1910).

Alyeska Pipeline Service Company. (2004, June 23). *Pipeline Facts, Crude Oil.* Retrieved September 28, 2011, from http://www.alyeska-pipe.com/Pipelinefacts/CrudeOil.html

Alyeska Pipeline Service Company. (2009, March 30). *Pipeline Facts.* Retrieved September 28, 2011, from SERVS: http://www.alyeska-pipe.com/Pipelinefacts/SERVS.html

Alyeska Pipeline Service Company. (2011, July 8). *About Us.* Retrieved September 28, 2011, from http://www.alyeska-pipe.com/about.html

Alyeska Pipeline Service Company. (2011b, February 3). *Pipeline Facts, Throughput.* Retrieved September 28, 2011, from Cumulative throughput at end of year & Throughput, total per year: http://www.alyeska-pipe.com/Pipelinefacts/Throughput.html

American Petroleum Institute. (2011, September 1). *About Oil and Natural Gas, Industry Sectors, Pipeline.* Retrieved September 27, 2011, from Pipeline: http://www.api.org/aboutoilgas/sectors/pipeline/index.cfm

Association of Oil Pipelines. (2011). *About Pipelines.* Retrieved September 27, 2011, from Why Pipelines: http://www.aopl.org/aboutPipelines/

Bandivadekar, A., Bodek, K., Cheah, L., Evans, C., Groode, T., Heywood, J., et al. (2008, July). *On the Road in 2035: Reducing Transportation's Petroleum Consumption and GHG Emissions.* Retrieved December 1, 2011, from Laboratory for Energy and the Environment, Massachusetts Institute of Technology: http://web.mit.edu/sloan-auto-lab/research/beforeh2/otr2035/On%20the%20Road%20in%202035_MIT_July%202008.pdf

BP. (2011). *Deepwater Horizon Accident.* Retrieved September 28, 2011, from http://www.bp.com/sectiongenericarticle800.do?categoryId=9036575&contentId=7067541

Center for Tankship Excellence. (2011). *CTX Glossary.* Retrieved September 28, 2011, from VLCC: http://www.c4tx.org/ctx/gen/glossary.html#lightweight

Chevron U.S.A. (2010). *Chevron Pascagoula Refinery.* Retrieved September 26, 2011, from Processing & Refining Crude Oil: http://pascagoula.chevron.com/home/abouttherefinery/whatwedo/processingandrefining.aspx

Chevron U.S.A. (2011). *What We Do: The Refining Process.* Retrieved September 26, 2011, from El Segundo Refinery: http://www.chevron.com/products/sitelets/elsegundo/about/what_we_do.aspx

Economist. (2006, January 21). No Questions Asked. *Economist, 378*(8461), 43-44.

Federal Energy Regulatory Commission. (2010, June 28). *About FERC, Office of Energy Market Regulation* . Retrieved September 27, 2011, from Pipeline Regulation: http://www.ferc.gov/about/about.asp

Government of Alberta. (2011, September 9). *About Oil Sands.* Retrieved September 28, 2011, from Facts and Statistics: http://www.energy.alberta.ca/OilSands/791.asp

Gulf Coast Ecosystem Restoration Task Force. (2011, April 19). *Last Fisheries Re-Opening Today.* Retrieved September 28, 2011, from NOAA: All Federal waters of the Gulf once closed to fishing due to spill now open: http://www.restorethegulf.gov/release/2011/04/19/last-fisheries-re-opening-today

Hamilton, J. D. (2007, October). Running Dry? *Atlantic Monthly, 300*(3), pp. 42-43.

Helman, C. (2010, July 9). *The World's Biggest Oil Companies*. Retrieved September 28, 2011, from http://www.forbes.com/2010/07/09/worlds-biggest-oil-companies-business-energy-big-oil.html

Hubbert, M. K. (1956). Nuclear energy and the fossil fuels. Presented before the spring meeting of the Southern District, Division of Production, American Petroleum Institute, San Antonio, Texas, March 7-8-9, 1956.

Hyne, N. J. (2001). *Nontechnical Guide to Petroleum Geology, Exploration, Drilling, and Production, 2nd Edition*. Tulsa: PennWell Corporation.

International Maritime Organization. (2011). *International Convention for the Prevention of Pollution from Ships (MARPOL)*. Retrieved September 28, 2011, from Adoption: 1973 (Convention), 1978 (1978 Protocol), 1997 (Protocol - Annex VI); Entry into force: 2 October 1983 (Annexes I and II).: http://www.imo.org/about/conventions/listofconventions/pages/international-convention-for-the-prevention-of-pollution-from-ships-(marpol).aspx

Kerr, R. A. (2012, February 3). Technology is Turning U.S. Oil Around But Not the World's. *Science*, 522-523.

Kinder Morgan. (2011). *Products Pipelines*. Retrieved September 27, 2011, from Transmix: http://www.kne.com/business/products_pipelines/transmix.cfm

King, N. (2008, May 5). Tracking Saudi Oil From Space. *Wall Street Journal - Eastern Edition, 251*(106), p. A15.

London Tanker Brokers' Panel. (2011). *Average Freight Rate Assessment (AFRA)*. Retrieved September 28, 2011, from http://www.ltbp.com/LTBP_AFRA.html

Louisiana Department of Transportation and Development. (2011). *LOOP Program*. Retrieved September 28, 2011, from All About LOOP: http://www.dotd.louisiana.gov/programs_grants/loop/

Maritime Connector. (2011). *World's Largest Ships*. Retrieved September 28, 2011, from Supertanker - Knock Nevis: http://www.maritime-connector.com/ContentDetails/1433/gcgid/191/lang/English/World-s-Largest-Ships.wshtml

Murkowski, S. (2011, March 18). *United States Senate Committee on Energy and Natural Resources*. Retrieved February 17, 2012, from Senator Murkowski on America's Tremendous Resource Potential : http://energy.senate.gov/public/index.cfm?FuseAction=PressReleases.Detail&PressRelease_id=91d1a907-f89a-4ec1-bf17-53e4394968fa&Month=3&Year=2011

National Commission on the BP Deepwater Horizon Oil Spill and Offshore Drilling. (2011, January). *Deepwater: The Gulf Oil Disaster and the Future of Offshore Drilling, Report to the President*. Retrieved September 29, 2011, from http://www.oilspillcommission.gov/sites/default/files/documents/DEEPWATER_ReporttothePresident_FINAL.pdf

National Highway Trafic Safety Administration. (2011). *CAFE Overview - Frequently Asked Questions*. Retrieved September 28, 2011, from For what years and at what levels have the passenger car CAFE standards been set?: http://www.nhtsa.dot.gov/cars/rules/cafe/overview.htm

National Oceanic and Atmospheric Administration. (2010, August 19). *Scientists Map Origin of Large, Underwater Hydrocarbon Plume in Gulf*. Retrieved September 29, 2011, from Plume detected 22 miles long and more than 3,000 feet below surface: http://www.noaanews.noaa.gov/stories2010/20100820_plume.html

National Oceanic and Atmospheric Administration. (2011, April 25). *Office of Response and Restoration*. Retrieved September 28, 2011, from Response to the Exxon Valdez Spill: http://response.restoration.noaa.gov/topic_subtopic_entry.php?RECORD_KEY%28entry_subtopic_topic%29=entry_id,subtopic_id,topic_id&entry_id(entry_subtopic_topic)=262&subtopic_id(entry_subtopic_topic)=2&topic_id(entry_subtopic_topic)=1

National Petrochemical & Refiners Association. (2011). *How a Refinery Works, Refining Process Overview*. Retrieved September 23, 2011, from http://www.npra.org/ourIndustry/refineryFacts/?fa=refineryWorks

Organization of Petroleum Exporting Countries. (2011). *About Us--Members*. Retrieved September 26, 2011, from http://www.opec.org/opec_web/en/17.htm

Plains All American Pipeline. (2010, October 1). *Pipeline Tariffs - Plains*. Retrieved September 28, 2011, from Proportional Tariff, Crude Petroleum, From Points Names in Louisiana and Mississippi to Point Named in Illinois: http://www.paalp.com/_filelib/FileCabinet/Tariffs/FERC/2010/FERC_No._87.0.0.pdf?FileName=FERC_No._87.0.0.pdf

Ruhl, C. (2010, March/April). Global Energy After the Crisis. *Foreign Affairs, 89*(2), 70.

Sandrea, I., & Sandrea, R. (2007, November 5). Global Oil Reserves – Recovery Factors Leave Vast Target for EOR Technologies. *Oil & Gas Journal, 105*(41), 44-48.

The White House--Office of the Press Secretary. (2009, May 19). *President Obama Announces National Fuel Efficiency Policy*. Retrieved September 28, 2011, from http://www.whitehouse.gov/the_press_office/President-Obama-Announces-National-Fuel-Efficiency-Policy/

Union of Concerned Scientists. (2010). *HybridCenter.org*. Retrieved September 28, 2011, from Hybrids Under the Hood (part 1): http://www.hybridcenter.org/hybrid-center-how-hybrid-cars-work-under-the-hood.html

United States Census Bureau. (2011, September 26). *Population Clicks*. Retrieved September 26, 2011, from http://www.census.gov/main/www/popclock.html

United States Congress, Office of Technology Assessment. (1989, September). *Polar Prospects: A Minerals Treaty forAntarctica, OTA-O-428*. Retrieved September 23, 2011, from http://www.princeton.edu/~ota/disk1/1989/8926/8926.PDF

United States Department of Energy. (2007, June 18). *DOE Office of Petroleum Reserves--Strategic Unconventional Fuels*. Retrieved September 28, 2011, from Fact Sheet: U.S. Oil Shale Resources: http://www.fossil.energy.gov/programs/reserves/npr/Oil_Shale_Resource_Fact_Sheet.pdf

United States Department of the Interior. (2011, 5 April). *Bureau of Land Management*. Retrieved September 28, 2011, from Details on the Oil Shale & Tar Sands PEIS: http://www.blm.gov/wo/st/en/prog/energy/oilshale_2/PEIS_details.html

United States Department of Transportation. (2010, May 3). *Research and Innovative Technology Administration (RITA), Bureau of Transportation Statistics*. Retrieved September 28, 2011, from Table 1-11: Number of U.S. Aircraft, Vehicles, Vessels, and Other Conveyances: http://www.bts.gov/publications/national_transportation_statistics/html/table_01_11.html

United States Energy Information Administration. (2010, December 20). *Independent Statsitics and Analysis*. Retrieved September 26, 2011, from Crude Oil Proved Reserves, Reserves Changes, and Production: http://www.eia.gov/dnav/pet/pet_crd_pres_dcu_NUS_a.htm

United States Energy Information Administration. (2011). *Independent Statistics and Analysis*. Retrieved September 23, 2011, from Petroleum Navigator, Definitions, Sources and Explanatory Notes: http://tonto.eia.doe.gov/dnav/pet/TblDefs/pet_pri_imc3_tbldef2.asp

United States Energy Information Administration. (2011b, September 21). *Independent Statistcs and Analysis*. Retrieved September 23, 2011, from Petroleum and Other Liquids, World Crude Oil Prices (Dollars per Barrel): http://www.eia.gov/dnav/pet/pet_pri_wco_k_w.htm

United States Energy Information Administration. (2011c, June 17). *Oil: Crude and Petroleum Products Explained.* Retrieved September 26, 2011, from http://tonto.eia.doe.gov/energyexplained/index.cfm?page=oil_home

United States Energy Information Administration. (2011d). *Independent Statistics and Analysis*. Retrieved September 26, 2011, from International Energy Statistics, Consumption: http://tonto.eia.doe.gov/cfapps/ipdbproject/iedindex3.cfm?tid=5&pid=54&aid=2&cid=US,&syid=2008&eyid=2008&unit=TBPD

United States Energy Information Administration. (2011e, July 28). *Independent Statistics and Analysis*. Retrieved September 26, 2011, from U.S. Imports by County of Origin: http://www.eia.gov/dnav/pet/pet_move_impcus_a2_nus_epc0_im0_mbblpd_a.htm

United States Energy Information Administration. (2011f, July 5). *Independent Statistics and Analysis, Oli: Crude and Petroleum Products Explained*. Retrieved September 26, 2011, from Petroleum Statistics: http://www.eia.gov/energyexplained/index.cfm?page=oil_home#tab2

United States Energy Information Administration. (2011g, July 28). *Independent Statistics and Analysis*. Retrieved September 26, 2011, from U.S. Field Production of Crude Oil (Thousand Barrels per Day): http://www.eia.gov/dnav/pet/hist/LeafHandler.ashx?n=PET&s=MCRFPUS2&f=A

United States Energy Information Administration. (2011h, July 28). *Independent Statistics and Analysis*. Retrieved Septemeber 26, 2011, from Crude Oil Production: http://www.eia.gov/dnav/pet/pet_crd_crpdn_adc_mbbl_a.htm

United States Energy Information Administration. (2011i, July 28). *Independent Statistics and Analysis*. Retrieved September 26, 2011, from U.S. Imports by Country of Origin, Crude Oil: http://www.eia.gov/dnav/pet/pet_move_impcus_a2_nus_epc0_im0_mbblpd_a.htm

United States Energy Information Administration. (2011j, July 5). *Independent statistics and Analysis*. Retrieved September 26, 2011, from Oil: Crude and Petroleum Products Explained, Petroleum Statistics, Total World Oil Production: http://www.eia.gov/energyexplained/index.cfm?page=oil_home#tab2

United States Energy Information Administration. (2011k, September 7). *Independent Statistics and Analysis*. Retrieved September 26, 2011, from STEO Table Browser, 3c: OPEC Crude Oil (excluding condensates) Supply: http://www.eia.gov/emeu/steo/pub/cf_tables/steotables.cfm?tableNumber=7&loadAction=Apply+Changes&periodType=Annual&startYear=2000&endYear=2012&startMonthChanged=false&startQuarterChanged=false&endMonthChanged=false&endQuarterChanged=false&noScroll=true

United States Energy Information Administration. (2011l, January). *Independent Statistics and Analysis*. Retrieved September 26, 2011, from Saudi Arabia: http://www.eia.gov/countries/cab.cfm?fips=SA

United States Energy Information Administration. (2011m, July 28). *Independent Statistics and Analysis*. Retrieved September 26, 2011, from U.S. Field Production of Crude Oil (Thousand Barrels): http://www.eia.gov/dnav/pet/hist/LeafHandler.ashx?n=pet&s=mcrfpus1&f=a

United States Energy Information Administration. (2011n, September). *Independent Statistics and Analysis*. Retrieved September 26, 2011, from International Energy Outlook 2011: http://www.eia.gov/forecasts/ieo/pdf/0484(2011).pdf

United States Energy Information Administration. (2011o, September 2). *Independent Statistics and Analysis*. Retrieved September 27, 2011, from Average Depth of Crude Oil and Natural Gas Wells: http://www.eia.gov/dnav/pet/PET_CRD_WELLDEP_S1_A.htm

United States Energy Information Administration. (2011p, January 31). *Independent Statistics and Analysis*. Retrieved September 28, 2011, from Voluntary Reporting of Greenhouse Gases Program (Voluntary Reporting of Greenhouse Gases Program Fuel Carbon Dioxide Emission Coefficients): http://www.eia.gov/oiaf/1605/coefficients.html

United States Energy Information Administration. (2011q). *Independent Statistics and Analysis*. Retrieved September 28, 2011, from CO2 Emissions from the Consumption of Petroleum (Million Metric Tons): http://www.eia.gov/cfapps/ipdbproject/iedindex3.cfm?tid=90&pid=5&aid=8&cid=&syid=2009&eyid=2009&unit=MMTCD

United States Energy Information Administration. (2012, February 20). *Independent Statistics and Analysis*. Retrieved February 20, 2012, from U.S. Field Production of Crude Oil (Thousand Barrels): http://www.eia.gov/dnav/pet/hist/LeafHandler.ashx?n=pet&s=mcrfpus1&f=a

United States Energy Information Administration. (2012b, February 3). *Independent Statistics and Analysis*. Retrieved February 20, 2012, from U.S. Average Depth of Crude Oil Exploratory and Developmental Wells Drilled (Feet per Well): http://www.eia.gov/dnav/pet/hist/LeafHandler.ashx?n=PET&s=E_ERTWO_XWD0_NUS_FW&f=A

United States Geolical Survey. (2011, August 4). *Energy, Bakken Formation*. Retrieved September 26, 2011, from Will the USGS update its assessment of the Bakken Formation?: http://www.usgs.gov/faq/index.php?sid=54684&lang=en&action=artikel&cat=21&id=1176&artlang=en

United States Geological Survey. (2008, April 20). *USGS Newsroom*. Retrieved September 26, 2011, from 3 to 4.3 Billion Barrels of Technically Recoverable Oil Assessed in North Dakota and Montana's Bakken Formation—25 Times More Than 1995 Estimate: http://www.usgs.gov/newsroom/article.asp?ID=1911

United States Government Accountability Office. (2007, February). *GAO-07-283: Crude Oil: Uncertainty About Future Oil Supply Makes it Important to Develop a Strategy for Addressing a Peak and Decline in Oil Production*. Retrieved September 26, 2011, from http://www.gao.gov/new.items/d07283.pdf

United States Senate Committee on Environment and Public Works. (2000, December 29). *Oil Pollution Act of 1990*. Retrieved September 28, 2011, from SEC. 4115: http://epw.senate.gov/opa90.pdf

Whitney, G., Behrens, C. E., & Glover, C. (2010, November 30). *Congressional Research Service*. Retrieved February 17, 2012, from U.S. Fossil Fuel Resources: Terminology, Reporting, and Summary: http://epw.senate.gov/public/index.cfm?FuseAction=Files.view&FileStore_id=04212e22-c1b3-41f2-b0ba-0da5eaead952

Wood, J. H., Long, G. R., & Morehouse, D. F. (2004, August 18). *Long-Term World Oil Supply Scenarios: The Future Is Neither as Bleak or Rosy as Some Assert*. Retrieved September 26, 2011, from United States Energy Information Administraion: http://www.eia.doe.gov/pub/oil_gas/petroleum/feature_articles/2004/worldoilsupply/oilsupply04.html

The 375 MW SGT5-8000H Gas Turbine, *Credit/Courtesy of Siemens*

Chapter 4
Methane

Natural Gas Production, Reserves & Consumption

Natural gas is primarily methane. Methane is the simplest hydrocarbon, having only 1 carbon atom and 4 hydrogen atoms—its molecular formula is CH_4. Reserves are commonly found in proximity to petroleum and coal deposits because they are all created with similar biologic and geologic origins. After studying the earth's interior by examining waves of energy traveling through various rock layers (this is called "seismic exploration"), production companies drill for natural gas just as they drill for oil. Rig operators use similar drilling and casing techniques as those previously explained in the petroleum chapter. In a producing well, natural gas flows from the rock, through the perforated cement and steel casing, and into the well. Since it is less dense than the air in the well, it naturally rises to the surface where it is directed into pipelines.

Wellhead natural gas is not ready for consumer use because it normally contains 10-15% impurities. These impurities include other important hydrocarbons, including ethane (C_2H_6), propane (C_3H_8) and butane (C_4H_{10}). These are useful gases—ethane is used as a feedstock for ethylene production, propane and butane are used for heating and cooking. Other impurities may include hydrocarbons with larger molecular structures, such as pentane (C_5H_{12}) and hexane (C_6H_{14}). Through a variety of methods these additional hydrocarbons are separated from the methane at gas processing plants and then sold. Since they are often liquids after processing, they are often called natural gas liquids (NGL), or, when present with methane at the wellhead, "wet natural gas".

"Pipeline quality" gas is nearly all methane and is ready to burn by consumers. To reach consumers, it is first pumped under very high pressure from processing plants through many thousands of miles of transmission pipelines. Before entering a home, the methane is pressurized to about 1.25 atm. This is necessary to gently push the gas out of the pipes and into various burners—be they in a water heater, on a stovetop or in any other appliance. Were it not for the slightly-higher-than atmospheric pressure, the methane would simply stop flowing through the pipes. As it stands, consumers are assured of a constant pressure (and thus constant flame size) when using this energy source.

Since methane is odorless, but highly flammable, it is doped with a mercaptan. Mercaptans are organic chemical compounds that contain sulfur and literally stink like rotten eggs or rotten cabbage. You well know this if you've ever experienced a leaking gas line or were unable to ignite a gas burner. It's not the methane you smelled (nobody can smell methane), but the mercaptan. This stinky compound is deliberately added to the methane as a safety precaution in order to warn people of natural gas leaks.

In the United States, natural gas production and consumption is measured in cubic feet (abbreviated as "cf"—most of the rest of the world uses cubic meters). It is also frequently reported in the number of Btu consumed during any given time frame. Since such enormous volumes are produced and consumed every year—indeed, every month—billion, trillion or even quadrillion are used as measurement prefixes. Thus one typically sees large scale data reported in Bcf (billion cubic feet), Tcf (trillion cubic feet), or billion, trillion or even quadrillion Btu. Adding yet another wrinkle to how natural gas is measured is the word "therm." **A therm is 100,000 Btu.** Therms are especially important when examining a typical consumer's monthly gas bill (this comes later in the chapter).

In 2009, 6,289 Tcf of natural gas reserves existed on the earth (United States Energy Information Administration, 2011). This is truly an enormous number—as seen when stated in these other manners: 6,289,000,000,000,000 cf, 6.2×10^{15} cf, or even, six quadrillion two hundred eighty nine trillion cubic feet. The world produced 106,471 Bcf of natural gas in 2009 (United States Energy Information Administration, 2011b). This is also a huge volume of natural gas. Remember that a "billion" is 10^9, so this number can also be expressed as 106,471,000,000,000, or as 1.06×10^{14}. Calculating the earth's natural gas R/P ratio is demonstrated in Calculation 4-1, and produces an answer of 58 years. So do we really have only 58 years left of natural gas left on the earth? No. This R/P ratio is just as problematic as those used to calculate how

Calculation 4-1, Calculating Earth's Natural Gas R/P Ratio

Calculating Earth's Natural Gas R/P Ratio

A) 6,289 Tcf of natural gas reserves existed in 2009 (6,289 Tcf = 6.2×10^{15} cf)
B) 106,471 Bcf of natural gas were produced in 2009 (106,471 Bcf = 1.06×10^{14})
C) Reserves / Production = $6.2 \times 10^{15} / 1.06 \times 10^{14} = 5.8 \times 10^1 = 58$ years.

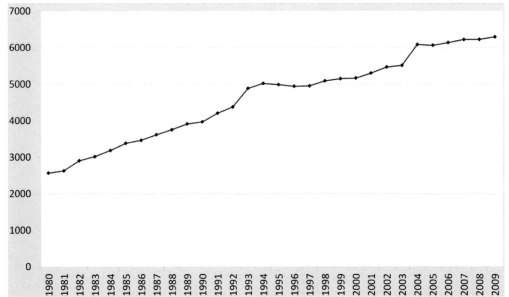

Figure 4-1, Worldwide Proven Reserves of Natural Gas, 1980-2009 (Tcf)

much coal and petroleum reserves are left. **World proven reserves of natural gas have risen over most of the last thirty years.**

Proven reserves of natural gas are unequally concentrated across the world's geologic and political landscapes. For example, Russia has by far the greatest proven reserves as of natural gas, measuring 1,680 Tcf in 2009. This means that as of 2009, about 27% of the earth's proven natural gas reserves were located in just one country—Russia. Countries holding other significant reserves in 2009 were Iran (992 Tcf) and Qatar (892 Tcf). The United States, by contrast, held 273 Tcf (United States Energy Information Administration, 2011c). This unbalanced concentration of natural gas is a concern for policy makers because of fears that one country—Russia in particular—might with-

hold it during an economic or political clash. This exact scenario unfolded in January of 2009 as Russia's Gazprom (the country's natural gas monopoly) began withholding natural gas piped to Ukraine after a prolonged pricing dispute. Not only did Ukraine receive inadequate natural gas supplies that winter, so too did much of Europe because the same pipelines also run west after first going through Ukraine. By by cutting supplies to Ukraine, Russia also cut supplies to millions of other consumers in western Europe (Kramer & Bennhold, 2009).

While small compared to Russia, Iran and Qatar, in a global context the proven reserves of natural gas within the United States are still quite robust. The United States produced 21,577 Bcf in 2010 (United States Energy Information Administration, 2011d), but it con-

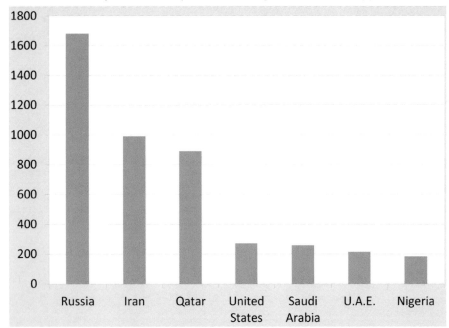

Figure 4-2, Countries With The Largest Proven Natural Gas Reserves, 2009 (Tcf)

Country	Bcf	Route
Canada	3,280	Pipeline
Trinidad	190	Liquefied Natural Gas
Egypt	73	Liquefied Natural Gas
Qatar	46	Liquefied Natural Gas
Nigeria	42	Liquefied Natural Gas
Yemen	39	Liquefied Natural Gas

Table 4-1, Largest Natural Gas Importers Into the United States, 2010 (Bcf)

sumed 24,133 Bcf (United States Energy Information Administration, 2011e). The difference was, of course, imported into the country via pipeline, primarily from Canada, and via liquefied natural gas (LNG), primarily from Trinidad, Egypt, Qatar, Nigeria and Yemen (United States Energy Information Administration, 2011f).

Somewhat surprisingly, the United States does in fact also export natural gas. In 2010, for example, the United States exported 1,137 Bcf (United States Energy Information Administration, 2011g). Why would a country that uses more natural gas than it produces actually export this energy source? This apparent contradiction is largely explained by geography. Natural gas, like any commodity, must be transported, and this of course costs money. Natural gas produced in the United States near the borders of Mexico and Canada may therefore be more economically marketed across the border rather than inside the country. There are

also important markets receiving LNG from the United States, such as Japan, South Korea and the United Kingdom. Thus, while the United States does indeed export some natural gas, the more important fact is that it is a net natural gas importer.

Natural gas consumers are sorted into one of the following five categories: residential, commercial, industrial, vehicle fuel and electric power (United States Energy Information Administration, 2011h). Residential consumers are private dwellings, such as single-family homes, condominiums and apartments, where natural gas is used primarily for cooking, space-heating, and water-heating. In 2010, residential consumers accounted for about 22% of the natural gas delivered to consumers. Residential consumers pay the most for natural gas because their distribution network is so vast. Each residence has its own dedicated pipeline, of which there can be many dozens in a single square

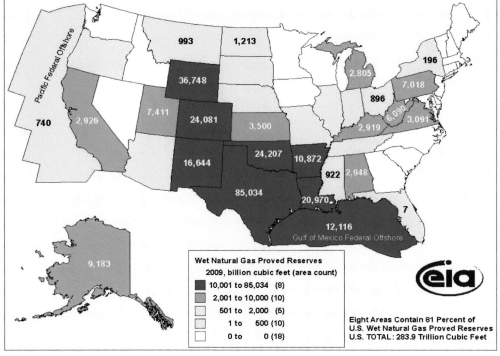

Figure 4-3, Proven Reserves of Natural Gas in the United States, 2009, *Credit/Courtesy of United States Energy Information Administration*

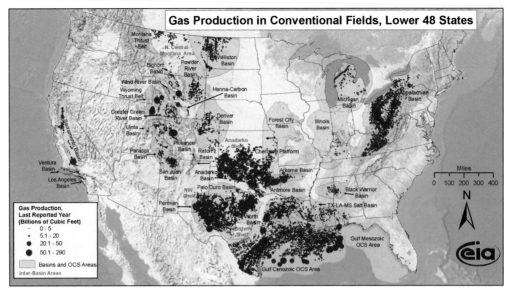

Figure 4-4, Natural Gas Field Production, 2009, *Credit/Courtesy of United States Energy Information Administration*

Figure 4-5, A Natural Gas Drilling Rig, *Credit/Courtesy of BP, © BP p.l.c*

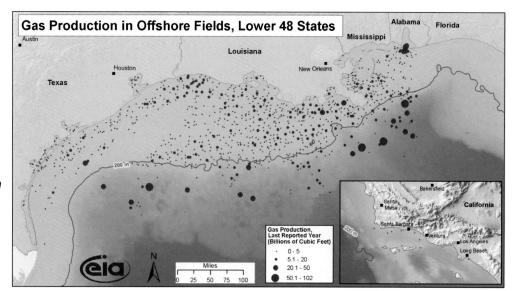

Figure 4-6, Natural Gas Offshore Production, 2009, *Credit/Courtesy of United States Energy Information Administration*

mile (especially in built-up, urban settings). In 2010, the average price paid by residential consumers in the United States per thousand cubic feet (Mcf) of natural gas was $11.21 (United States Energy Information Administration, 2011i).

Figure 4-7, The Goodwyn Platform, Northwest Shelf Project, Western Australia, *Credit/Courtesy of Woodside Energy Ltd.*

1,000 cubic feet of natural gas is abbreviated "Mcf". This is often confusing because we normally think "million" when we see "M." In the context of natural gas however, the "M" stands for the Roman numeral for the number 1,000. In this vein, 1 MMcf is 1 million cubic feet, because 1,000 x 1,000 = 1,000,000. Keep in mind that $11.21 was the average price paid by residential consumers in 2010, and thus some unexpected and surprising values occur for specific states. For example, the average residential price paid in Hawaii in 2010 was $44.62 per Mcf (United States Energy Information Administration, 2011j). Having no natural gas of its own (the state is geologically too young), The Gas Company in Hawaii manufactures synthetic natural gas (SNG) from imported petroleum.

Contrasting the high prices paid by residential consumers is the comparatively low average price paid by electric power generators in 2010. These consumers accounted for about 33% of the delivered natural gas that year. As the category name suggests, electric power generators use natural gas to generate electricity in power plants. In 2010, the average price paid by electric power producers in the United States per Mcf of natural gas was $5.26. This inexpensive price (compared to residential consumers) is explained by two facts: 1) over 10% more gas was consumed by this group, and 2) there are relatively few gas-powered electricity producers compared to the millions of residential consumers, and thus a comparatively small amount of administrative and physical infrastructure serves this gas-hungry sector.

As one might discern from the above paragraphs, two main components determine the price of natural gas. The first and most obvious is the gas itself. In this respect it is a commodity like any other and traded as

Figure 4-8, **The Angel Platform, Northwest Shelf Project, Western Australia,** *Credit/ Courtesy of Woodside Energy Ltd.*

such (think coal, petroleum, pork bellies, soybeans, etc.). The price of this energy commodity rises and falls according to supply and demand, with the free market determining the price. The second variable determining the price of natural gas is the cost to transport it to consumers. Most natural gas consumers live far away from the gas production fields, and thus this energy commodity must be transported to them via pipeline. Pipelines in the United States are privately owned and publically regulated companies that charge transportation fees to natural gas buyers. These two main cost components then, as well as taxes and other charges, are passed on to consumers in the form of a monthly natural gas bill.

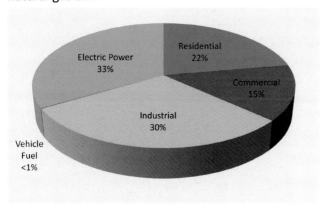

Figure 4-9, United States Natural Gas Consuming Sectors, 2010

Proven natural gas reserves in the United States rose 63% between 1999 and 2009—from 167 Tcf to 273 Tcf (United States Energy Information Administration, 2011k). This remarkable growth is driven largely by the development of unconventional natural gas reservoirs—specifically shale gas. **Shale gas is methane trapped within the fine pore spaces of the sedimentary rock shale.** Shale gas "plays" are large bodies

of this rock that have recently become economically viable for gas production due to advances in drilling technology and extraction techniques. Examples of shale plays that are often in the news include the Barnett play in Texas, the Haynesville-Bossier play in Texas and Louisiana, and the Marcellus play in the eastern United States. Well over a dozen unique plays exist within the United States, boosting the probable methane reserves of the United States to a whopping 2,552 Tcf (United States Energy Information Administration, 2011l), and prompting some analysts to call the United States the Saudi Arabia of shale gas. Considerable optimism now exists about the ability of the United States to rely almost wholly on domestic supplies of natural gas well into the future, perhaps even for as long as a century.

Until recently, accessing commercially-viable quantities of methane in shale was too expensive. Because the grains in shale are very tightly packed they impede any trapped gas from migrating through it and towards a well. Traditional, vertically-situated wells have a very limited production zone within shale beds and therefore run dry very quickly. Directional drilling, by contrast, is a significant technological advancement that is frequently called a "game changer" within the gas industry. **Directional drilling allows crews to drill horizontally through shale beds for many thousands of feet and thereby create horizontally-situated wells with very extensive production zones.** This is accomplished using curve drilling assemblies (abbreviated "CDA") which contain articulating motors, inclination sensors and directional tracking systems. Curve radii vary depending on reservoir conditions and can be as small as 20 feet or as large as 3,000 feet (Grand Directions, pp. 5-6). Horizontal wells are cased like other

Figure 4-10, Shale Gas Plays in the United States, *Credit/Courtesy of United States Energy Information Administration*

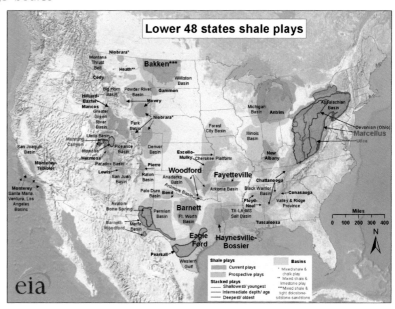

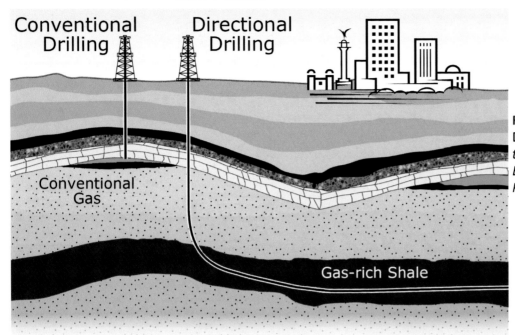

Figure 4-11, Directional Drilling, *Credit/Courtesy of National Energy Education Development Project (adapted)*

wells, and production zones are opened using firing charges similar to those already discussed in Chapter 3.

Coupled with directional drilling is an extraction technique called hydraulic fracturing—often termed "fracking". **Hydraulic fracturing involves pumping high-pressure water, sand and chemicals into a well's production zone in order to generate cracks in reservoir rock and thereby facilitate the flow of hydrocarbons into the well** (National Energy Technology Laboratory, 2009, p. 4). Sand is an important component of the fracturing recipe because it penetrates the newly-opened cracks and props open hydrocarbon flow routes that would otherwise collapse. Compared to the water and sand, relatively small amounts—between 1% and 2% by volume—of chemicals are used. The chemicals can include agents to control foaming, pH and friction, as well as others that dissolve minerals, kill bacteria and prevent corrosion. As one might imagine, pumping chemicals underground is a controversial technique that has recently garnered the gas industry a lot of unwanted attention. This issue is explored more deeply later in this chapter when the vices and virtues of natural gas are examined.

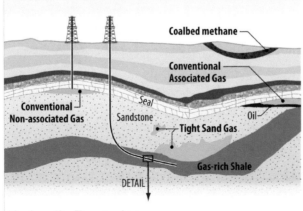

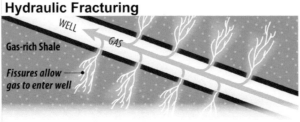

Figure 4-12, Hydraulic Fracturing in Gas-Rich Shale, *Credit/Courtesy of National Energy Education Development Project*

Calculating the Natural Gas Charge for a Typical Home

Calculation 4-2,
Calculating the Natural
Gas Charge for a Typical
Home

A) Natural gas consumed over 28 days = 92 ccf
B) Recall that 1 ccf = 100 cf
C) 92 x 100 = 9,200 cf
D) Btu factor = 1,011 Btu/cf
E) 9,200 x 1,011 = 9,301,200 total Btu consumed
F) Recall that 1 therm = 100,000 Btu
G) 9,301,200 / 100,000 = 93.012 total therms consumed
H) 1 therm = $0.43
I) Natural gas charge = 93.012 x 0.43 = $39.96

Your Confusing Natural Gas Bill

When consumers purchase electricity or gasoline, it is relatively easy to understand the bill. For example, when we buy a certain number of kWh of electricity, we are charged per kWh for that electricity. We fill up our vehicles with a certain number of gallons of gasoline, and are then charged per gallon for that gasoline. Natural gas pricing is unfortunately more complicated. This increased complexity is due to three issues. The first issue concerns how our gas consumption is reported. We burn a certain number of cubic feet (cf) every month, but when our bill arrives, the total is reported in "ccf" units, instead of "cf" units. **A ccf of natural gas is 100 cubic feet.** Like the "M" in Mcf, the "c" in ccf is a Roman numeral—this time for 100.

The second confusing issue with natural gas pricing is that every cf of natural gas contains slightly more or less energy than every other cf. The energy, usually re-ferred to as the "heat content" or "Btu factor" of the gas, is measured in Btu/cf. Over the past 38 years, the average Btu/cf delivered to residential consumers has varied from a low of 1,016 Btu/cf in 1978, to a high of 1,033 Btu/cf in 1998 (United States Energy Information Administration, 2011m). The gas industry blends natural gas in order to make the Btu factor as consistent as possible from year to year, and thereby to ensure the safety of consumers and their appliances.

A third potentially confusing part about a typical gas bill is that we are charged by the number of therms consumed within a billing cycle, not by the total number of cubic feet (recall that 1 therm is 100,000 Btu). This makes sense, however, when considering that the heat content of natural gas varies, so in one month a single cf might contain 1016 Btu, and the next it might contain 1,033 Btu. Since gas companies are in the business of selling energy, not just volumes of "stuff", they charge for Btu consumed, not cubic feet.

Figure 4-13,
A Typical Natural Gas Bill

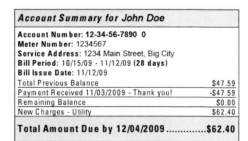

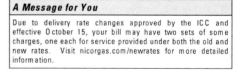

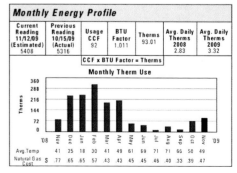

Consider a natural gas bill for a typical single-family-detached home in Illinois for the 28 day period ranging from October 15 through November 12. The salient parts of the bill are: 1) 92 ccf, 2) Btu factor = 1,011, 3) 93.01 therms. We interpret these three issues as follows: Over the 28 day period 9,200 cf of natural gas was consumed. Each cf contained 1,011 Btu, so a total of 9,301,200 Btu were consumed. 9,301,200 Btu calculates into 93.01 therms, which is the energy—not the cubic feet—consumed during the billing period. If each therm costs $0.43, our natural gas charge is $39.99. Calculation 4-2 illustrates the steps to understanding how consumers are charged for natural gas.

From the previous discussion about natural gas prices, we know that $39.99 is not in fact the total bill for the above home. Pipeline transportation costs (often called "delivery charges") are added, as well as (possibly) franchise costs, efficiency program costs, and environmental program costs. Municipal, state and utility taxes are also tacked on to the bill. In the end, a total bill of $62.00 may arrive, of which about 65% is for the actual gas, and 35% is for delivery charges, taxes and other expenses.

Pipelines & Underground Natural Gas Storage

Natural gas must be transported from wellheads to processing plants. Processing plants separate the methane from other gases, including ethane, propane, butane and water vapor. After processing, pipeline quality natural gas is then transported to consumers. A complex system of intrastate and interstate pipelines exists to satisfy these transportation needs. **Intrastate pipelines operate within the borders of one state, while interstate pipelines cross state borders.** As one might expect, pipelines come in different sizes and types that accommodate different needs, from producers to consumers. **"Gathering pipelines" serve producers and transport natural gas from wellheads to processing plants.** These pipelines have a comparatively small diameter, beginning at 6 inches.

Trunklines, also called mainlines or transmission lines, serve local distribution companies (LDCs) that pur-

Figure 4-14, The Karratha Gas Plant, North West Shelf Project, Western Australia, *Credit/Courtesy of Woodside Energy Ltd.*

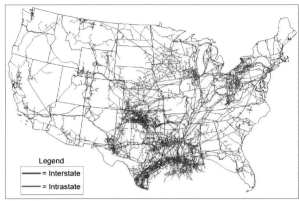

Figure 4-15, United States Natural Gas Pipeline Network, 2009, *Credit/Courtesy of United States Energy Information Administration*

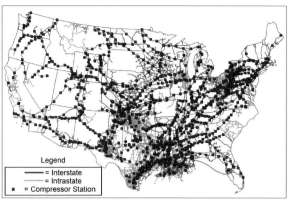

Figure 4-16, United States Natural Gas Pipeline Compressor Stations, 2008, *Credit/Courtesy of United States Energy Information Administration*

chase natural gas with the intent of selling it to consumers. **Trunklines are the largest diameter pipes, generally measuring between 16 to 56 inches wide.** The largest pipes therefore measure over 4 ½ feet wide, large enough for most people to walk through while crouching! Trunklines transport pipeline quality natural gas long distances—sometimes over a thousand miles—from processing plants to LDCs.

Distribution pipelines carry natural gas to consumers. These are comparatively small diameter pipes, and shrink even further when reaching a typical home where the smallest-diameter pipes exist, measuring about ½ inch in diameter. In all, there are over 1.5 million miles of natural gas pipelines in the United States (United States Energy Information Administration, 2011n), with over 305,000 of those miles being trunkline diameter pipe (United States Energy Information Administration, 2011o).

Compressor stations run the length of pipelines—on average about one station every 70 miles—to ensure the continued flow of natural gas. You may have unknowingly passed one of these stations while driving—look for unmanned, fenced-off areas with large-diameter white pipes coming out of the ground. These stations are needed because pipeline pressure drops due to friction (against the inside of the pipe), consumption, and leakage. **Compressor stations boost the pressure at regular intervals along interstate pipelines, providing an average ramp-up of 250 psi, and discharging natural gas back into a trunkline at the preferred pressure of 1,500 to 1,750 psi.** In 2006, 1,201 compressor stations existed along United States trunklines, with each station having an average of 4 compressors (United States Energy Information Administration, 2007, p. 2). The compressors are either turbine or piston-driven engines that are fueled either by natural gas taken directly from the pipeline, or by

Figure 4-17, A Natural Gas Pipeline in Texas, *Credit/ Courtesy of BP, © BP p.l.c.*

electricity served up by the local grid (United States Energy Information Administration, 2011p). Other important functions of compressor stations include filtering out any particulate matter that may have accumulated (due to internal pipeline corrosion), as well as scrubbing out any liquids that may have condensed (NaturalGas.org, 2011).

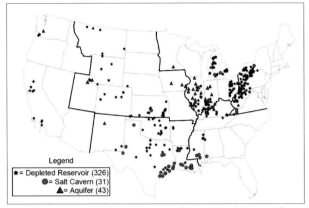

Figure 4-18, United States Underground Natural Gas Storage Facilities, 2007, *Credit/Courtesy of United States Energy Information Administration*

Compressor stations are also situated at underground natural gas storage facilities and are used to inject and withdraw natural gas in conjunction with supply and demand. As of April 2009, the United States had 3,889 Bcf of underground natural gas storage capacity, an increase of 100 Bcf from a year earlier (United States Energy Information Administration, 2009, p. 1). **Underground storage facilities ensure uninterrupted supplies of natural gas.** As many readers know, natu-

ral gas consumption is seasonal, with greater use occurring during winter months. Underground storage allows an LDC to purchase natural gas when prices are low and store this energy source until prices are high. This does not mean that LDCs are gouging natural gas consumers. LDCs are under intense public oversight and often must charge consumers the same price they paid for natural gas. By using underground storage facilities, natural gas companies can better plan their finances and thereby provide consumers a reasonably priced commodity whenever it is demanded, and at whatever volume.

There are three types of underground natural gas storage facilities—depleted fields, aquifers, and salt caverns. In 2009, 409 underground storage facilities existed in the United States (United States Energy Information Administration, 2011q). Depleted fields include oil and gas reservoirs, and are the most common type of underground storage facility. In 2009, for example, 331 of the 409 facilities were depleted fields, 43 were aquifers, and 35 were salt caverns. Widespread use of depleted fields for natural gas storage is due to their extensive availability, especially in formerly productive areas of western Pennsylvania, eastern Ohio and West Virginia. By contrast, storing natural gas in aquifers requires that the water-saturated rock be overlaid by an impermeable cap-rock, lest the natural gas leak up to the earth's surface. Geology therefore dictates that the Midwest—especially the states of Illinois and Indiana—is the most geologically suitable place for aquifer facilities.

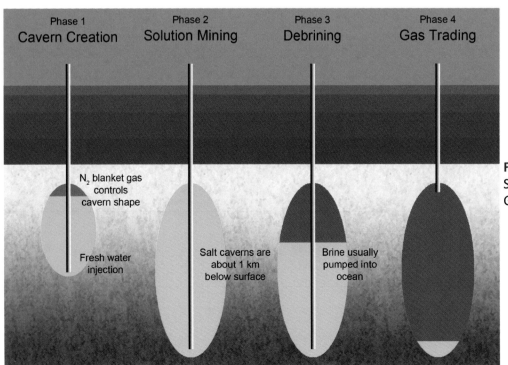

Figure 4-19, Creating a Salt Cavern for Natural Gas Storage

Local geology is also a critical variable when storing natural gas in salt caverns (also called "salt domes"), because massive salt deposits occur only in select areas of the country. Geologically-suitable deposits are found in Texas, Louisiana and Mississippi, and are normally over one thousand feet deep and measure hundreds of feet thick. Storage facilities are created by leaching a cavity into a salt bed. A borehole is first drilled into an existing salt formation, after which fresh water is injected into the borehole. The water dissolves the salt and is then pumped out of the borehole and replaced with more fresh water. Eventually a cavity emerges that, unlike many aquifers and depleted fields, enables high injection and withdrawal rates and thereby makes possible multiple gas injections and withdrawals in a single year (United States Energy Information Administration, 2002, p. 1).

FERC Order 636: The Restructuring Rule (1992)

Interstate natural gas pipelines in the United States operate much like trains—merely providing transportation services to commodity buyers and sellers. This was not always the case. Prior to 1994, pipeline companies actually bought and sold natural gas in a regulated marketplace that disadvantaged just about every player, from natural gas producers, to transporters, to buyers. Imagine Union Pacific Railroad actually buying the vegetables, computers, lumber, vehicles and anything else it moves, and then selling those commodities after the train arrives at its destination. This ridiculous scenario would encumber the transportation company with unnecessary costs and thus divert resources from its specialized (and thus most efficient) task—transporting commodities. Prior to 1994, natural gas pipelines wrestled with an analogous scenario, buying natural gas from producers, moving it, and then selling it to LDCs (but only at cost, as mandated by the Federal Energy Regulatory Commission at that time). The effect was a regulated system riddled with inefficiencies that hurt producers, transporters and consumers (Federal Energy Regulatory Comission, 1992, pp. 36-38).

In April, 1992 FERC passed Order Number 636 to rectify the inefficiencies in the natural gas marketplace. At its heart the order mandated that natural gas pipeline companies "unbundle"—that is, separate, sell or divest—any sales services from their transportation services. This eliminated their transportation monopoly, released them from FERC regulations requiring them

to sell natural gas at cost, and increased competition amongst natural gas producers and LDCs (United States Energy Information Administration, 2011r).

Pipeline companies were given until November of 1993 (19 months) to comply with Order 636. Since this date then, natural gas pipeline companies in the United States merely transport the commodity instead of buying and selling the natural gas. This does not mean, however, that they are free of FERC regulations. FERC still asserts its authority over this industry in the following ways, including:

- *Regulates the transmission and sale of natural gas for resale in interstate commerce.*

- *Approves the siting and abandonment of interstate natural gas pipelines and storage facilities.*

- *Ensures the safe operation and reliability of proposed and operating LNG terminals.*

- *Enforces FERC regulatory requirements through imposition of civil penalties and other means.*

- *Oversees environmental matters related to natural gas projects.*

- *Administers accounting and financial reporting regulations and conduct of regulated companies.*

(Federal Energy Regulatory Commission, 2010)

Liquefied Natural Gas—LNG

Stranded natural gas is located in places without existing pipelines or transportation networks. The other two other fossil fuels—coal and petroleum (a solid and a liquid)—are readily transported across land and oceans with comparatively little logistical, technical and financial challenges. CH_4, however, exists as a gas at the earth's surface temperature and pressure, and if moved in this state would constitute an incredibly bulky and therefore exceptionally expensive commodity to transport. LNG carriers resolve this issue by lowering the temperature of natural gas to -162.2 °C (-260 °F), which changes its state from gas to liquid, and thereby reduces its volume by a factor of 600. The reduced volume makes transportation economically feasible, and thus connects previously stranded gas fields with the international energy marketplace.

Figure 4-20, The North West Shelf LNG Plant in West-ern Australia, *Credit/Courtesy of BP, © BP p.l.c.*

LNG carriers are as large as oil tankers—the largest stretch over 1,130 feet in length—but have very different designs due to the challenges of transporting large quantities of extremely cold liquid. Two basic types exist, the spherical and the membrane, both of which were initially developed in the 1970s. Spherical design LNG carriers are perhaps the most visually striking be-

Figure 4-21, A Spherical Design LNG Carrier Delivers Natural Gas to the United States, *Credit/Courtesy of State of Connecticut, Department of Utility Control*

cause of the 4 or 5 giant spheres that protrude many feet above a ship's deck. Each insulated sphere is situated on individual supports which are welded into the hull of the carrier. The sphere's shape limits sloshing during transport (important for ship stability), and any expansion or contraction of the sphere is transferred to the supports instead of the ship's hull. A disadvantage of this design is that the spheres do not conform to the ship's hull, and thus space that could otherwise be used to carry LNG is simply wasted. Membrane de-

Figure 4-22, The LNG Northwest Seaeagle, A Spherical Design LNG Carrier, *Credit/Courtesy of Woodside Energy Ltd.*

Figure 4-23, A Docked LNG Carrier, *Credit/Courtesy of Personnel of NOAA Ship Thomas Jefferson*

Figure 4-24, The Northwest Shearwater, A Spherical Design LNG Carrier, *Credit/Courtesy of BP, © BP p.l.c.*

Figure 4-25, The British Trader, A Membrane Design LNG Carrier,
Credit/Courtesy of BP, © BP p.l.c.

Figure 4-26, The British Merchant, A Membrane Design
LNG Carrier, *Credit/Courtesy of BP, © BP p.l.c.*

sign LNG carriers, on the other hand, integrate heavily-insulated tanks directly into a ship's hull (United States Department of Energy, 2005, p. 12). This design maximizes cargo space, making membrane carriers more efficient than spherical carriers.

The majority of LNG carriers have a capacity between 100,000 and 150,000 m³ (GlobalSecurity.org, 2011). Remember that this volume expands 600 times upon regasification. Thus, an average LNG tanker carrying 125,000 m³ of LNG will contribute 75,000,000 m³ of dry natural gas to a pipeline (125,000 x 600 = 75,000,000). Since 1 m³ = 35.3 ft³, 75,000,000 m³ = 2,647,500,000 ft³ (75,000,000 x 35.3 = 2,647,500,000). To get some perspective about this enormous quantity, consider this example: A typical single-family home in northern Illinois consumes about 650 ccf of natural gas in a year, which is 65,000 ft³ (650 x 100 = 65,000). Thus, in a single shipment, this average LNG carrier satisfies the natural gas requirements of 40,731 homes in northern Illinois for 1 year (2,647,500,000 ft³ / 65,000 ft³ = 40,731). This is an impressive energy cargo for just one average-sized LNG carrier.

The largest class of LNG carrier is called the Q-Max. The "Q" stands for Qatar, the tiny Persian Gulf country holding 892 Tcf of natural gas reserves in 2009. The "Max" in Q-Max stands for the maximum berthing size at Qatar's LNG export facilities in Ras Laffan Industrial City, about 45 miles north-northeast of Doha. This facility, operated by Qatargas Operating Company Limited, processes natural gas extracted from beneath the nearby seabed—Qatar's North Field. Qatargas is the largest LNG producer in the world, and has a production capacity at Ras Laffan Industrial City of 42 million metric tons of LNG per year (Qatargas, 2011). Massive carriers regularly dock here to load LNG for destina-

Figure 4-28, LNG Storage Tank, *Credit/Courtesy of BP, © BP p.l.c.*

tions in Europe, Asia and North America. The Q-Max class has a 266,000 m³ capacity (Business Wire, 2008), the first of which was named "Mozah" in honor of Her Highness Sheikha Mozah Bint Nasser Al-Misnad, a wife of Sheikh Hamad bin Khalifa Al Thani, the Emir of Qatar since 1995 (Qatargas Operating Company Limited, 2008, p. 4). **A Q-Max LNG carrier satisfies the natural gas requirements of 86,675 homes in northern Illinois for 1 year with a single shipment.**

LNG must be "regasified" at LNG import terminals. The United States has 10 LNG import/export facilities—1 export facility in Alaska, 6 import facilities in the Gulf of Mexico, and 3 import facilities along the east coast (Federal Energy Regulatory Commission, 2011). Import facilities pump LNG from docked carriers into double-walled insulated tanks where it is stored until regasification prior to pipeline injection. Each terminal normally has enough storage capacity for 2 to 3 LNG carrier loads, and typical offloading time is about 12 hours (United States Energy Information Administration, 2004, pp. 3-4). **In 2010, the United States received the majority (44%) of its LNG from the Caribbean country of the Republic of Trinidad**

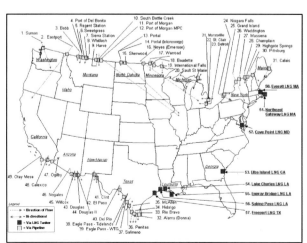

Figure 4-27, U.S. Natural Gas Import/Export Locations, as of the end of 2008, *Credit/Courtesy of United States Energy Information Administration*

and **Tobago** (United States Energy Information Administration, 2011s).

LNG projects are traditionally expensive because the liquefaction plant, fleet of LNG carriers, and regasification plant are planned and constructed for near-simultaneous delivery. Sales and purchase agreements are coordinated amongst a variety of shareholders, including publically-traded oil and gas firms, sovereign governments, municipal LNG buyers, as well as landowners. No single firm or government typically accepts all the risk. The effect is that long-term buying and selling contracts are established before a project ever begins. The $15 billion LNG project in Papua New Guinea currently under construction, for example, has the following 6 major shareholders: Exxon Mobile (33.2%), Oil Search (29%), the Papua New Guinea government (16.6%), Santos Ltd. (13.5%), Nippon Oil Corp. (4.7%), and landowners (2.8%). As these shareholders coordinated their various commitments, Tokyo Electric agreed to buy 1.8 million tons of LNG annually over the next 20 years, while China Petroleum & Chemical agreed to buy 2 million tons each year. Other Asia- Pacific buyers are also at the table, committing to purchase the project's estimated 6.6 million metric tons of annual LNG production (Sharples & Paton, 2009).

As consumers use ever-increasing volumes of natural gas, more import/export terminals must be built, along with more LNG carriers. This reality is eroding the fairly static long-term LNG sales and purchase agreements that have defined the industry for many decades. **Today, private entrepreneurs can purchase and sell LNG quickly, independently, and without any long-term commitments (this is called the LNG "spot market").** Buying and selling LNG on the spot market is expanding the reach of natural gas as a fuel source and knitting together a previously demarcated and less-efficient marketplace (Rühl, 2010).

Henry Hub and the Futures Marketplace

When natural gas arrives (or is produced) in the United States, it travels via pipeline to various hubs where it is bought and sold (perhaps many times over). **The most important natural gas hub in the United States is named Henry Hub, located in Erath, Louisiana.** This distribution and delivery point—owned by Sabine Pipe Line LLC.—connects 9 interstate and 4 intrastate pipelines so that natural gas producers and buyers can make their exchanges at a single, well-connected place. Henry Hub can handle 1.8 Bcf/d of gas transfer (Sabine Pipe Line LLC, 2011). For perspective on this massive transfer capacity, consider that over the entire year of 2010, the United States consumed about 66 Bcf/d.

A single location for natural gas distribution and delivery facilitates the creation of a futures marketplace. Thus, in early 1990, natural gas contracts began trading on the New York Mercantile Exchange (NYMEX). Contracts trade in units of MMBtu, which you already

Figure 4-29, LNG Storage Tank in Trinidad and Tobago, *Credit/Courtesy of BP,* © *BP p.l.c.*

Table 4-2,
Daily Settlements for
Henry Hub Natural Gas
Futures, October 11, 2011
($/MMBtu)

Month	Open	High	Low	Last	Change	Settle	Estimated Volume	Prior Day Open interest
Nov 11	3.562	3.627	3.513	3.625	+.075	3.616	208,362	180,055
Dec 11	3.844	3.867	3.784	3.851	+.023	3.859	108,326	99,184
Jan 12	4.010	4.024	3.940	4.000	+.002	4.008	132,913	199,647
Feb 12	4.042	4.042	3.964	4.020	UNCH	4.028	25,195	66,014
Mar 12	4.009	4.009	3.928	3.991	+.005	3.999	23,202	73,918
Apr 12	3.995	3.995	3.920	3.978	+.006	3.987	24,451	85,845
May 12	3.990	4.024	3.950	4.013	+.006	4.021	6,291	25,004

know means 1,000,000 Btu (M = 1,000, thus 1,000 x 1,000 = 1,000,000). **The futures marketplace exists primarily to mitigate risk.** As a natural gas *producer*, you may want the financial assurance of selling your commodity at a set price on a future date (usually no more than 18 months into the future), regardless of what the spot price of natural gas may be at that time. As a natural gas *buyer*, you may also need the financial certainty of a set price on a future date. Producers and buyers can thus "meet" on the natural gas futures exchange and agree that, for example, 1 MMBtu will be sold for $5.59 many months out from the present date. This satisfies the financial certainty requirements of both producers and buyers, and allows them to plan ahead without worrying about the daily volatility of natural gas prices.

Natural Gas's Virtues and Vices

Natural gas is the cleanest burning fossil fuel. This statement is made by natural gas advocates, and is indeed true, but what does it really mean? To begin, recall that the other fossil fuels are coal and petroleum. These energy resources are made of many hundreds of molecules, some of which are incompletely burned during combustion. Emissions include nitrogen oxides, unburned hydrocarbons, sulfur dioxide as well as particulate matter. These emissions are partially controlled with emission-control systems, including scrubbers (in coal-fired power plants) and catalytic converters (in automobiles) that exist primarily due to a host of state and federal pollution mitigation policies that mandate their use. Pipeline quality natural gas, on the other hand, contains carbon, hydrogen and very few nitrogen and/or sulfur compounds. **When pipeline quality CH_4 burns, it produces heat, water vapor (H_2O), carbon dioxide (CO_2) and almost nothing else.**

Carbon dioxide (CO_2) emissions, however, are a problem associated with the combustion of all fossil fuels because they all contain carbon. **By definition, burning carbon-bearing energy sources produces carbon dioxide and/or carbon monoxide.** Burning natural gas, however, emits comparatively less CO_2 than coal or petroleum. Since coal is a solid, petroleum is a liquid, and natural gas is, of course, a gas, comparing these three forms of energy requires using a common base. Thus we return to the Btu—specifically, how many pounds of CO_2 are emitted per million Btu

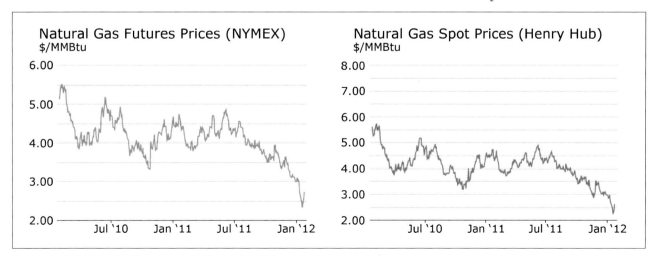

Figure 4-30, NYMEX Futures and Henry Hub Spot Prices, *Credit/ Courtesy of United States Energy Information Administration*

produced. **216 pounds of CO_2 are emitted per million Btu of average coal, 157 pounds of CO_2 are emitted per million Btu of motor gasoline, and 118 pounds of CO_2 are emitted per million Btu of pipeline natural gas** (United States Energy Information Administration, 2011t). Thus natural gas is "cleaner" than the other two fossil fuels, but not "clean" in the sense that there are no undesirable combustion byproducts.

As noted earlier in this chapter, world proven reserves of natural gas have risen over most of the past thirty years. This increasing-volume trend is also seen in the United States. For example, every year since 1998 there has been an increase in dry natural gas proven reserves, from 164,041 Bcf in 1998 to 272,509 Bcf in 2009 (United States Energy Information Administration, 2010). Thus, natural gas has an impressive growth record as a domestically-available energy resource. **Most analysts believe that domestic production of natural gas in the United States will continue growing through 2035.** This growth will largely come from unconventional natural gas sources, specifically shale gas and to a lesser extent coalbed methane (CBM) (United States Energy Information Administration, 2011u, pp. 2-4).

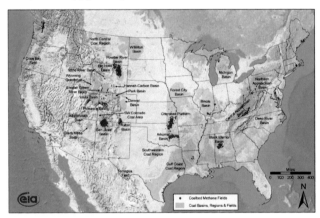

Figure 4-31, Coalbed Methane Fields, 2009, *Credit/ Courtesy of United States Energy Information Administration*

As a result of directional drilling technology opening previously inaccessible shale gas plays, many analysts assert that the United States is now at the cusp of a gas glut. Natural gas is indeed cheap, averaging $4.21/ MMBtu for the first 9 months of 2011 (United States Energy Information Administration, 2011v). Some experts even use the term "game changer" in reference to the United States and its natural gas reserves. These assertions are understood by keeping the following issues in mind: 1) the latest combined cycle power plants operate at a remarkable 60% efficiency, 2) burning methane to generate electricity results in

significantly less CO_2 emissions compared to burning coal or petroleum, 3) natural gas is cheap now and will likely remain inexpensive well into the foreseeable future, and 4) abundant quantities of natural gas are domestically available. Given these realities, the potential for natural gas to radically disadvantage the renewable and nuclear energy sectors is a very real possibility (Maize, Butcher, & Peltier, 2011).

While the energy industry is quite optimistic about the future of domestic natural gas production, considerable public anxiety exists about extracting natural gas from unconventional sources—specifically shale gas plays. This apprehension is documented in the movie *Gasland*, which alleges harmful health and environmental effects due to the chemicals used in hydraulic fracturing (Gasland, 2010). While debate exists about the film's veracity, it has nonetheless gained a popular following, much to the vexation of energy companies as they scramble to mollify burgeoning concerns about hydraulic fracturing. People are most worried about the safety of their drinking water and have voiced their concerns loudly enough to prompt investigations by congressional committees (Committee on Energy and Commerce, 2011) and the Environmental Protection Agency (United States Environmental Protection Agency, 2011). One can argue, therefore, that methane is suffering a black-eye due to the negative publicity surrounding hydraulic fracturing, and this publicity has somewhat dampened otherwise enthusiastic attitudes about this clean, plentiful and domestically-available fossil fuel.

Important Ideas:

- Natural gas is primarily methane.

- "Pipeline quality" gas is nearly all methane and is ready to burn by consumers.

- A therm is 100,000 Btu.

- World proven reserves of natural gas have risen over most of the last thirty years.

- Proven reserves of natural gas are unequally concentrated across the world's geologic and political landscapes.

- 1,000 cubic feet of natural gas is abbreviated "Mcf".

- Proven natural gas reserves in the United States rose 63% between 1999 and 2009— from 167 Tcf to 273 Tcf.

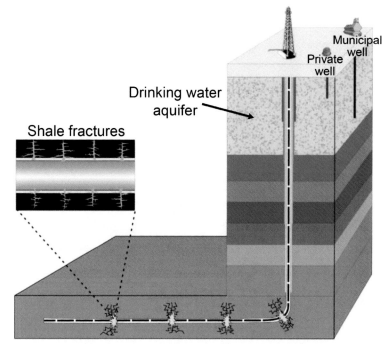

Figure 4-32, Model of Hydraulic Fracturing and Drinking Water Aquifer, *Credit/Courtesy of United States Environmental Protection Agency*

Chemical Compound(s)	Purpose	Common Use
Hydrochloric acid	Dissolve minerals & initiate cracks in rock	Swimming pool chemical & cleaner
Glutaraldehyde	Eliminates bacteria that produce corrosive byproducts	Disinfectant; sterilize medical & dental equipment
Ammonium persulfate	Delayed break-down of gel polymer chains	Bleaching agent, manufacture of household plastics
N, N-Dimethylformamide	Prevents pipe corrosion	Used in pharmaceuticals, acrylic fibers & plastics
Borate salts	Maintains fluid viscosity as temperature increases	Laundry detergents, hand soaps & cosmetics
Polyacrylamide	Minimizes friction between fluid and the pipe	Water treatment, soil conditioner
Mineral oil	Minimizes friction between fluid and the pipe	Make-up remover, laxatives, & candy
Guar gum	Thickens the water to suspend sand	Cosmetics, toothpaste, sauces, baked goods, ice cream
Citric acid	Prevents precipitation of metal oxides	Food additive, lemon juice is ~7% citric acid
Potassium chloride	Creates a brine carrier fluid	Table salt substitute
Ammonium bisulfite	Removes oxygen from water to prevent pipe corrosion	Cosmetics, food & beverage processing, water treatment
Sodium or potassium carbonate	Maintains effectiveness of other compounds	Detergent, soap, water softener, glass & ceramics
Ethylene glycol	Prevents scale deposits in pipe	Automotive antifreeze, cleansers & deicing agent
Isopropanol	Increases viscosity of fracture fluid	Glass cleaner, antiperspirant, & hair color

Table 4-3, Common Hydraulic Fracturing Chemicals

- Shale gas is methane trapped within the fine pore spaces of the sedimentary rock shale.

- Directional drilling allows crews to drill horizontally through resource-bearing rock for many thousands of feet and thereby create horizontally-situated wells with very extensive production zones.

- Hydraulic fracturing involves pumping high-pressure water, sand and chemicals into a well's production zone in order to generate cracks in reservoir rock and thereby facilitate the flow of hydrocarbons into the well.

- A ccf of natural gas is 100 cubic feet.

- Intrastate pipelines operate within the borders of one state, while interstate pipelines cross state borders.

- "Gathering pipelines" serve producers and transport natural gas from wellheads to processing plants.

- Trunklines are the largest diameter pipes, generally measuring between 16 to 56 inches wide.

- Distribution pipelines carry natural gas to consumers.

- Compressor stations boost the pressure at regular intervals along interstate pipelines, providing an average ramp-up of 250 psi, and discharging natural gas back into a trunkline at the preferred pressure of 1,500 to 1,750 psi.

- Underground storage facilities ensure uninterrupted supplies of natural gas.

- There are three types of underground natural gas storage facilities—depleted fields, aquifers, and salt caverns.

- Interstate natural gas pipelines in the United States operate much like trains—merely providing transportation services to commodity buyers and sellers.

- Stranded natural gas is located in places without existing pipelines or transportation networks.

- A Q-Max LNG carrier satisfies the natural gas requirements of 86,675 homes in northern Illinois for 1 year with a single shipment.

- In 2010, the United States received the majority (44%) of its LNG from the Caribbean country of the Republic of Trinidad and Tobago.

- LNG projects are traditionally expensive because the liquefaction plant, fleet of LNG carriers, and regasification plant are planned and constructed for near-simultaneous delivery.

- Today, private entrepreneurs can purchase and sell LNG quickly, independently, and without any long-term commitments (this is called the LNG "spot market").

- The most important natural gas hub in the United States is named Henry Hub, located in Erath, Louisiana.

- The futures market exists primarily to mitigate risk.

- Natural gas is the cleanest burning fossil fuel.

- When pipeline quality CH_4 burns, it produces heat, water vapor (H_2O), carbon dioxide (CO_2) and almost nothing else.

- By definition, burning carbon-bearing energy sources produces carbon dioxide and/or carbon monoxide.

- 216 pounds of CO_2 are emitted per million Btu of average coal, 157 pounds of CO_2 are emitted per million Btu of motor gasoline, and 118 pounds of CO_2 are emitted per million Btu of pipeline natural gas.

- Most analysts believe that domestic production of natural gas in the United States will continue growing through 2035.

Works Cited

Business Wire. (2008, December 17). *ExxonMobil Technology Yields World's Largest LNG Carrier: Technological Advancements Result in the Globalization of a Once-Regional Energy Resource*. Retrieved October 10, 2011, from http://www.businesswire.com/portal/site/home/permalink/?ndmViewId=news_view&newsId=20081217005080&newsLang=en

Committee on Energy and Commerce. (2011, May 26). *Democrats Press for Committee Hearing on Hydraulic Fracturing*. Retrieved October 11, 2011, from http://democrats.energycommerce.house.gov/index.php?q=news/democrats-press-for-committee-hearing-on-hydraulic-fracturing

Federal Energy Regulatory Comission. (1992, April 8). *Pipeline Service Obligations and Revisions to Regulations Governing Self-Implementing Transportation Under Part 284 ofthe Commission's Regulations, Docket No. RM91-11-000; Regulation of Natural Gas Pipelines After Partial Wellhead Decontrol*. Retrieved October 10, 2011, from Order Number 636: http://www.ferc.gov/legal/maj-ord-reg/land-docs/rm91-11-000.txt

Federal Energy Regulatory Commission. (2010, December 3). *About FERC*. Retrieved October 10, 2011, from What FERC Does: http://www.ferc.gov/about/ferc-does.asp

Federal Energy Regulatory Commission. (2011, June 20). *Existing FERC Jurisdictional LNG Import/Export Terminals*. Retrieved October 10, 2011, from http://ferc.gov/industries/gas/indus-act/lng/exist-term.asp

Gasland. (2010). *About the Film*. Retrieved June 27, 2011, from http://www.gaslandthemovie.com/about-the-film/

GlobalSecurity.org. (2011, July 7). *LNG Tanker History*. Retrieved October 10, 2011, from http://www.globalsecurity.org/military/systems/ship/tanker-lng-history.htm

Grand Directions. (n.d.). *Horizontal Drilling, Chapter 5*. Retrieved October 7, 2011, from Oklahoma Geological Survey, Highlights Files: http://www.ogs.ou.edu/highlightsfiles/HDRILLChap5HorizDrill.ppt

Kramer, A. E., & Bennhold, K. (2009, January 6). Gazprom Reduces Deliveries of Gas Through Ukraine. *The New York Times*, p. 8.

Maize, K., Butcher, C., & Peltier, R. (2011, September 1). *Power Magazine*. Retrieved October 10, 2011, from Global Gas Glut: http://www.powermag.com/business/3974.html

National Energy Technology Laboratory. (2009, April). *Modern Shale Gas Development in the United States: A Primer*. Retrieved October 7, 2011, from U.S. Department of Energy, Office of Fossil Energy: http://www.netl.doe.gov/technologies/oil-gas/publications/EPreports/Shale_Gas_Primer_2009.pdf

NaturalGas.org. (2011). *Natural Gas - From Wellhead to Burner Tip*. Retrieved October 10, 2011, from The Transportation of Natural Gas: http://www.naturalgas.org/naturalgas/transport.asp

Qatargas. (2011). *About Us*. Retrieved October 10, 2011, from Current Operations: http://www.qatargas.com/AboutUs.aspx?id=62

Qatargas Operating Company Limited. (2008, July-August). *The Pioneer: The Magazine of Qatargas Operating Company Limited*. Retrieved October 10, 2011, from Issue No. 120: http://www.qatargas.com/uploadedFiles/QatarGas/Media_Center/Publications/Pioneer_Jul-Aug_2008-English.pdf

Rühl, C. (2010, March/April). Global Energy After the Crisis. *Foreign Affairs*, pp. 72-73.

Sabine Pipe Line LLC. (2011). *The Henry Hub*. Retrieved October 10, 2011, from http://www.sabinepipeline.com/Home/Report/tabid/241/default.aspx?ID=52

Sharples, B., & Paton, J. (2009, December 9). Exxon to Build $15 Billion Gas Project for Asia. *BusinessWeek Online*, p. 22.

United States Department of Energy. (2005, August). *Office of Fossil Energy*. Retrieved October 10, 2011, from Liquified Natural Gas: Understanding the Basics: http://www.fe.doe.gov/programs/oilgas/publications/lng/LNG_primerupd.pdf

United States Energy Information Administration. (2002, June). *The Basics of Unerground Natural Gas Storage*. Retrieved October 10, 2011, from http://tonto.eia.doe.gov/ftproot/natgas/storagebasics.pdf

United States Energy Information Administration. (2004, June). *U.S. LNG Markets and Uses: June 2004 Update*. Retrieved October 10, 2011, from http://www.eia.doe.gov/pub/oil_gas/natural_gas/feature_articles/2004/lng/lng2004.pdf

United States Energy Information Administration. (2007, November). *Office of Oil and Gas*. Retrieved October 10, 2011, from Natural Gas Compressor Stations on the Interstate Pipeline Network: Developments Since 1996, p. 2: http://www.eia.doe.gov/pub/oil_gas/natural_gas/analysis_publications/ngcompressor/ngcompressor.pdf

United States Energy Information Administration. (2009, April). *Independent Statistics and Analysis.* Retrieved October 10, 2011, from Estimates of Peak Underground Working Gas Storage Capacity in the United States, 2009 Update: http://www.eia.doe.gov/pub/oil_gas/natural_gas/feature_articles/2009/ngpeakstorage/ngpeakstorage.pdf

United States Energy Information Administration. (2010, December 30). *Independent Statistics and Analysis.* Retrieved October 11, 2011, from Natural Gas Navigator: http://www.eia.gov/dnav/ng/hist/rngr11nus_1a.htm

United States Energy Information Administration. (2011). *Independent Statistics and Analysis.* Retrieved October 5, 2011, from International Energy Statistics--Natural Gas Reserves: http://www.eia.gov/cfapps/ipdbproject/iedindex3.cfm?tid=3&pid=3&aid=6&cid=&syid=1980&eyid=2009&unit=TCF

United States Energy Information Administration. (2011b). *Independent Statistics and Analysis.* Retrieved October 5, 2011, from International Energy Statistics--Natural Gas Production: http://www.eia.gov/cfapps/ipdbproject/iedindex3.cfm?tid=3&pid=26&aid=1&cid=regions&syid=1980&eyid=2009&unit=BCF

United States Energy Information Administration. (2011c). *Independent Statistics and Analysis.* Retrieved October 6, 2011, from International Energy Statistics--Natural Gas Reserves: http://www.eia.gov/cfapps/ipdbproject/iedindex3.cfm?tid=3&pid=3&aid=6&cid=&syid=2009&eyid=2009&unit=TCF

United States Energy Information Administration. (2011d). *Independent Statistics and Analysis.* Retrieved October 6, 2011, from International Energy Statistics, Dry Natural Gas Production 2010: http://www.eia.gov/cfapps/ipdbproject/iedindex3.cfm?tid=3&pid=26&aid=1&cid=US,&syid=2010&eyid=2010&unit=BCF

United States Energy Information Administration. (2011e). *Independent Statistics and Analysis.* Retrieved October 6, 2011, from International Energy Statistics, Dry Natural Gas Consumption: http://www.eia.gov/cfapps/ipdbproject/iedindex3.cfm?tid=3&pid=26&aid=2&cid=US,&syid=2010&eyid=2010&unit=BCF

United States Energy Information Administration. (2011f, September 29). *Independent Statistics and Analysis.* Retrieved October 6, 2011, from U.S. Natural Gas Imports by Country, 2005-2010: http://www.eia.gov/dnav/ng/ng_move_impc_s1_a.htm

United States Energy Information Administration. (2011g, September 29). *Independent Statistics and Analysis.* Retrieved October 6, 2011, from U.S. Natural Gas Exports by Country, 2005-2010: http://www.eia.gov/dnav/ng/ng_move_expc_s1_a.htm

United States Energy Information Administration. (2011h, September 29). *Independent Statistics and Analysis.* Retrieved October 6, 2011, from Natural Gas Consumption by End Use: http://www.eia.gov/dnav/ng/ng_cons_sum_dcu_nus_a.htm

United States Energy Information Administration. (2011i, September 29). *Independent Statistics and Analysis.* Retrieved October 6, 2011, from Natural Gas Prices, Dollars per Thousand Cubic Feet: http://www.eia.gov/dnav/ng/ng_pri_sum_dcu_nus_a.htm

United States Energy Information Administration. (2011j, September 29). *Independent Statistics and Analysis.* Retrieved October 6, 2011, from Natural Gas Prices for Hawaii: http://www.eia.gov/dnav/ng/ng_pri_sum_dcu_SHI_a.htm

United States Energy Information Administration. (2011k). *Independent Statistics and Analysis.* Retrieved October 7, 2011, from International Energy Statistics, United States Natural Gas Proved Reserves, 1999-2009: http://www.eia.gov/cfapps/ipdbproject/iedindex3.cfm?tid=3&pid=3&aid=6&cid=US,&syid=1999&eyid=2009&unit=TCF

United States Energy Information Administration. (2011l, August 3). *Independent Statistics and Analysis.* Retrieved October 7, 2011, from Frequently Asked Questions: How much shale gas does the United States have?: http://www.eia.gov/tools/faqs/faq.cfm?id=58&t=8

United States Energy Information Administration. (2011m, September). *Monthly Energy Review, September 2011.* Retrieved October 7, 2011, from Table A4. Approximate Heat Content of Natural Gas (Btu per cubic foot): http://www.eia.doe.gov/emeu/mer/pdf/pages/sec13_4.pdf

United States Energy Information Administration. (2011n, June 9). *Independent Statistics and Analysis.* Retrieved October 10, 2011, from Natural Gas Explained--Natural Gas Pipelines: http://www.eia.gov/energyexplained/index.cfm?page=natural_gas_pipelines

United States Energy Information Administration. (2011o). *Independent Statistics and Analysis.* Retrieved October 10, 2011, from About U.S. Natural Gas Pipelines--Transporting Natural Gas--Estimated Natural Gas Pipeline Mileage in the Lower 48 States, Close of 2008: http://www.eia.doe.gov/pub/oil_gas/natural_gas/analysis_publications/ngpipeline/mileage.html

United States Energy Information Administration. (2011p). *Independent Statistics and Analysis.* Retrieved October 10, 2011, from About U.S. Natural Gas Pipelines - Transporting Natural Gas: http://www.eia.doe.gov/pub/oil_gas/natural_gas/analysis_publications/ngpipeline/process.html

United States Energy Information Administration. (2011q, September 29). *Independent Statistics and Analysis*. Retrieved October 10, 2011, from Natural Gas, Underground Natural Gas Storage Capacity, Total Number of Active Fields: http://www.eia.gov/dnav/ng/ng_stor_cap_a_EPG0_SAD_Count_a.htm

United States Energy Information Administration. (2011r). *Independent Statistics and Analysis.* Retrieved October 10, 2011, from Major Legislative and Regulatory Actions, FERC Order 636: The Restructuring Rule (1992): http://www.eia.doe.gov/oil_gas/natural_gas/analysis_publications/ngmajorleg/ferc636.html

United States Energy Information Administration. (2011s, September 29). *Independent Statistics and Analysis*. Retrieved October 10, 2011, from U.S. Natural Gas Imports by Country, Annual, : http://www.eia.gov/dnav/ng/ng_move_impc_s1_a.htm

United States Energy Information Administration. (2011t, January 31). *Independent Statistics and Analysis*. Retrieved October 10, 2011, from Voluntary Reporting of Greenhouse Gases Program (Fuel and Energy Source Codes and Emission Coefficients): http://www.eia.doe.gov/oiaf/1605/coefficients.html

United States Energy Information Administration. (2011u, April). *Independent Statistics and Analysis.* Retrieved October 11, 2011, from Annual Energy Outlook 2011: http://www.eia.gov/forecasts/aeo/pdf/0383(2011).pdf

United States Energy Information Administration. (2011v, October 5). *Independent Statistics and Analysis*. Retrieved October 10, 2011, from Natural Gas Futures Contract 1 (Dollars/Mil. BTUs): http://www.eia.gov/dnav/ng/hist/rngc1M.htm

United States Environmental Protection Agency. (2011, October 6). *Hydraulic Fracturing*. Retrieved October 11, 2011, from EPA's Draft Hydraulic Fracturing Study Plan: http://water.epa.gov/type/groundwater/uic/class2/hydraulicfracturing/index.cfm

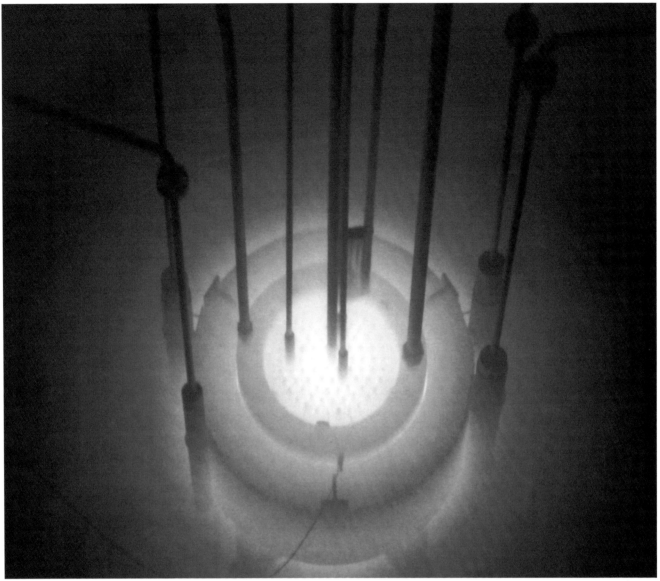

The Cerenkov Effect in a Nuclear Reactor, *Credit/ Courtesy of United States Nuclear Regulatory Commission*

Chapter 5
Uranium

Atoms, Isotopes & Fission

The previous chapters, about coal, petroleum and natural gas, consider non-renewable fossil fuels that we ignite in order to release their energy. Uranium is different because it is not a fossil fuel, and it is not burned in order to release its stored energy. Recall that fossil fuels originate with plant and animal matter that lived around 300 million years ago. Uranium, on the other hand, originates from the cataclysmic death of a massive star—a supernova—when lighter atomic elements fuse into heavier atomic elements. A supernova event took place in our celestial neighborhood over 6 *billion* years ago and produced all the naturally-occurring uranium that will ever exist on our planet. **Uranium is not a fossil fuel, but it is a non-renewable energy source because it cannot be replaced in a short time span.**

Releasing the energy contained in uranium is a lot more complicated than working with coal, petroleum and natural gas. The energy contained within fossil fuels is liberated by combustion—a relatively simple process that, to put it bluntly, simply requires lighting a match to break a variety of chemical bonds. By contrast, the energy contained within uranium is released by fission—a comparatively complex process that requires particularly behaved neutrons, a particular form of uranium (usually), and a variety of additional compounds and elements that are able to moderate and sustain a chain reaction.

Understanding fission requires examining atoms, and a logical place to start is inside an atom's nucleus where protons are located. Protons are positively charged particles, and their number within a nucleus is literally the atomic number of that element. You are probably familiar with the atomic number of elements because any physical science class you've ever been in probably had the Periodic Table of the Elements hanging somewhere on a wall. Each of the elements on this table has a unique atomic number (hydrogen = 1, helium = 2, lithium = 3, etc.). The atomic number is literally the number of protons in the nucleus of the atom. The number of protons determines an element's chemical properties and name. For example, carbon always has

Figure 5-1, The Periodic Table of the Elements, *Credit/Courtesy of Los Alamos National Lab*

[Periodic Table of the Elements]

element names in **blue** are liquids at room temperature
element names in red are gases at room temperature
element names in black are solids at room temperature

6 protons in its nucleus, iron always has 26 protons in its nucleus, and uranium always has 92 protons in its nucleus. Changing the number of protons in a nucleus literally changes the atom into a different element with different chemical properties.

All atomic nuclei except a particular form of hydrogen (H-1) also contain neutrons, which add mass but no charge to an atom. Atomic mass is crudely understood as the number of protons plus the number of neutrons in an atomic nucleus. For example, if carbon, with its 6 protons, also contains 6 neutrons, its atomic mass is commonly said to be 12 (6 + 6 = 12). If, on the other hand, the nucleus contains 8 neutrons, its atomic mass is commonly said to be 14 (6 + 8 = 14). As straightforward as this seems, it is too rough a measure to understand why uranium is an energy resource. This simple addition exercise reveals only the number of nucleons (protons and neutrons) inside an atomic nucleus.

Look closely at the atomic mass values of many elements on the Periodic Table of the Elements and you will often see numerals reported with decimal places. For example, carbon has an atomic mass value 12.01, magnesium, with 12 protons in its nucleus has an atomic mass value of 24.31, and silicon, with 14 protons in its nucleus has an atomic mass value of 28.09. These finer measures more accurately report the atomic mass of an element and reveal that this value is not simply the sum of an atom's nucleons. Rather, the atomic mass of an atom is scaled in reference to the mass of a particular form of carbon, and thus fractions of whole values are common. This discussion of atomic mass is not simply an academic exercise. In order to understand why uranium is an energy source, this more accurate measure of mass is essential because the whole point of a nuclear power plant is to transform a very tiny portion of this mass into energy.

From the above discussion it is apparent that while two atoms of the same element *must* have the same number of protons, they need *not* have the same number of neutrons. Different forms of the same element naturally exist. **Isotopes are forms of the same chemical element that differ by the number of neutrons in their nucleus, and thereby have different physical and nuclear properties.** For example, uranium-235 (U-235) is the most important isotope of uranium for the electrical generation industry. Because U-235 is uranium, it *must* have 92 protons in its nucleus. The "235" comes from adding these 92 protons to 143 neutrons (92 + 143 = 235). This isotope of uranium is quite rare. A much more common isotope is U-238. Because U-238 is uranium, it too *must* have 92 pro-

tons in its nucleus, but this form has 146 neutrons (92 + 146 = 238). **Naturally occurring uranium is about 99.3% U-238, and about 0.7% U-235.**

You have undoubtedly heard the aphorism "opposites attract." Perhaps you have also played with magnets and found that while opposite magnetic poles attract each other, similar magnetic poles repel. Try "sticking" two similar poles of two magnets together and you easily experience a force called electrostatic repulsion. With this in mind, now consider that inside an atomic nucleus the protons, each with a positive electric charge, are bound tightly together and strongly *resist* breaking apart. How is this possible? Shouldn't the protons repel each other? Well, they do, at least initially, until they are forced into close enough proximity and bind together in a process called fusion (very different than fission). During fusion, protons overcome their considerable electrostatic repulsion and join together with the substantially greater nuclear binding energy (Brookhaven National Lab, 2011). **The "strong nuclear force" binds together protons and neutrons inside an atomic nucleus.**

Fission is a process where a large atom splits into smaller atoms. These smaller atoms are literally fragments of the larger, initial atom. This is accomplished when the atomic nucleus of U-235 absorbs a deliberately-slowed neutron. These neutrons are called "moderated" or "thermal" neutrons, and they are purposely slowed from a velocity of 10^9 cm/second to about 10^5 cm/second so that they have a higher probability of interacting with U-235 and thereby inducing fission (World Nuclear Association, 2010). Neutrons are often moderated with large quantities of "light" water (just plain old H_2O), which is very different from "heavy" water, which is discussed later.

The atomic structure of the hydrogen (H-1) in light water makes interaction with fast-moving neutrons likely, and thereby produces thermal (slowed) neutrons. Slowing the velocity of neutrons is essential because 1) if they move too fast they simply ignore the U-235 and no fission occurs, and 2) the velocity of fission fragments is dependent upon the velocity of the thermal neutron. If the thermal neutron is too fast or too slow, fission is less likely *and* less than optimal fission fragment velocity is produced when fission does happen to occur (Madland, 2006, p. 119). This is very important for the nuclear power industry because fission fragment velocity constitutes about 85% of the energy liberated during the fission process.

When the nucleus of a U-235 atom absorbs a thermal neutron, it becomes highly unstable U-236. The strong

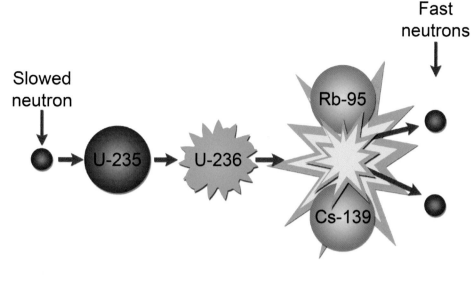

Figure 5-2, Fission Model & Equation

$$n + {}^{235}U \rightarrow [{}^{236}U] \rightarrow {}^{95}Rb + {}^{139}Cs + 2n + energy$$

nuclear force within the nucleus simply cannot accommodate the additional neutron, and the atom literally breaks apart as the previously-overcome electrostatic repulsion forces take over (Lawrence Berkeley National Laboratory, 2007). The U-236 atom then splits into 2 lighter pieces called fission fragments. Fission fragments are unique chemical elements with atomic numbers about half that of U-236. There are hundreds of possible, random combinations of fission fragments that result from a U-236 atom undergoing fission. 2 or 3 fast–moving neutrons are also released during fission, along with a burst of gamma rays (gamma rays are a form of electromagnetic radiation). One possible fission reaction is symbolized as: $n + {}^{235}U \rightarrow [{}^{236}U] \rightarrow {}^{95}Rb + {}^{139}Cs + 2n + energy$

For our purposes—understanding why uranium is an energy source—we need to consider the combined mass of the U-235 atom and the neutron that has entered its nucleus. These are called "reactants" and they have an atomic mass of 236.05. If we compare this mass to the mass of the fission fragments and neutrons (these are called "products"), what we find is really quite amazing. The products have a combined mass of 235.865, so subtracting the product

mass from the reactant mass yields 0.19 *missing* mass (CANTEACH, 2011, p. 53). **The combined mass of its fission fragments and neutrons is *less* than the mass of the U-236 atom.**

Where did the "missing" mass go? One of the fundamental laws of nature is that matter and energy cannot be created or destroyed, they can only change form. Another way to consider this is that matter is energy, and energy is matter—they are simply different forms of the same thing. Since we can't violate any laws of nature, there is in fact no "missing" mass. Instead, this mass has been transformed into energy. **Uranium is an energy source because fission transforms mass into energy.**

When the fission fragments split away from the atomic nucleus of U-236 they have an enormous amount of kinetic energy. This kinetic energy is a large portion of the "missing" mass of the U-236 atom. Since the fuel in nuclear reactors is solid, this motion is immediately stopped by adjacent material (Hore-Lacy, 2006, p. 49). The effect is that the kinetic energy of the fission fragments is transformed into thermal energy—heat—which in turn boils water to spin a turbine. Essentially,

Calculation 5-1, Transformation of Mass to Energy from Fission

Transformation of Mass to Energy from Fission	
A) Mass of Reactants:	236.05
B) Mass of Products:	235.865
C) Mass Converted to Energy (subtract):	0.19

a nuclear power plant transforms a tiny bit of mass into energy in order to 1) boil water to 2) spin a turbine to 3) spin a generator.

This seems like a heck of a lot of effort, and it is, but it's deemed worthwhile because of the enormous quantities of energy released during the fission process. The energy released from fission is often quantified in a unit called an electron volt (eV). An eV is the amount of energy a single electron acquires when moving across an electric potential difference of 1 volt. This definition can be tricky to get one's brain around, so consider this more down-to earth description: "...if you get a 9 volt battery and build a little particle accelerator out of it, you would be able to give an electron 9 volts of energy" (Cox & Forshaw, 2009, p. 148). A single eV is really a tiny amount of energy because an electron is so small. Consider, for example, that "it would take 6.2×10^{20} eV/sec to light a 100 watt light bulb" (National Aeronautics and Space Administration, 2011).

A single atom of U-236 that undergoes fission releases an average of 200 MeV of energy (recall that "M" means million, so 200 MeV = 200,000,000 eV), and trillions of reactions occur inside a nuclear reactor at any particular moment. These are very large numbers and can therefore be a difficult to grasp, so consider the following comparisons: A single carbon atom in a fossil fuel that undergoes combustion releases about 5 eV of energy (CANTEACH, 2011, p. 51). The energy liberated from a single gram of uranium is equivalent to the energy liberated from 6,000 pounds of coal (Lawrence Berkeley National Laboratory, 2000).

From Ore to Pellets

Uranium is an abundant element that typically occurs in very small amounts throughout most of the soil, rock and water that exists on the earth. Its quantity is often measured in parts per million (ppm), with normal concentrations around 2 to 4 ppm (World Nuclear Association, 2006). This literally means that of 1 million units of normal soil, rock and water, a maximum of 4 are typically uranium. If, for example, 1,000,000 kilograms is mined, a maximum of about 4 kilograms will be uranium. Since naturally occurring uranium is 99.3% U-238 and about 0.7% U-235, about 3.972 kilograms will likely be U-238 and just 28 grams will likely be U-235.

Such a small amount of U-235 is not sufficient to sustain a commercial mining operation, so the industry seeks out ore (ore is rock that contains valuable minerals) with uranium concentrations greater than 1,000 ppm. Higher grade ores contain greater amounts of uranium and are thus likely to be more profitable to mine. **The world's largest, high-grade uranium deposit is the McArthur River Operation in Northern Saskatchewan, Canada.** It contains 729,200 tons of ore with >200,000 ppm and will eventually yield 332,600,000 pounds of uranium oxide (U_3O_8) (Cameco Corporation, 2009). To put this in another perspective, consider that this ore is over 20% uranium.

Uranium is mined around the world in a variety of countries with diverse economic and political characteristics. While it is indeed a non-renewable energy source, there is no sense of an impending uranium

Figure 5-3, McArthur River Operation Uranium Mine, *Credit/Courtesy of Cameco Corporation*

Figure 5-4, Rabbit Lake Uranium Mine, *Credit/ Courtesy of Cameco Corporation*

shortage. New investments in exploration are presently yielding enough uranium deposits to readily satisfy current world demand for the next 100 years (International Atomic Energy Agency, 2008). In 2008, a total of 17 countries mined uranium, producing 113.87 million (113,870,000) pounds of U_3O_8, with Canada and Australia accounting for nearly 40% (45.42 million pounds) of that total. Other major producers in 2008 included Kazakhstan (22.15 million pounds), Namibia (11.35 million pounds) and Russia (9.15 million pounds) (Ux Consulting Company LLC., 2009). The United States, by contrast, produced 3.9 million pounds of U_3O_8 in 2008 and 3.7 million pounds in 2009 (United States Energy Information Administration, 2010).

Uranium oxide is not traded like other, better-know commodities, because no open and transparent marketplace exists in which U_3O_8 units might be publically exchanged. Instead, buying and selling are privately negotiated. A futures marketplace does exist, however, which is based on an agreement between the New York Mercantile Exchange (NYMEX) and weekly price data provided by The Ux Consulting Company, LLC (New York Mercantile Exchange--NYMEX, 2011). Two major independent marketing consultant companies—Ux Consulting Company LLC and TradeTech—publish weekly futures prices that private industry and governments use for budgeting and investment decisions. On August 26, 2011, for example, TradeTech listed 1 pound of U_3O_8 selling for $48.85 (TradeTech, 2011). On August 29, 2011, by contrast, Ux Consulting Company LLC listed 1 pound of U_3O_8 selling for $49.00 (Ux Consulting Company, LLC., 2011).

Figure 5-5, Uranium Ore, *Credit/Courtesy of United States Geological Survey*

Uranium is often mined in traditional underground mines. Unlike coal, however, many steps must take place before the uranium is ready to serve as a fuel. In the dry conversion process, the ore is first heated to drive off impurities, and then agglomerated and milled into a fine powder. This powder is often yellow or dark-khaki and commonly called "yellowcake". **Yellowcake is a powdery solid that is about 85% pure uranium oxide (U_3O_8).**

Traditional mining creates mill tailings, the sandy debris from the milling and extraction process that is pumped as slurry into a tailings dam. These dams are also called "impounds," and can look like shallow, man-made ponds. Tailings contain the element radium, Ra-226, that decays to form radon, R-222, a radioactive gas and public health hazard. **To protect the public from the harmful effects of tailings, the United**

Figure 5-6, Yellowcake, *Credit/Courtesy of Cameco Corporation*

States Congress enacted the Uranium Mill Tailings Radiation Control Act of 1978 (UMTRCA). The Act's two programs are called UMTRCA Title I and Title II. Title I legislated remedial actions at abandoned tailing sites, while Title II legislated various United States Nuclear Regulatory Commission (NRC) and United States Environmental Protection Agency (EPA) authorities over existing and future tailing sites (United States Nuclear

Regulatory Commission, 2011s). Since 1978, about 40 million cubic yards of low-level radioactive material has been encapsulated in 19 engineered disposal cells from 22 inactive uranium ore-processing sites in the United States (United States Department of Energy, 2011).

Uranium is also mined using in-situ recovery (ISR) (sometimes called in-situ leaching), where the ore remains beneath the surface of the earth, literally unexcavated, while the valuable uranium is withdrawn. ISR is a viable option when uranium ore is low-grade (near 1,000 ppm) and/or environmental or economic issues make traditional mining and milling unfeasible. ISR requires a liquid called "lixiviant". **Lixiviant is a solution of water, hydrogen peroxide, and sodium carbonate or carbon dioxide** (United States Nuclear Regulatory Commission, 2011). Injection wells first pump lixiviant down into uranium-bearing ore. The solution dissolves the uranium oxide, after which recovery wells pump the solution back up to the surface. The uranium-rich solution is concentrated and then evaporated, again yielding yellowcake, but this time without the mining scar or mill tailings. While environmentally less destructive, ISR produces liquid wastes (and some waste pipes, pumps and other processing equipment). The liquid waste is usually pumped back into the ground via deep disposal wells (United States Nuclear Regulatory Commission, 2011b).

In 2010, 4 underground and 4 ISR uranium mines existed in the United States. Together that year they produced 4.2 million pounds of U_3O_8 (United States Energy Information Administration, 2011). This is roughly close to the amount produced each year in the United

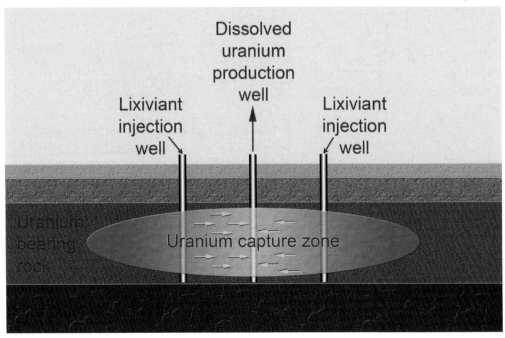

Figure 5-7, In-Situ Recovery (ISR) Uranium Mining

States since 2006. While moderately valuable, U_3O_8 is not immediately useful for the nuclear power industry in the United States because it does not contain enough of the fissile isotope U-235.

Yellowcake has the same isotopic ratio as naturally occurring uranium (99.3% U-238 and 0.7% U-235). For light-water reactors (these are the type in the United States) yellowcake must be enriched in order to boost the amount of U-235, the fissile isotope. **Light water nuclear reactors require a U-235 isotope concentration of between 3 to 5%.** The yellowcake cannot be enriched in its current—solid—state, so a conversion plant processes it into uranium hexafluoride (UF_6). Uranium hexafluoride is often called "hex" because of its 6 fluorine atoms. This compound is readily transformed among all three states—solid, liquid and gas—using well-established industrial processes, and thus facilitates uranium enrichment and shipment procedures (Argonne National Laboratory, 2011).

The United States has 1 uranium hexafluoride conversion facility—the Metropolis Works Plant (MTW)—located in Metropolis, Illinois and owned by Honeywell International, Inc. After its 2007 expansion, this facility annually processes up to 15,000 metric tons of yellowcake into UF_6 (ConverDyn, 2011). Conversion of yellowcake to UF_6 requires several steps. First, the yellowcake is chemically reduced in a kiln containing hydrogen to UO_2. Second, the UO_2 is reacted in another kiln with gaseous hydrogen fluoride (HF) to form uranium tetrafluoride (UF_4). This step is called hydrofluorination. Finally, the UF_4 is reacted with gaseous fluorine (F_2) to produce uranium hexafluoride (UF_6) (World Nuclear Association, 2011). At this stage in the conversion process the UF_6 is a gas. It is cooled and thereby

U_3O_8 to UF_6 Conversion

1. Reduction: $U_3O_8 + 2H_2 \longrightarrow 2H_2O + 3UO_2$

2. Hydrofluorination: $UO_2 + 4HF \longrightarrow UF_4 + 2H_2O$

3. Fluorination: $UF_4 + F_2 \longrightarrow UF_6$

Figure 5-8, U_3O_8 to UF_6 Conversion

condensed into a liquid, after which it is poured into 14 ton steel cylinders and allowed to cool for about 5 days until it solidifies into solid white crystals (ConverDyn, 2011b).

The UF_6—now solid and encased in steel—is then shipped across the Ohio River to Paducah, Kentucky, where the single operating uranium enrichment facility in the United States currently exists (additional facilities are in standby mode, under planning or review, or under construction). The facility is owned by the United States Enrichment Corporation (a public company listed on the New York Stock Exchange under the ticker "USU"). **In Paducah, Kentucky, a gaseous diffusion plant (GDP) boosts the U-235 isotope concentration in UF_6 from 0.7% to between 3 to 5%, thereby creating low-enriched uranium (LEU).**

Gaseous diffusion is one of three ways to enrich UF_6. Another method employs gas centrifuges, and the latest technique (still being developed and tested) is laser separation. All three processes rely on the fact that U-235 has a slightly different mass than U-238 (recall that they differ by 3 neutrons). All three processes also first change the solid "feed material" (the UF_6) back into a gas in order to accomplish enrichment.

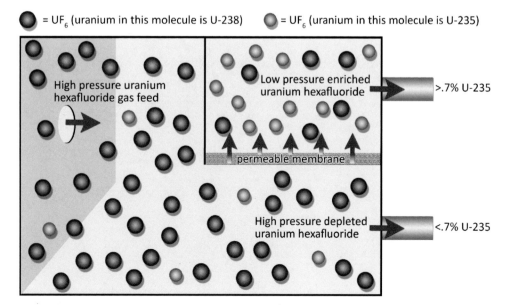

Figure 5-9, Enriching UF_6 with Gaseous Diffusion

Figure 5-10, Paducah Gaseous Diffusion Plant, Paducah, Kentucky, *Credit/Courtesy of USEC Inc.*

Thus, at the GDP in Kentucky, newly arrived LEU, in the form of UF$_6$, is slowly warmed and then forced into a series of gaseous diffusion converters. Porous membranes (called "barriers") inside the converters allow lighter U-235 molecules of UF$_6$ to diffuse more readily than heavier U-238 UF$_6$ molecules. Since the process is not very efficient, the gaseous feed material must be passed through many hundreds of diffusion converters (a series of converters is called a "cascade") before a high-enough concentration of U-235 isotopes exist within the enriched UF$_6$ stream (United States Nuclear Regulatory Commission, 2011c). When the desired amount of enrichment is achieved, the gaseous UF$_6$ is cooled and condensed back into a liquid. The liquid is then poured into 2.5 ton steel "product cylinders" and allowed to cool further into a solid state before shipment to a uranium fuel fabrication facility (United States Enrichment Corporation, 2011).

While not currently operating in the United States, the gas centrifuge enrichment method is well-established and more efficient than gaseous diffusion. The process relies upon a cascade of hundreds of vertically-oriented, rotating cylinders. Non-enriched UF$_6$ is forced into these cylinders, which rotate at high-speeds and thus create strong centrifugal forces. Heavier U-238 molecules of UF$_6$ collect near the sides, while lighter U-235 molecules of UF$_6$ remain in the center. The enriched UF$_6$ is repeatedly passed into successive centrifuges until the desired U-235 isotope ratio is achieved.

There are 6 uranium fuel fabrication facilities in the United States (United States Nuclear Regulatory Commission, 2011d). At these sites, product cylinders arrive via train or truck. They are gradually warmed up in order to convert the enriched UF$_6$ back into a gaseous state for chemical transformation into low enriched uranium dioxide (UO$_2$). This is accomplished with several chemical conversion steps that ultimately produce a fine a black powder of UO$_2$. The UO$_2$ powder is then pressed into small ceramic pellets and baked at about 1400 °C in a process called "sintering". **Uranium dioxide pellets, each measuring about 2 centimeters long and 1.5 centimeters in diameter, serve as the fuel for most nuclear power plants.**

The uranium dioxide pellets are stacked, end-to-end, inside hollow fuel rods made from zirconium alloy, often simply called "zircaloy" (zirconium is a metal, and an "alloy" is a mixture of metals). **Zircaloy is used for fuel rods because it is both corrosion resistant and does not readily absorb the neutrons that sustain the**

Figure 5-11, Moving a UF$_6$ cylinder at the Paducah Gaseous Diffusion Plant, Paducah, Kentucky, *Credit/Courtesy of USEC Inc.*

Figure 5-12, A Cylinder Hauler Moves UF$_6$ cylinders at the Paducah Gaseous Diffusion Plant, Paducah, Kentucky, *Credit/ Courtesy of USEC Inc.*

Figure 5-13, Enriching UF$_6$ with Gas Centrifuges, *Credit/Courtesy of United States Department of Energy*

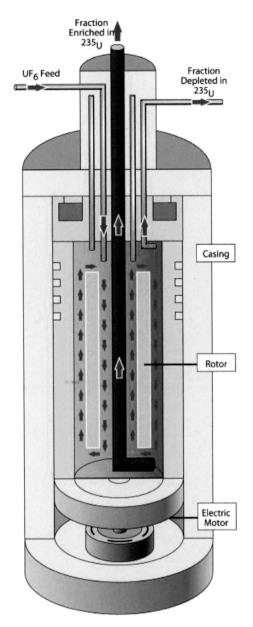

Figure 5-14, Model of Gas Centrifuge, *Credit/Courtesy of United States Nuclear Regulatory Commission*

Figure 5-15, Uranium Enrichment Facility, *Credit/Courtesy of United States Nuclear Regulatory Commission*

Figure 5-16, Uranium Dioxide Pellets, the Fuel in Most Nuclear Reactors, *Credit/Courtesy of United States Nuclear Regulatory Commission*

chain reaction inside a nuclear reactor. These rods are quite long, often measuring 4 meters long (over 13 feet high). Since 4 meters = 400 centimeters, about 200 individual uranium dioxide pellets fit inside each rod. Each rod is then bundled with many other rods to complete a nuclear fuel assembly (Areva, 2011). **Fuel assemblies can measure about 30 centimeters on each side and hold up to 264 rods total** (United States Nuclear Regulatory Commission, 2011e).

Reactors

In 2010, the 104 nuclear power reactors that existed in the United States had a total capacity of 101,004 MW (101 GW) and produced 806,968,301 MWh of electricity (United States Energy Information Administration, 2011b). **The United States generates far more electricity from nuclear power than any other country.** France, well-known for its nuclear power, places a distant second in MWh of nuclear power generation compared to the United States (it generates a bit more than 50% as many MWh from nuclear power during any particular recent year). Total electricity generation in the United States in 2010 from all sources was 4,120,028,000 MWh (United States Energy Informa-

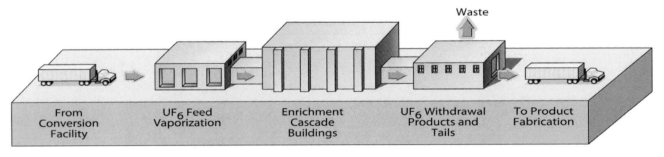

Figure 5-17, Fuel Rod Manufacturing with Uranium Dioxide Pellets, *Credit/Courtesy of United States Nuclear Regulatory Commission*

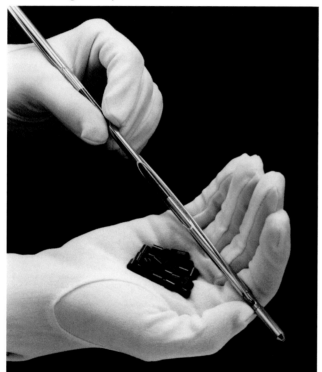

Figure 5-18, Uranium Dioxide Pellets Stacked Inside A Fuel Rod, *Credit/Courtesy of United States Department of Energy*

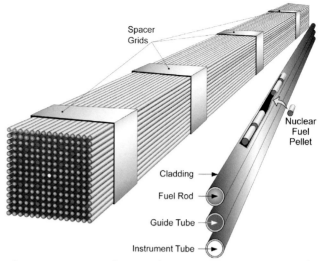

Figure 5-19, A Nuclear Fuel Assembly Model, *Credit/Courtesy of United States Department of Energy*

Figure 5-20, A Nuclear Fuel Assembly, *Credit/Courtesy of United States Nuclear Regulatory Commission*

tion Administration, 2012b). **Nuclear power accounted for 19.5% of the electricity generated in the United States in 2010.**

To put such numbers in perspective, consider that at the start of 2011, 444 operating nuclear power reactors existed in the world and had a total nameplate capacity of 378,313 MW (American Nuclear Society, 2011, p. 67). Thus, at the start of 2011, the United States contained 23% of all the world's nuclear reactors and 26% of the world's nuclear generation capacity. Also consider that while in any given year nuclear power generates approximately 20% of the electricity in the United States, about 76% of France's total

Figure 5-21, Reactor Head, *Credit/Courtesy of United States Nuclear Regulatory Commission*

electricity production comes from nuclear power (United States Energy Information Administration, 2012). France is more dependent on nuclear power, but generates far fewer MWh, than the United States. This of course makes sense given that France's population and economy are a fraction of those of the United States.

In the United States, 35 reactors are Boiling Water Reactors (BWRs), 69 are Pressurized Water Reactors (PWRs), for a total of 104 nuclear reactors (United States Nuclear Regulatory Commission, 2011f). Other reactor types exist throughout the world, but PWRs and BWRs dominate global reactor designs. Both essentially do the same thing—turn liquid water into steam to spin a turbine, to spin a generator, to produce electricity. Whether BWR or PWR, the essential design consists of a reactor core filled with normal water (H_2O—this is called a "light" water reactor) holding numerous fuel assemblies. BWRs contain between 370-800 fuel assemblies (United States Nuclear Regulatory Commission, 2011g), while PWRs contain between 150-200 fuel assemblies (United States Nuclear Regulatory Commission, 2011h). Water in the reactor core serves the dual purpose of neutron moderator and coolant. **As neutron moderator, reactor water slows down high-velocity neutrons, thereby initiating and sustaining fission. As coolant, reactor water transfers energy (in the form of steam) to a turbine.**

As fission begins in a reactor core, water both sustains the chain reaction and transfers energy. The essential difference between reactor designs is that water in a BWR literally boils and is the *same water* (in the form of steam) that spins the turbine. By contrast, the water in a PWR does not boil (it is hot, but since the pressure is very high, it is unable to boil). This hot, pressurized and radioactive water is routed to a heat exchanger. The heat exchanger is made up of thousands of tubes through which the water flows. The water heats the tubes, and the tubes heat water, which flashes to

Figure 5-22, Reactor Core, *Credit /Courtesy of United States Nuclear Regulatory Commission*

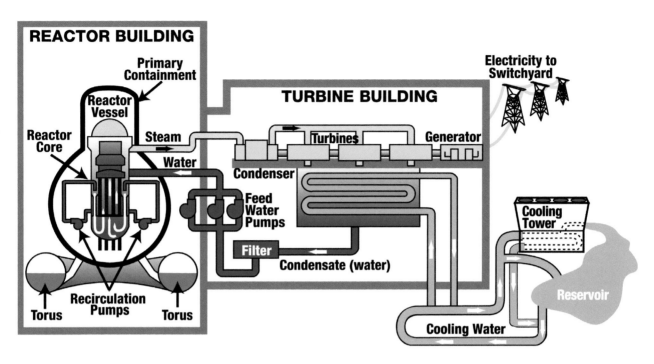

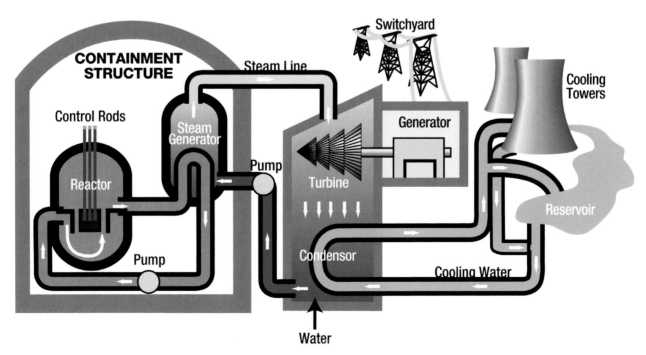

steam, which in turn spins a turbine. Steam in a PWR *never touches* the reactor core. An advantage of this design is that fewer parts of a power plant are exposed, and thus contaminated, by radioactive water.

Heavy water is used in some reactors as neutron moderator (think "neutron slowing material"). A heavy water molecule is made not from 2 atoms of H-1 hydrogen (the "2" subscript in H_2O), but rather 2 atoms of deuterium (H-2). Deuterium is simply an isotope of hydrogen, a hydrogen atom with a neutron in its nu-

Figure 5-23, A Boiling Water Reactor (BWR) Model, *Credit/Courtesy of Tennessee Valley Authority*

cleus. The most common form of hydrogen, H-1, contains only a proton in its nucleus and no neutrons. The addition of a neutron literally makes H-2 heavier than H-1. Heavy water's chemical formula can be written as ^2H_2O or D_2O, and a vessel holding heavy water literally contains a higher-than-normal concentration of these

Figure 5-24, A Pressurized Water Reactor (PWR) Model, *Credit/Courtesy of Tennessee Valley Authority*

molecules. Since deuterium already contains a neutron, it is less likely to absorb neutrons created from the fission process. This allows naturally-occurring, un-enriched UO_2 to be used as reactor fuel, thereby eliminating the need to invest in an enrichment facility.

In addition to heavy water reactors, other nuclear power plant designs exist, such as those using carbon dioxide and liquid sodium as a reactor coolant (think "heat transferring material"). These designs can have varying advantages compared to those using light water, including higher efficiencies and cheaper costs.

Besides fuel assemblies and coolant, a nuclear reactor must also contain control rods. **Control rods absorb neutrons and thus control the rate of fission inside a nuclear reactor.** Control rods are corrosion and radiation-resistant stainless steel tubes into which cadmium, hafnium and/or boron carbide (B_4C) powder or sintered pins are placed (Westinghouse Electric Company, 2011). Control rods are inserted and/or withdrawn from a reactor core as needed. Since neutrons are responsible for sustaining the fission chain reaction, by absorbing some of these neutrons the chain reaction can be slowed (with partially inserted control rods) or even stopped (with fully inserted control rods) (United States Department of Energy, 1993, pp. 48-50).

Each fission occurrence renders one less atom of U-235 inside the uranium dioxide (UO_2) pellets situated in a reactor's fuel assemblies. Over time, fission fragments accumulate which cannot themselves undergo fission and are thus useless as reactor fuel. Recall that UO_2 pellets are composed of LEU (between 3 to 5% U-235). When less than 1% of U-235 remains, the fuel assembly must be pulled and replaced with a fresh assembly. **Normally about 1/3 of a reactor's fuel assemblies are replaced during any regularly scheduled refueling outage, with outages occurring about every 18 months.** In the meantime, a typical reactor runs day-and-night, serving-up baseload power to the electric grid. Examples abound of scheduled refueling outages—just check the news headlines. For instance, in April of 2010, the Salem Nuclear Generating Station, Unit 1 in New Jersey was shut down for about 1 month to replace 1/3 of its fuel assemblies (New Jersey On-Line LLC, 2010). In another case, also in April of 2010, the Byron 2 reactor in Illinois was shut down, also for about 1 month, to perform maintenance activities and replace about 1/3 of the reactor's fuel assemblies (Reuters, 2010).

Spent Nuclear Fuel (SNF)

When depleted fuel assemblies are pulled from a nuclear reactor they are literally hot, and emit dangerous levels of radioactivity from all the accumulated fission fragments. Both of these issues are temporarily resolved by initially placing the fuel assembly into a spent fuel pool. **A spent fuel pool is a large concrete basin of water that absorbs heat and shields workers from radiation.** Federal regulations in the United States require that the SNF assemblies be under at least 20 feet of water to ensure worker safety (United States Nuclear Regulatory Commission, 2011i). Typical spent fuel pools are therefore about 40 feet deep, have of 4 to 6 feet thick concrete walls, and contain a 0.5 inch stainless steel inner liner. The spent fuel assemblies are normally submerged under 27 feet of water (Entergy Corporation, 2011). This amount of water is more-than sufficient to shield workers from any radiation. Outside the United States, SNF assemblies are normally submerged under at least 3 meters (almost 10 feet) of water (World Nuclear Association, 2011b).

Every operating nuclear power plant in the United States has SNF stored in on-site spent fuel pools. These pools (never intended to be permanent repositories) eventually fill up, however, thus mandating a second on-site storage method for those plants that have sufficiently cooled SNF but have run out of "wet" storage capacity (Idaho National Laboratory, 2011). **A typical nuclear power plant in the United States produces about 23 short tons of SNF each year, and the country as a whole has produced about 71,870 short tons of SNF over the past 40 years** (Nuclear Energy Institute, 2011). This waste is currently stored at or nearby the nuclear power plant that generated the waste.

After between 1 to 5 years (depending on the fuel assembly design), when the SNF has cooled enough to be extracted from the spent fuel pool, it can be placed in "dry cask storage". **In dry cask storage, SNF is surrounded by inert gas inside a leak-tight, steel and concrete cylinder that shields workers from radiation** (United States Nuclear Regulatory Commission, 2011j). Only some nuclear power plants in the United States (those with at-capacity spent fuel pools) employ dry cask storage for their SNF. Casks are vertically situated on above-ground concrete pads, or horizontally inside above-ground concrete silos. **Independent Spent Fuel Storage Installations (ISFSI) are sites where dry cask SNF storage occurs.** These sites are adjacent to the nuclear power plant or at a nearby location where truck or rail can readily transport the casks. As of late

Figure 5-25, A Spent Fuel Pool, *Credit/Courtesy of United States Nuclear Regulatory Commission*

2011, 33 states had at least one ISFSI (United States Nuclear Regulatory Commission, 2011k).

SNF can be either be reprocessed or stored indefinitely. Reprocessing essentially recycles the SNF by re-enriching the uranium that remains in the spent fuel. Recall that when fuel assemblies are pulled from reactors, the valuable fissile isotope of uranium (U-235) is reduced to less than 1%. This does not mean, however, that the spent uranium dioxide (UO$_2$) pellets are valueless. **SNF still contains about 96% uranium and holds enough of the fissile U-235 isotope to make re-enrichment possible.** Reprocessing involves literally chopping up fuel assemblies and soaking the material

in nitric acid. The acid dissolves the fuel but not the zircaloy, so the initial product of reprocessing is a fuel-rich, nitric acid solution that can be drawn-off, precipitated and then re-enriched (Nuclear Energy Agency, Organisation for Economic Co-operation and Development, 1994). Several countries in the world currently reprocess SNF, including the United Kingdom, Russia, France, India and Japan. These countries have what is called a "closed" fuel cycle because very little of the SNF is committed to permanent disposal (International Atomic Energy Agency, 2008b, pp. 79-141).

An "open" or "once-through" fuel cycle, on the other hand, describes indefinitely storing SNF without any

Figure 5-26, Map of Used Nuclear Fuel in Storage (Metric Tons, End of 2010), *Credit/Courtesy of Nuclear Energy Institute*

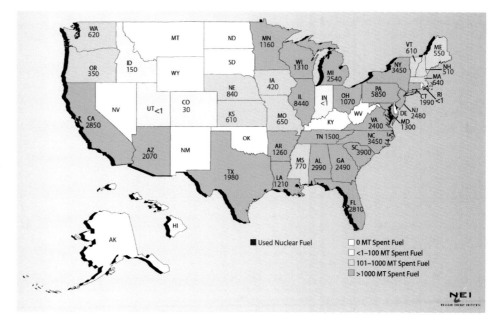

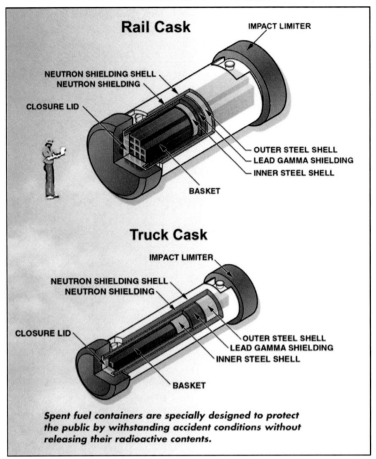

Figure 5-27, Models of Transportation Casks, *Credit/Courtesy of United States Nuclear Regulatory Commission*

Figure 5-28, Model of Truck Transportation Cask, *Credit/Courtesy of United States Department of Energy*

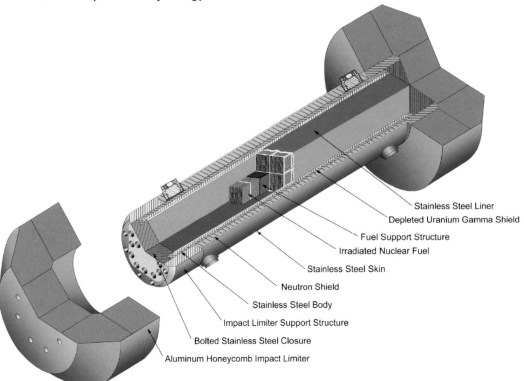

Figure 5-29, Rail & Truck Transport Casks, *Credit/ Courtesy of United States Department of Energy*

Figure 5-30, Rail Casks on Train, *Credit/Courtesy of United States Department of Energy*

Figure 5-31, A Truck Cask With Lid Removed, *Credit/ Courtesy of United States Department of Energy*

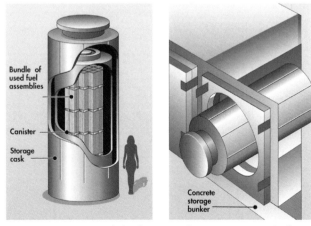

Figure 5-32, A Model of Dry Cask Storage, *Credit/ Courtesy of United States Nuclear Regulatory Commission*

Figure 5-33, Dry-Cask Transporter, *Credit/Courtesy of United States Nuclear Regulatory Commission*

Figure 5-34, Dry Cask Storage on Concrete Pad, *Credit/Courtesy of National Nuclear Security Administration*

Figure 5-35, Concrete Silos for Dry Cask Storage, *Credit/Courtesy of Idaho National Laboratory*

reprocessing. Several countries in the world currently do this, including the United States, Canada and Sweden. Prior to 1976, the United States did reprocess its SNF. The decision to temporarily halt reprocessing was initiated in late 1976 by President Ford, citing concerns about plutonium and nuclear weapons proliferations. In 1977 President Carter extended this temporary ban, which stood until 1981 when it was overturned by President Reagan. Market forces, however, including cheap and available uranium as well as the perceived high financial risk associated with nuclear power plants, preclude the establishment of any civilian reprocessing facilities in the United States to this date (Hylko, James M, 2008).

Nuclear power plants store their SNF on-site because 1) the United States does not reprocess civilian SNF and 2) no permanent centralized repository exists in the United States. Most countries with nuclear power plants do not have centralized repositories, including Belgium, Canada, China, Finland, France, Japan and the United Kingdom (United States Nuclear Waste Technical Review Board, 2009). **Under the Nuclear Waste Policy Act (NWPA)—initially passed in 1982 and later amended—the federal government of the United States is legally required to take full title of SNF and dispose of it in a permanent geological repository** (United States Nuclear Regulatory Commission, 2011l). This obligation exists because the NWPA established the Nuclear Waste Fund (NWF), into which electricity consumers pay one-tenth of a cent

(1/1,000[th] of a dollar, also sometimes written as 1.0 mil) for every kilowatt-hour of electricity generated via nuclear power (Cornell University Law School, 2010). About $750 million is collected annually which, coupled with interest, now totals around $24 billion (Sapien, 2011).

From 1987 to 2010 (23 years) the United States spent about $8.6 billion studying and licensing the now infamous Yucca Mountain site. Located on federal land about 90 miles northwest of Las Vegas, this was to become the country's sole long-term nuclear waste geological repository (Hylko, James M; Peltier, Robert, 2010). Environmental, political and legal controversy mired the project from the start. On March 3, 2010, the United States Department of Energy formally gave up the fight to develop the location by issuing a motion to the Nuclear Regulatory Commission to "withdraw its pending license application for a permanent geologic repository at Yucca Mountain, Nevada" (United States Department of Energy, 2010).

While giving up on Yucca Mountain, however, the federal government reiterated its commitment to stand by its long-term obligations to take title of and store the nation's SNF and nuclear waste, as outlined in the NWPA. In the meantime (as they have for decades), electrical utilities with operating nuclear power plants are preparing, paying for, and storing their own SNF at nearby, "temporary" sites. **Unable to reprocess and having no permanent repository, United States util-**

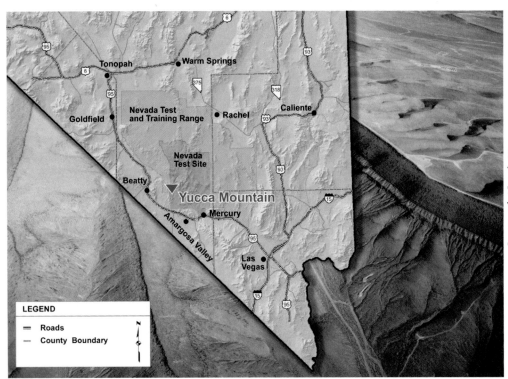

Figure 5-36, Location of Yucca Mountain, About 90 Miles Northwest of Las Vegas, *Credit/Courtesy of United States Department of Energy*

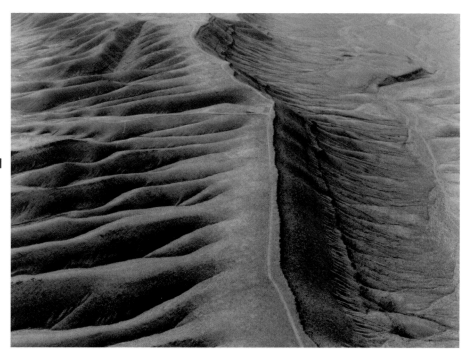

Figure 5-37, Yucca Mountain Arial View, *Credit/Courtesy of United States Department of Energy*

Figure 5-38, Yucca Mountain Tunnel Borer, *Credit/ Courtesy of United States Nuclear Regulatory Commission*

Figure 5-40, Yucca Mountain Tunnel Curve, *Credit/ Courtesy of United States Department of Energy*

Figure 5-39, Yucca Mountain Tunnel, *Credit/Courtesy of United States Nuclear Regulatory Commission*

Figure 5-41, Inside Yucca Mountain, *Credit/Courtesy of United States Nuclear Regulatory Commission*

ity companies use **Independent Spent Fuel Storage Installations (ISFSI) to store their SNF until the United States government takes possession.** This essentially means that the "temporary" dry cask storage scheme is becoming the de facto permanent solution to the SNF in the United States (Smith, 2010).

Since the billions of dollars that currently exist in the NWF cannot legally be used to offset "temporary" storage costs, utilities must absorb these costs. One ought to expect litigation to arise in such circumstances, and that's exactly what is happening. Just over a month after the motion to withdraw Yucca Mountain's pending license was filed, over a dozen utility companies across the United States, as well as their trade association—the National Association of Regulatory Utility Commissioners—sued the United States Department of Energy for continuing to collect fees as outlined in the NWF (Wald, 2010). The lawsuit claims that the United States government is failing to live up to its obligation to move and store SNF, and therefore electrical utility ratepayers should not continue to be charged under the NWF (National Association of Regulatory Utility Commissioners, 2010).

Figure 5-42, Independent Spent Fuel Storage Installations (ISFSI) Sites, *Credit/Courtesy of United States Nuclear Regulatory Commission*

Uranium, Fission and Radiation

Naturally occurring uranium, the LEU used in nuclear reactors, as well as the fission fragments produced in a nuclear reactor, are radioactive. **A radioactive material spontaneously emits electromagnetic radiation and/or particles.** The two major categories of radiation are ionizing and non-ionizing radiation. **Ionizing radiation has enough energy to break molecular bonds, or strip electrons from atoms (thus creating an ion—a charged atom), and therefore damage living cells** (United States Environmental Protection Agency, 2010). Examples of ionizing radiation include gamma rays, x-rays, alpha particles, beta particles and neutrons emitted during fission (World Health Organization, 2011). In our daily lives, the most common form of ionizing radiation we come across is radon gas, Rn-222.

Non-ionizing radiation does not contain enough energy to break molecular bonds or remove electrons from atoms. Examples of non-ionizing radiation include radio waves, microwaves, infrared light, visible light, and low frequency ultraviolet light (high frequency ultraviolet light is ionizing). Non-ionizing radiation is commonly used for communications, heating and seeing (without visible light radiation, our eyes would not see anything).

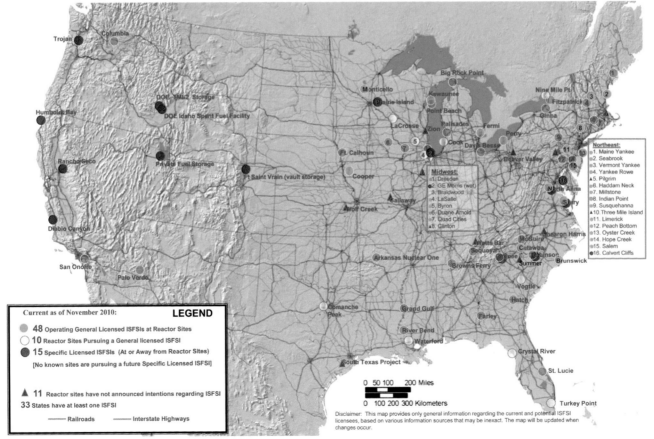

Uranium, whether naturally occurring or enriched, emits alpha particles (abbreviated with the small Greek letter α) from its atomic nucleus. Alpha particles are relatively large and heavy, composed of two protons and two neutrons, and have an approximate velocity of one-twentieth the speed of light (that's about 60,000 kilometers per second) (United States Environmental Protection Agency, 2011). Alpha particles are stopped by several sheets of paper, the outer layer of dead human skin, or a plastic bag. This is why a nuclear power plant employee might briefly handle a uranium dioxide (UO_2) pellet with no more protection than plastic safety glasses and not worry about acquiring radiation burns or sickness. Remember, however, that alpha particles are ionizing and can therefore damage living cells. **Inhaling or ingesting material that emits alpha particles is dangerous because atoms in living cells might become ionized and thereby undergo malignant, cancerous transformations.**

Recall that when fission occurs in a nuclear power plant, 2 or 3 fast–moving neutrons are released, along with a burst of gamma rays. Recall also that the reactor moderator (often water) slows the fast-moving neutrons so that they have a higher probability of interacting with the atomic nuclei of U-235 atoms and thus inducing fission. Thermal neutrons have considerable penetrating distance and can readily interact with atoms and molecules within your body. **Neutrons are a form of ionizing radiation because they can break molecular bonds, or strip electrons from atoms, if they collide with living cells.** Therefore, protection—neutron shielding—is required whenever fission occurs.

Water (H_2O) provides some of the best shielding because neutrons are 1) likely to interact with the water's hydrogen and 2) experience high energy loss when they do collide with the water's hydrogen (United States Department of Energy, 1993b, p. 18). This is one of the reasons that water is normally used in nuclear reactors, the other being heat transfer, as discussed earlier. About 103 centimeters of light water is required to absorb about 90% of thermal neutrons (Gunnerson, 2011, p. 23). Since 103 centimeters is a little over three feet, the requirement that SNF assemblies be under at least 20 feet of water is more than adequate neutron shielding for anyone in a nuclear power plant. In addition to water, several feet of concrete mixed with a high (7% or greater) percentage weight of water is an adequate neutron shield, again because of the hydrogen in the water molecule (United States Department of Energy, 1993b, p. 19).

The gamma rays (abbreviated with the small Greek letter γ) released during fission are not particles (like alpha particles and neutrons), but rather a form of electromagnetic radiation. Gamma rays have no mass, travel at the speed of light (300,000 kilometers per second) and have a very short wavelength. A wavelength is the distance from one wave crest to the next. The light you see the world with, for example, is called "visible light" and has a wavelength of between about 400 and 700 nanometers. A "nanometer" (nm) is one billionth of a meter. Imagine a meter stick divided into 1 billion equal slices. One of these slices is 1 nm. Now stack 400 of these slices together—this is about the smallest wavelength of electromagnetic radiation that our eyes can see, and is perceived as the color violet. By contrast, stacking 700 of these little slices together is about the largest wavelength of electromagnetic radiation that our eyes can see, and is perceived as the color red. Gamma rays, on the other hand, have the shortest wavelengths of electromagnetic radiation—far beyond the capacity of the human eye to perceive. Gamma wavelengths are measured in fractions of nanometers (take 1 nm and slice it up into one thousand equal pieces), for example from .03 to .003 nm (United States Environmental Protection Agency, 2011b).

Since gamma rays are ionizing radiation and contain a lot of energy, exposure can breakdown living cells. **Dense metals (such as lead or depleted uranium) provide the best shielding from gamma rays.** For example, lead is a good shielding material because it carries about 82 electrons for each atom and is also very dense. Therefore, if we have a sheet of lead 10 centimeters thick, an incoming gamma ray will 1) have a high likelihood of interaction, and 2) have its energy dissipated due to lead's high density. If lead is unavailable, or a different shielding material is needed, several feet of concrete is an appropriate substitute, and has the additional benefits of structural adaptability and strength (United States Department of Energy, 1993b, p. 20).

The fission fragments produced in a nuclear power plant are about half the size of a uranium atom. These fragments can include many different radioactive isotopes, some commonly noted examples being barium (Ba), strontium (Sr), cesium (Cs), iodine (I), krypton (Kr) and xenon (Xe) (Hore-Lacy, 2006, p. 76). The isotopes Sr-90 and Cs-137 in particular are well known because of their high beta particle radioactivity, their ability to enter the food supply, and the fact that they will remain radioactive for hundreds of years. **A beta particle is an electron, emitted from a neutron, within the nucleus of an atom.**

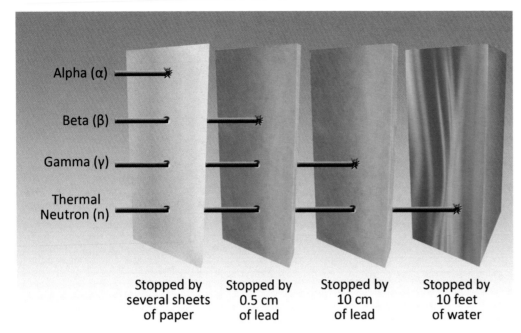

Alpha (α)

Beta (β)

Gamma (γ)

Thermal
Neutron (n)

Stopped by
several sheets
of paper

Stopped by
0.5 cm
of lead

Stopped by
10 cm
of lead

Stopped by
10 feet
of water

Figure 5-43, Penetrating Ability of Radiation

Beta particles (abbreviated with the small Greek letter β) are smaller, faster and more penetrating than alpha particles, but not nearly as penetrating as the neutrons and gamma rays. Beta particles can penetrate past the outer layers of dead human skin and into live tissue, and thus pose a greater threat than alpha particles. Like alpha particles, beta emitters mostly pose significant cellular risk if ingested (United States Environmental Protection Agency, 2011c). Beta particles are readily absorbed by a sheet of lead about 0.5 cm thick. Other objects of varying thicknesses and densities can also stop beta particles, such as "a thin sheet of metal or plastic or a block of wood" (United States Nuclear Regulatory Commission, 2011m).

Recall that the number of protons in the nucleus of an atom determines the element. When an atom releases an alpha or beta particle, the number of protons in the atomic nucleus literally changes, which necessarily changes the element. This is why the word "decay" is often used to describe radioactive materials. **Radioactive decay is the process of an atom emitting radiation, and in doing so changing into a different element.** For example, if an alpha (α) particle is released, an atomic nucleus *loses* 2 protons. If a beta (β) particle is released, an atomic nucleus *gains* 1 proton. This is because the beta particle—likely an electron—comes from a neutron within the atomic in the nucleus, and when this electron leaves, the neutron is no longer a neutron but is transformed into a proton. For example, by spontaneously emitting a specific series of alpha and beta particles, U-238 eventually decays into lead, Pb-206 (Argonne National Laboratory, EVS, 2005). Pb-206 is a stable isotope of lead and the last

stop in the decay series for U-238. **Stable isotopes are not radioactive.**

Radiation is measured in several different ways, which, while important, adds an unfortunate amount of confusion to any discussions about uranium, nuclear power and radiation. For our purposes let's concentrate on what people are most concerned about. **A "dose equivalent" is the amount of radiation a person might absorb and the resulting biological damage** (United States Nuclear Regulatory Commission, 2011n). Dose equivalent is measured in different ways—a "rem" (this is an abbreviation of "roentgen equivalent man") and a "sievert" (named after a Swedish researcher). (Note that the two measures can be equated as follows: 100 rem = 1 sievert). Since 1 rem, or 1 sievert, is an exceptionally large dose of radiation to receive at any one time, and thus very uncommon, much smaller units are normally used to discuss radiation exposure and resulting biological damage. Thus, 1/1000th (.001) of a rem, and 1/1000th (.001) of a sievert—respectively called a "millirem" and "millisievert"—are often used.

Our universe, our planet, the foods we eat, and even our own bodies, are radioactive. Every single person is constantly exposed to naturally-occurring, ionizing radiation. This naturally-occurring radiation comes primarily from radon, Rn-222 (United States Nuclear Regulatory Commission, 2011o). Radon is an invisible, odorless and tasteless gas that comes from the uranium that naturally exists in the rock and soil beneath our feet (United States Environmental Protection Agency, 2010b). Manmade sources of ionizing radiation are primarily from medical and dental x-rays,

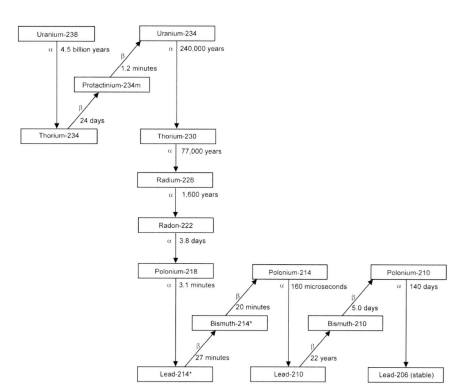

Figure 5-44, U-238 Decay Series, *Credit/Courtesy of Argonne National Laboratory (adapted)*

CT scans, and nuclear medicine. **In 2006, the average person in the United States received a typical annual dose of 3.11 millisievert (311 millirem) of naturally-occurring, ionizing, ubiquitous background radiation, as well as another 3.138 millisievert (313.8 millirem) from manmade, ionizing radiation sources, for a total of 6.248 millisievert (624.8 millirem) per person per year** (National Council on Radiation Protection and Measurements, 2009, pp. 236-237). Health effects from radiation exposure vary and are dependent upon the amount, type, tissue exposed, and length of exposure time.

Acute Radiation Syndrome (ARS) is the most feared health effect and comes from receiving a high dose of penetrating radiation over most of one's body in a short time span. **ARS symptoms can include fatigue, fever, nausea, vomiting, diarrhea, seizures, coma and death** (Centers for Disease Control and Prevention, 2006). Readers wishing to calculate their own annual

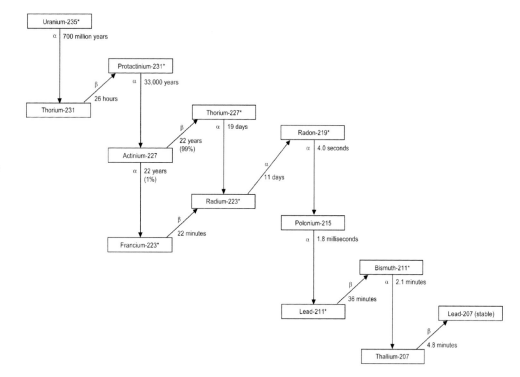

Figure 5-45, U-235 Decay Series, *Credit/ Courtesy of Argonne National Laboratory (adapted)*

Exposure Category		Millisievert	Millirem
Ubiquitous Background			
	Internal (radon & thoron inhalation)	2.28	228
	External (space)	0.33	33
	Internal (ingestion)	0.29	29
	External (terrestrial)	0.21	21
Total Ubiquitous Background		**3.11**	**311**
Medical			
	Computed Tomography	1.47	147
	Nuclear Medicine	0.77	77
	Interventional Fluoroscopy	0.43	43
	Conventional Radiography & Fluoroscopy	0.33	33
Total Medical		**3**	**300**
Consumer		**0.13**	**13**
Industrial		**0.003**	**.3**
Occupational		**0.005**	**.5**
Total		6.248	624.8

Table 5-1, Effective 2006 Radiation Dose per Individual In the United States

radiation dose will find convenient on-line calculators at The American Nuclear Society (http://www.ans.org/pi/resources/dosechart/) and The United States Nuclear Regulatory Commission (http://www.nrc.gov/about-nrc/radiation/around-us/calculator.html).

Regardless of background and medical radiation, people worry most about radiation exposure from nuclear power plants. The United States Nuclear Regulatory Commission comments on this concern quite succinctly, stating that any additional radiation exposure that comes from living next to a nuclear power plant has never been shown to cause any harm to people, most likely because the amount is so negligible:

> "...a person who spends a full year at the boundary of a nuclear power plant site would receive an additional radiation exposure of less than 1 percent of the radiation that everyone receives from natural background sources."

(United States Nuclear Regulatory Commission, 2011p)

Uranium's Virtues and Vices

High-profile nuclear accidents in Ukraine and Japan, in addition to unsettled SNF disposal problems, have stymied nuclear power expansion in the United States. However, faced with concerns about global warming, an over-reliance on foreign sources of energy, and the need for jobs, as well as the nuclear power industry's exemplary safety record, some public perceptions are shifting in favor of this technology (Helman, Schoenberger, & Wherry, 2005). This is perhaps best demonstrated with the fact that, after a 30 year hiatus, construction of two new nuclear power reactors is currently underway at the Vogtle power plant near Waynesboro, Georgia. Southern Company, the owner of the Vogtle plant, already has two operating reactors at the site. The addition of reactor units 3 and 4, at a cost of about $14 billion dollars, will increase the plant's nameplate capacity by 2,200 MW (Southern Company, 2012) to a total of 4,436 MW. The new reactors are scheduled to come online in 2016 and 2017.

As an energy source, uranium has a lot to offer. Recall that in any given year nuclear power generates approximately 20% of the electricity in the United States. Also impressive is the raw amount of energy that this

Figure 5-46, Aerial View of Vogtle 3 and 4 Construction Site, *Credit/Courtesy of Southern Company*

Figure 5-47, Vogtle Unit 4 Circulating Water System Pipes and Nuclear Island, *Credit/Courtesy of Southern Company*

Figure 5-48, Assembly of Vogtle Unit 3 Containment Vessel Bottom Head, *Credit/Courtesy of Southern Company*

fissile metal offers. **A single uranium dioxide (UO$_2$) fuel pellet (measuring about 2 centimeters long and 1.5 centimeters in diameter) contains as much energy as 149 gallons of oil, or 1,780 pounds of coal, or 17,000 cf of natural gas** (Nuclear Energy Institute, 2011b).

Uranium is an emission-free energy source because nuclear power plants do not emit any carbon dioxide, nitrogen or sulfur compounds while generating electricity. This fact can radically lower a country's annual carbon emissions. For example, recall from Chapter Two that a single, typical 500 MW coal-fired power plant produces about 4 million short tons of CO$_2$ in a single year. If a nuclear power plant replaced that coal-fired power plant, all that carbon would remain trapped in the ground and not released into the atmosphere. Of course, other energy sources also generate CO$_2$ free, or CO$_2$ neutral electricity, such as biomass, hydroelectric, wind, geothermal and sunlight. In 2011, however, these sources *combined* accounted for 478,289,000 MWh of electricity generation in the United States while nuclear power *alone* generated 718,388,000 MWh (United States Energy Information Administration, 2012b, p. 18). **Nuclear power generates the vast majority of emission free electricity in the United States.** This emission-freedom is convincing some people who once opposed nuclear power

to reconsider their position and tentatively support policy initiatives promoting nuclear power.

Nuclear power does produce SNF, which presents many challenges and is perhaps the most important "vice" of this energy source. Statisticians, however, often point out that the *volume* of the waste is quite small—especially when compared to the volume of waste produced from comparably-sized fossil fuel power plants. **In one year, a 1,000 MW nuclear power plant produces about 23 short tons of SNF waste, while a 1,000 MW coal-fired power plant produces about 8,000,000 short tons of CO$_2$ waste.** Neither waste-stream is desirable, of course, and nuclear waste presents unique disposal challenges, but the fact remains that there is simply a lot less waste to contend with when nuclear power plants are compared to fossil fuel power plants. Nuclear advocates also point out that while there does not currently exist a permanent geologic repository for SNF in the United States, there is no storage-space crisis at ISFSI sites across the country. While SNF pools are indeed filling up, there is still abundant room for dry cask storage across the country.

Nuclear power advocates also point to the fact that plutonium-239 (Pu-239) is created inside nuclear reactors when non-fissile but "fertile" U-238 absorbs a neutron and undergoes a specific radioactive decay series (this is called "breeding" Plutonium-239 by "transmuting" U-238). Plutonium is a "transuranic" element, meaning it has an atomic number greater than uranium's atomic number (92 on the Periodic Table of the Elements). This decay series is described in the following manner: n→U-238 → U-239 β→ Np-239 β→ Pu-239. Pu-239 is fissile, and thus can serve as nuclear fuel in mixed-oxide (called "MOX") fuel assemblies, provided that SNF assemblies are first reprocessed to separate and then concentrate the Pu-239. **MOX fuel assemblies contain the fissile isotopes of U-235 and Pu-239.** The transmutation of U-238 into Pu-239 has been well-understood since 1941 when Enrico Fermi worked on isolating this transuranic element at the University of Chicago's Metallurgical Laboratory (Frost, 1996).

Since the United States does not reprocess SNF, no Pu-239 is currently used in MOX fuel assemblies at any commercial nuclear power plant in the country. This practice is under pressure to change, however. In May 2010 an Argonne National Laboratory representative testified before the United States House of Representatives Committee on Science and Technology and argued for the development of a SNF recycling demonstration program that would close the nuclear fuel cycle in the United States by re-enriching spent fuel and extracting fissile Pu-239, while satisfying proliferation, environmental, health and safety concerns (Peters, 2010). By demonstrating that reprocessing is both technologically and economically viability, the project hopes to incentivize and entice private capital into this under-developed area of nuclear power generation in the United States.

Coupled with the virtues of uranium as an energy resource, significant vices also exist. Most damning is the dual fact that SNF is toxic and some SNF fission fragments literally take thousands of years to decay into stable, non-radioactive elements. Describing the decay series over time requires getting our brains around two issues: the term "half-life", and realizing that radionuclides decay spontaneously and randomly, but with a constant probability. **A half-life is the time it takes for half the radionuclides of a specific radioactive isotope in a sample to decay into a more stable state.**

Consider this illustration: A small rock sample contains 1 million atoms of U-238. U-238, as we know from the previous discussion, is unstable, with each atom emitting an alpha particle and thereby transforming itself into thorium-234 (Th-234). Not all 1 million atoms of U-238 emit their alpha particle same moment, however. Instead, a single atom of U-238 will win the decay "lottery" and spontaneously decay first, leaving all 999,999 others remaining intact. Our sample now, literally, has 1 Th-234 atom and 999,999 U-238 atoms. More time elapses, and then another U-238 atom spontaneously decays into Th-234. Our sample now has only 999,998 atoms of U-238. This process continues, and after 4.5 billion years only 500,000 U-238 atoms (half of the original 1 million) remain in our sample. A single half-life has thus elapsed. After another 4.5 billion years, a second half-life elapses and we are left with only 250,000 U-238 atoms. After 7 half-lives, less than 1% of the 1 million U-238 atoms remains in the sample.

Figure 5-49, Half-Life Graph

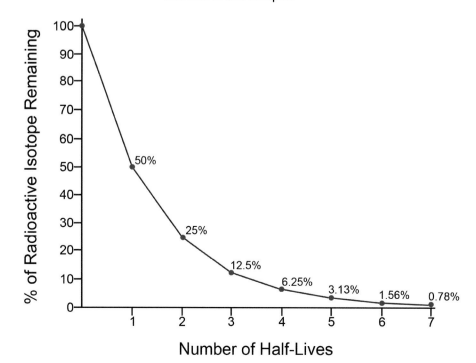

Every radioactive isotope has a unique half-life — some are only seconds long (Radon-219 has a half-life of 4 seconds) while others last millions of years (Uranium-235 has a half-life of 700 million years). SNF, as we already know, contains both fissile fragments and transuranic elements, all of which have unique decay series and half-lives. For example, Pu-239, previously discussed as a "virtue" of uranium, has a half-life of 24,100 years. Placed into either an ISFSI, or in some future geologically isolated repository, this alpha-emitting element will last 168,700 years before < 1% of its atoms remain in any particular SNF assembly (24,100 years x 7 half-lives = 168,700 years). While plutonium is not inherently dangerous unless inhaled (Centers for Disease Control and Prevention, 2006), nuclear critics contend that waiting over 168,000 years to ensure that no Pu-239 becomes airborne dust or particulate matter in a water supply is simply too long for any society to remain vigilant about the effects that this waste might inflict on future generations or civilizations. **Long-term, passive control plans for the now scrapped Yucca Mountain site, called for various monuments, warning markers and radiation symbols designed to last 10,000 years** (United States Department of Energy, 2008, p. 136), **far short of the expected lifetime of many radionuclides.**

Nuclear critics also point to government insurance subsidies as evidence that nuclear power is a failed technology that cannot exist on its own because no company is willing to accept the inherent risk (Public Citizen, 2004, p. 2). Particularly egregious, they claim, is the Price-Anderson Act. **As of 2011, the Price-Anderson Act limits the offsite liability of a nuclear power plant to $12 billion** (United States Nuclear Regulatory Commission, 2011q). This coverage is paid for by nuclear power plant operators through a combination of insurance premiums and required assessments across the industry if a nuclear accident occurs. Who pays the rest of the bill, however, if a catastrophic nuclear accident occurs and the liability exceeds $12 billion? You do, along with everyone else in the United States, because congress will then allocate taxpayer funds to cover any additional costs. The Price-Anderson Act, originally passed with the Atomic Energy Act of 1954, was recently extended through 2025 (United States Nuclear Regulatory Commission, 2011r, p. 118).

The mere possibility of a catastrophic nuclear accident also constitutes an important vice of uranium. No other energy source conjures up images of vaporized cities with millions of people dying in mere seconds from searing heat and colossal blast forces. While it is impossible for a nuclear power plant to explode like a nuclear bomb, this matters little in popular imagination where the word "nuclear" is habitually associated with the word "weapon". Simply put, people often fear nuclear power. **People often fear nuclear power because the technology is complicated and because accidents, while improbable, can be catastrophic.**

While cataclysmic mushroom clouds are impossible from a nuclear power plant mishap, an accident can release plumes of radioactive gasses and particles which can then be inhaled and ingested (United States Department of Homeland Security, Federal Emergency Management Agency, 2011). In 1986, for example, due to human error, the nuclear plant in Chernobyl, Ukraine, contaminated 155,000 km^2 (a land area a bit larger than the state of Georgia)with over 520 different radionuclides, including strontium-90 (Sr-90) and cesium-137 (Cs-137) (United Nations Office for the Coordination of Humanitarian Affairs, 2004). Still today, over 25 years after the accident, the concrete sarcophagus surrounding the damaged plant contains 216 tons of uranium and plutonium at a temperature of over 200 °C (400 °F) (Maize, 2011).

The 2011 Fukushima Daiichi nuclear power plant disaster in Japan offers a second example of a large-scale accident. An earthquake-triggered tsunami destroyed the plant's diesel back-up generators, thereby making it impossible to circulate cool water through the spent fuel pools. Heat and pressure rose to unsustainable levels until explosions ripped through several buildings, releasing successive plumes of radioactive dust that included Iodine-131 (I-131), cesium-134 (Cs-134), cesium-137 (Cs-137) and gamma radiation, and prompted Japanese officials to mandate an evacuation zone of 30 kilometers around the nuclear power plant (International Atomic Energy Agency, 2011). Cleanup efforts continue still today, with Japan's Nuclear and Industrial Safety Agency (NISA) recently stating that the explosions of units 1-4 at Fukushima Daiichi released about seven times more radiation than the Three Mile Island incident, but only 15% of that released from Chernobyl (Patel, 2011).

Important Ideas:

- Uranium is not a fossil fuel, but it is a non-renewable energy source because it cannot be replaced in a short time span.

- Isotopes are forms of the same chemical element that differ by the number of neutrons in their nucleus, and thereby have different physical and nuclear properties.

- Naturally occurring uranium is about 99.3% U-238, and about 0.7% U-235.

- The "strong nuclear force" binds together protons and neutrons inside an atomic nucleus.

- Fission is a process where a large atom splits into smaller atoms.

- The combined mass of its fission fragments and neutrons is *less* than the mass of the U-236 atom.

- Uranium is an energy source because fission transforms mass into energy.

- The world's largest, high-grade uranium deposit is the McArthur River Operation in Northern Saskatchewan, Canada.

- Yellowcake is a precipitated, powdery solid that is about 85% pure uranium oxide (U_3O_8).

- To protect the public from the harmful effects of tailings, the United States Congress enacted the Uranium Mill Tailings Radiation Control Act of 1978 (UMTRCA).

- Lixiviant is a solution of water, hydrogen peroxide, and sodium carbonate or carbon dioxide.

- U_3O_8 (yellowcake) has the same isotopic ratio as naturally occurring uranium (99.3% U-238 and 0.7% U-235).

- Light water nuclear reactors require a U-235 isotope concentration of between 3 to 5%.

- The United States has 1 uranium hexafluoride conversion facility—the Metropolis Works Plant (MTW)—located in Metropolis, Illinois and owned by Honeywell International, Inc.

- In Paducah, Kentucky, a gaseous diffusion plant (GDP) boosts the U-235 isotope concentration in UF_6 from 0.7% to between 3 to 5%, thereby creating low-enriched uranium (LEU).

- Uranium dioxide pellets, each measuring about 2 centimeters long and 1.5 centimeters in diameter, serve as the fuel for most nuclear power plants.

- Zircaloy is used for fuel rods because it is both corrosion resistant and does not readily absorb the neutrons that sustain the chain reaction inside a nuclear reactor.

- Fuel assemblies can measure about 30 centimeters on each side and hold up to 264 rods total.

- The United States generates far more electricity from nuclear power than any other country.

- Nuclear power accounted for 19.5% of the electricity generated in the United States in 2010

- In the United States, 35 reactors are Boiling Water Reactors (BWRs), 69 are Pressurized Water Reactors (PWRs), for a total of 104 nuclear reactors.

- As neutron moderator, reactor water slows down high-velocity neutrons, thereby initiating and sustaining fission.

- As coolant, reactor water transfers energy (in the form of steam) to a turbine.

- Control rods absorb neutrons and thus control the rate of fission inside a nuclear reactor.

- Normally about 1/3 of a reactor's fuel assemblies are replaced during any regularly scheduled refueling outage, with outages occurring about every 18 months.

- A spent fuel pool is a large concrete basin of water that absorbs heat and shields workers from radiation.

- A typical nuclear power plant in the United States produces about 23 short tons of SNF each year, and the country as a whole has produced about 71,870 short tons of SNF over the past 40 years.

- In dry cask storage, SNF is surrounded by inert gas inside a leak-tight, steel and concrete cylinder that shields workers from radiation.

- Independent Spent Fuel Storage Installations (ISFSI) are sites where dry cask SNF storage occurs.

- SNF still contains about 96% uranium and holds enough of the fissile U-235 isotope to make re-enrichment possible.

- Nuclear power plants store their SNF on-site because 1) the United States does not reprocess SNF and 2) no permanent centralized repository exists in the United States.

- Under the Nuclear Waste Policy Act (NWPA)—initially passed in 1982 and later amended—the federal government of the United States is legally required to take full title of SNF and dispose of it in a permanent geological repository.

- Unable to reprocess and having no permanent repository, United States utility companies use Independent Spent Fuel Storage Installations (ISFSI) to store their SNF until the United States government takes possession.

- A radioactive material spontaneously emits electromagnetic radiation and/or particles.

- Ionizing radiation has enough energy to break molecular bonds and strip electrons from atoms (thus creating an ion—a charged atom), and therefore damage living cells.

- Non-ionizing radiation does not contain enough energy to break molecular bonds or remove electrons from atoms.

- Inhaling or ingesting material that emits alpha particles is dangerous because atoms in living cells might become ionized and thereby undergo malignant, cancerous transformations.

- Neutrons are a form of ionizing radiation because they can break molecular bonds, or strip electrons from atoms, if they collide with living cells.

- Dense metals (such as lead or depleted uranium) provide the best shielding from gamma rays.

- A beta particle is an electron, emitted from a neutron, within the nucleus of an atom.

- Radioactive decay is the process of an atom emitting radiation, and in doing so changing into a different element.

- Stable isotopes are not radioactive.

- A "dose equivalent" is the amount of radiation a person might absorb and the resulting biological damage.

- Our universe, our planet, the foods we eat, and even our own bodies, are radioactive.

- In 2006, the average person in the United States received a typical annual dose of 3.11 millisievert (311 millirem) of naturally-occurring, ionizing, ubiquitous background radiation, as well as another 3.138 millisievert (313.8 millirem) from manmade, ionizing radiation sources, for a total of 6.248 millisievert (624.8 millirem) per person per year.

- A single uranium dioxide (UO_2) fuel pellet (measuring about 2 centimeters long and 1.5 centimeters in diameter) contains as much energy as 149 gallons of oil, or 1,780 pounds of coal, or 17,000 cf of natural gas.

- Uranium is an emission-free energy source because nuclear power plants do not emit any carbon dioxide, nitrogen or sulfur compounds while generating electricity.

- Nuclear power generates the vast majority of emission free electricity in the United States.

- In one year, a 1,000 MW nuclear power plant produces about 23 short tons of SNF waste, while a 1,000 MW coal-fired power plant produces about 8,000,000 short tons of CO_2 waste.

- MOX fuel assemblies contain the fissile isotopes of U-235 and Pu-239.

- A half-life is the time it takes for half the radionuclides of a specific radioactive isotope in a sample to decay into a more stable state.

- Long-term, passive control plans for the now scrapped Yucca Mountain site, called for various monuments, warning markers and radiation symbols designed to last 10,000 years, far short of the expected lifetime of many radionuclides.

- As of 2011, the Price-Anderson Act limits the offsite liability of a nuclear power plant to $12 billion.

- People often fear nuclear power because the technology is complicated and because accidents, while improbable, can be catastrophic.

- ARS symptoms can include fatigue, fever, nausea, vomiting, diarrhea, seizures, coma and death.

Works Cited

American Nuclear Society. (2011, March). World List of Nuclear Power Plants: Operatble, Under Construction, or on Order as of December 31, 2010. *Nuclear News*, p. 67.

Areva. (2011). *Operations, Front End Fuel Fabrication*. Retrieved September 2, 2011, from What is a fuel assembly? Stages of fuel assembly production: http://www.areva.com/EN/operations-807/fuel-production-integrated-expertise-from-a-to-z.html

Argonne National Laboratory. (2011). *Depleted UF6 Management Information Network*. Retrieved September 2, 2011, from What is uranium hexafluoride (UF6)?: http://web.ead.anl.gov/uranium/faq/uf6properties/faq8.cfm

Argonne National Laboratory, EVS. (2005, August). *Human Health Fact Sheet*. Retrieved September 7, 2011, from Natural Decay Series: Uranium, Radium, and Thorium: http://www.ead.anl.gov/pub/doc/natural-decay-series.pdf

Brookhaven National Lab. (2011). *Supernovas: Making Astronomical History*. Retrieved December 13, 2011, from Nuclear Energy, Fission and Fusion: http://snews.bnl.gov/popsci/nuclear-energy.html

Cameco Corporation. (2009, February 16). *McArthur River Operation, Northern Saskatchewan, Canada. National Instrument 43-101. Technical Report*. Retrieved August 30, 2011, from p. 6: http://www.cameco.com/common/pdf/investors/financial_reporting/technical_report/McArthur_Technical_Report_2009.pdf

CANTEACH. (2011). *CANDU Fundamentals*. Retrieved Decwember 15, 2011, from Chapter 7, Fission: http://can-teach.candu.org/library/20040707.pdf

Centers for Disease Control and Prevention. (2006, May 10). *Acute Radiation Syndrome (ARS): A Fact Sheet for the Public*. Retrieved September 7, 2011, from http://emergency.cdc.gov/radiation/ars.asp

Centers for Disease Control and Prevention. (2006, May 10). *Radiation*. Retrieved September 6, 2011, from Radioisotope Brief: Plutonium: http://www.bt.cdc.gov/radiation/isotopes/plutonium.asp

ConverDyn. (2011). *Metropolis Works History*. Retrieved September 2, 2011, from http://www.converdyn.com/metropolis/mtwhistory.html

ConverDyn. (2011b). *Honeywell Dry Fluoride Volatility Conversion Process*. Retrieved September 2, 2011, from http://www.converdyn.com/product/conversion.html

Cornell University Law School. (2010, February 1). *Title 42 > Chapter 108 > Subchapter III > Section 10222*. Retrieved September 6, 2011, from Nuclear Waste Fund: http://www.law.cornell.edu/uscode/42/usc_sec_42_00010222----000-.html

Cox, B., & Forshaw, J. (2009). *Why Does E=mc2?: (And Why Should We Care?) [Kindle Edition]*. Cambridge, MA: Da Capo Press.

Entergy Corporation. (2011). *Indian Point Energy Center*. Retrieved December 9, 2011, from Spent Fuel, Highly Effective Barriers: http://www.safesecurevital.org/safe-secure-vital/spent-fuel.html

Frost, B. R. (1996, September 3). *A Brief History of Materials R&D at Argonne National Laboratory from the Met Lab to Circa 1995*. Retrieved September 6, 2011, from Argonne National Laboratory: http://www.anl.gov/Science_and_Technology/History/Frosthist.html

Gunnerson, F. S. (2011). *Interaction of Radiation and Matter*. Retrieved December 9, 2011, from ME443-543, LectureNotes: http://www.if.uidaho.edu/~gunner/ME443-543/LectureNotes/radandmatter.pdf

Helman, C., Schoenberger, C. R., & Wherry, R. (2005, January 31). The Silence of the Nuke Protesters. *Forbes, 175*(2), pp. 84-92.

Hore-Lacy, I. (2006). Nuclear Energy in the 21st Century. London, United Kingdom: World nuclear University Press.

Hylko, James M. (2008, August 15). *Power Magazine*. Retrieved September 6, 2011, from How to Solve the Used Nuclear Fuel Storage Problem: http://www.powermag.com/issues/features/How-to-solve-the-used-nuclear-fuel-storage-problem_1388.html

Hylko, James M; Peltier, Robert. (2010, May 1). *The U.S. Spent Nuclear Fuel Policy: Road to Nowhere*. Retrieved September 6, 2011, from POWER Magazine: http://www.powermag.com/issues/cover_stories/2651.html

Idaho National Laboratory. (2011). *What is Spent Nuclear Fuel?* Retrieved September 2, 2011, from How is SNF handled and stored?: https://inlportal.inl.gov/portal/server.pt/community/national_spent_nuclear_fuel/389/national_spent_nuclear_fuel_-_what_is_snf_

International Atomic Energy Agency. (2008, June 3). *Uranium Report: Plenty More Where That Came From*. Retrieved June 2, 2010, from Supply Sufficient for Next Century Amid Robust Demand Growth: http://www.iaea.org/NewsCenter/News/2008/uraniumreport.html

International Atomic Energy Agency. (2008b, August). *Spent Fuel Reprocessing Options, Annex III, Country Reports*. Retrieved September 6, 2011, from IAEA-TECDOC-1587: http://www-pub.iaea.org/MTCD/publications/PDF/te_1587_web.pdf

International Atomic Energy Agency. (2011, June 2). *Radiological Monitoring and Consequences of Fukushima Nuclear Accident*. Retrieved September 7, 2011, from http://www.slideshare.net/iaea/radiological-monitoring-and-consequences-of-fukushima-nuclear-accident-2-june-2011

Lawrence Berkeley National Laboratory. (2000, August 9). *Guide to the Nuclear Wallchart*. Retrieved December 14, 2012, from Nuclear Fission Energy: http://www.lbl.gov/abc/wallchart/chapters/14/1.html

Lawrence Berkeley National Laboratory. (2007, March 30). *ABC's of Nuclear Science*. Retrieved December 13, 2011, from Fission: http://www.lbl.gov/abc/Basic.html

Madland, D. G. (2006). Total Prompt Energy Release in the Neutron-Induced Fission of 235 U, 238 U, and 239 Pu. *Nuclear Physics A*(772), 113-137.

Maize, K. (2011, May 1). *Power Magazine*. Retrieved September 7, 2011, from Chernobyl: Twenty-Five Years of Wormwood: http://www.powermag.com/nuclear/Chernobyl-Twenty-Five-Years-of-Wormwood_3633.html

National Aeronautics and Space Administration. (2011, April 6). *Science News*. Retrieved December 15, 2011, from Definition of an Electron Volt: http://science.nasa.gov/science-news/science-at-nasa/2001/comment2_ast15jan_1/

National Association of Regulatory Utility Commissioners. (2010, April 2). *State Regulators Go to Court with DOE over Nuclear Waste Fees*. Retrieved September 6, 2011, from http://www.naruc.org/News/default.cfm?pr=193

National Council on Radiation Protection and Measurements. (2009). *NCRP Report No. 160 - Ionizing Radiation Exposure of the Population of the United States.* Bethesda, MD: National Council on Radiation Protection and Measurements.

New Jersey On-Line LLC. (2010, April 6). *New Jersey Business News.* Retrieved June 4, 2010, from PSEG shuts Salem unit for scheduled outage: http://www.nj.com/business/index.ssf/2010/04/pseg_shuts_salem_unit_for_sche.html

New York Mercantile Exchange--NYMEX. (2011). *NYMEX UxC Uranium U3O8 Futures Contract frequently Asked Questions.* Retrieved August 31, 2011, from http://www.uxc.com/data/nymex/Uranium_FAQsPRESS.pdf

Nuclear Energy Agency, Organisation for Economic Co-operation and Development. (1994). *The Economics of the Nuclear Fuel Cycle.* Retrieved September 6, 2011, from 3.4 The back-end of the fuel cycle, 3.4.2 Reprocessing option: http://www.nea.fr/ndd/reports/efc/efc02.pdf

Nuclear Energy Institute. (2011). *Nuclear Waste: Amounts and On-Site Storage*. Retrieved September 2, 2011, from Used Nuclear Fuel and High-Level Radioactive Waste: http://www.nei.org/resourcesandstats/nuclear_statistics/nuclearwasteamountsandonsitestorage/

Nuclear Energy Institute. (2011b). *How it Works*. Retrieved September 7, 2011, from Nuclear Power Plant Fuel: http://www.nei.org/howitworks/nuclearpowerplantfuel/

Patel, S. (2011, July 1). *Power Magazine*. Retrieved September 7, 2011, from TEPCO: Most Fuel at Daiichi 1 Melted: http://www.powermag.com/nuclear/TEPCO-Most-Fuel-at-Daiichi-1-Melted_3786.html

Peters, M. T. (2010, May 19). *Charting the Course for American Nuclear Technology: Evaluating the Department of Energy's Nuclear Energy Research and Development Roadmap*. Retrieved September 6, 2011, from Testimony to U.S. House of Representatives Committee on Science and Technology: http://www.anl.gov/Media_Center/News/2010/PetersHouseScienceTestimonyMay2010.pdf

Public Citizen. (2004, September). *Price-Anderson Act: The Billion Dollar Bailout for Nuclear Power Mishaps.* Retrieved September 6, 2011, from http://www.citizen.org/documents/Price%20Anderson%20Factsheet.pdf

Reuters. (2010, April 19). *UPDATE 2-Exelon shuts Ill. Byron 2 reactor for refuel.* Retrieved June 4, 2010, from Traders see unit back in about a month: http://www.reuters.com/article/idUSN1920772120100419

Sapien, J. (2011, March 30). *$24 Billion In Nuclear Waste Fund Remains Untouched In U.S.* Retrieved September 6, 2011, from Huffington Post: http://www.huffingtonpost.com/2011/03/30/nuclear-waste-fund-us-24-billion_n_842762.html

Smith, R. (2010, June 2). 'Temporary' Home Lasts Decades for Nuclear Waste. *The Wall Street Journal, 255*(127), pp. A1-A18.

Southern Company. (2012). *Plant Vogtle Units 3 and 4 Background*. Retrieved February 29, 2012, from Vogtle Units 3 and 4 Benefits, Key Facts: http://www.southerncompany.com/nuclearenergy/presskit/docs/GTF_onePager_Vog_3_4_benefits_changes_v2.pdf

TradeTech. (2011, August 26). *Uranium Spot Price Indicator*. Retrieved August 31, 2011, from http://www.uranium.info/

United Nations Office for the Coordination of Humanitarian Affairs. (2004). *The United Nations and Chernobyl*. Retrieved September 7, 2011, from History of the United Nations and Chernobyl: http://www.un.org/ha/chernobyl/history.html

United States Department of Energy. (1993, January). *DOE Fundamentals Handbook , DOE-HDBK-1019/2-93*. Retrieved September 6, 2011, from Nuclear Physics and Reactor Theory, Volume 2 of 2: http://www.hss.doe.gov/nuclearsafety/ns/techstds/docs/handbook/h1019v2.pdf

United States Department of Energy. (1993b, January). *DOE Fundamentals Handbook, DOE-HDBK-1017/2-93* . Retrieved September 6, 2011, from Material Science, Volume 2 of 2, Module 5, Plant Material, Shielding Materials, Neutron Radiation: http://www.artikel-software.com/file/mats-v2.pdf

United States Department of Energy. (2008, October). *United States of America, Third National Report for the Joint Convention on the Safety of Spent Fuel Management and on the Safety of Radioactive Waste Management*. Retrieved September 7, 2011, from http://www.em.doe.gov/pdfs/3rd%20US%20Rpt%20on%20SNF%20JC--%20COMPLETE%20REPORT%20-%2010%2013%2008.pdf

United States Department of Energy. (2010, March 3). *Unitied States of America Nuclear Regulatory Commission*. Retrieved September 6, 2011, from U.S. Department of Energy's Motion to Withdraw: http://www.doe.gov/sites/prod/files/edg/media/DOE_Motion_to_Withdraw.pdf

United States Department of Energy. (2011, August 2). *Regulatory Framework*. Retrieved August 31, 2011, from UMTRCA Title I Disposal and Processing Sites: http://www.lm.doe.gov/pro_doc/references/framework.htm

United States Department of Homeland Security, Federal Emergency Management Agency. (2011, March 30). *Are You Ready?* Retrieved September 7, 2011, from Nuclear Power Plants: http://www.fema.gov/areyouready/nuclear_power_plants.shtm

United States Energy Information Administration. (2010, May 12). *Independent Statistics and Analysis*. Retrieved June 2, 2010, from Total Production of Uranium Concentrate in the United States: http://www.eia.doe.gov/cneaf/nuclear/dupr/qupd_tbl1.html

United States Energy Information Administration. (2011, June 15). *Independent Statistics and Analysis*. Retrieved August 31, 2011, from U.S. Uranium Mine Production and Number of Mines and Sources: http://www.eia.gov/uranium/production/annual/html/umine.html

United States Energy Information Administration. (2011b, August 18). *Independent Statistics and Analysis, U.S. Nuclear Generation of Electricity*. Retrieved September 2, 2011, from U.S. Nuclear Generation and Generating Capacity, 2010 Capacity and Generation: Final: http://www.eia.gov/cneaf/nuclear/page/nuc_generation/gensum.html

United States Energy Information Administration. (2012, February). *Independent Statistcis and Analysis*. Retrieved February 29, 2012, from Nuclear Explained, Data and Statistics: http://www.eia.gov/energyexplained/index.cfm?page=nuclear_home#tab2

United States Energy Information Administration. (2012b, January). *Independent Statistics and Analysis*. Retrieved February 29, 2012, from Electric Power Monthly, Net Generation by Energy Source: Total (All Sectors): http://www.eia.doe.gov/cneaf/electricity/epm/table1_1.html

United States Enrichment Corporation. (2011). *Nuclear Fuel Cycle*. Retrieved September 2, 2011, from Uranium Enrichment: http://www.usec.com/nuclearfuelcycle.htm

United States Environmental Protection Agency. (2010, July 19). *RadTown USA*. Retrieved September 6, 2011, from What is Radiation?: http://www.epa.gov/radtown/basic.html#what

United States Environmental Protection Agency. (2010b, October 12). *A Citizen's Guide to Radon*. Retrieved September 7, 2011, from Radon is a cancer-causing, radioactive gas: http://www.epa.gov/radon/pubs/citguide.html

United States Environmental Protection Agency. (2011, July 8). *Alpha Particles*. Retrieved September 6, 2011, from What are the properties of an alpha particle?: http://www.epa.gov/radiation/understand/alpha.html#properties

United States Environmental Protection Agency. (2011b, July 8). *Gamma Rays.* Retrieved September 7, 2011, from What are the properties of gamma radiation?: http://www.epa.gov/radiation/understand/gamma.html#properties

United States Environmental Protection Agency. (2011c, July 8). *Beta Particles*. Retrieved September 7, 2011, from What are the properties of beta particle?: http://www.epa.gov/radiation/understand/beta.html#emitters

United States Nuclear Regulatory Commission. (2011, August 1). *Lixiviant*. Retrieved September 2, 2011, from http://www.nrc.gov/reading-rm/basic-ref/glossary/lixiviant.html

United States Nuclear Regulatory Commission. (2011b, March 31). *Comparison of Conventional Mill, Heap Leach, and In Situ Recovery Facilities*. Retrieved September 2, 2011, from http://www.nrc.gov/materials/uranium-recovery/extraction-methods/comparison.html

United States Nuclear Regulatory Commission. (2011c, April 7). *Uranium Enrichment*. Retrieved September 2, 2011, from Gaseous Diffusion: http://www.nrc.gov/materials/fuel-cycle-fac/ur-enrichment.html#2

United States Nuclear Regulatory Commission. (2011d, April 13). *Locations of Major U.S. Fuel Cycle Facilities*. Retrieved September 2, 2011, from http://www.nrc.gov/info-finder/materials/fuel-cycle/

United States Nuclear Regulatory Commission. (2011e, March 31). *Fuel Fabrication*. Retrieved September 2, 2011, from Light Water Reactor Low-Enriched Uranium Fuel: http://www.nrc.gov/materials/fuel-cycle-fac/fuel-fab.html

United States Nuclear Regulatory Commission. (2011f, March 12). *Power Reactors*. Retrieved September 2, 2011, from http://www.nrc.gov/reactors/power.html

United States Nuclear Regulatory Commission. (2011g, April 29). *Boiling Water Reactors*. Retrieved September 2, 2011, from http://www.nrc.gov/reactors/bwrs.html

United States Nuclear Regulatory Commission. (2011h, April 21). *Pressurized Water Reactors*. Retrieved September 2, 2011, from http://www.nrc.gov/reactors/pwrs.html

United States Nuclear Regulatory Commission. (2011i, March 4). *Spent Fuel Pools*. Retrieved September 2, 2011, from http://www.nrc.gov/waste/spent-fuel-storage/pools.html

United States Nuclear Regulatory Commission. (2011j, April 18). *Dry Cask Storage*. Retrieved September 6, 2011, from http://www.nrc.gov/waste/spent-fuel-storage/dry-cask-storage.html

United States Nuclear Regulatory Commission. (2011k, March 31). *Locations of Independent Spent Fuel Storage Installations*. Retrieved September 6, 2011, from http://www.nrc.gov/waste/spent-fuel-storage/locations.html

United States Nuclear Regulatory Commission. (2011l, February 4). *Backgrounder on Radioactive Waste.* Retrieved September 6, 2011, from Storage and Disposal: http://www.nrc.gov/reading-rm/doc-collections/fact-sheets/radwaste.html

United States Nuclear Regulatory Commission. (2011m, March 31). *Radiation Basics.* Retrieved September 7, 2011, from Beta Particles: http://www.nrc.gov/about-nrc/radiation/health-effects/radiation-basics.html#beta

United States Nuclear Regulatory Commission. (2011n, March 31). *Measuring Radiation.* Retrieved September 7, 2011, from http://www.nrc.gov/about-nrc/radiation/health-effects/measuring-radiation.html

United States Nuclear Regulatory Commission. (2011o, March 31). *Sources of Radiation--Natural Background Sources.* Retrieved September 7, 2011, from Terrestrial Radiation: http://www.nrc.gov/about-nrc/radiation/around-us/sources/nat-bg-sources.html#terr

United States Nuclear Regulatory Commission. (2011p, March 31). *Uses of Radiation.* Retrieved September 7, 2011, from Nuclear Power Plants: http://www.nrc.gov/about-nrc/radiation/around-us/uses-radiation.html#npp

United States Nuclear Regulatory Commission. (2011q, June 9). *Fact Sheet on Nuclear Insurance and Disaster Relief Funds.* Retrieved September 7, 2011, from Nuclear Insurance: Price-Anderson Act: http://www.nrc.gov/reading-rm/doc-collections/fact-sheets/funds-fs.html

United States Nuclear Regulatory Commission. (2011r, January). *Office of the General Counsel.* Retrieved September 7, 2011, from Nuclear Regulatory Legislation, 111th Congress; 2nd Session, NUREG-0980, Vol. 1, No. 9: http://www.nrc.gov/reading-rm/doc-collections/nuregs/staff/sr0980/v1/sr0980v1.pdf#pagemode=bookmarks&page=5

United States Nuclear Regulatory Commission. (2011s, February 4). *Fact Sheet on Uranium Mill Tailings.* Retrieved September 7, 2011, from http://www.nrc.gov/reading-rm/doc-collections/fact-sheets/mill-tailings.html

United States Nuclear Waste Technical Review Board. (2009, October). *Survey of National Programs for Managing High-Level Radioactive Waste and Spent Nuclear Fuel.* Retrieved September 6, 2011, from A Report to Congress and the Secretary of Energy, P.6: http://www.nwtrb.gov/reports/nwtrb%20sept%2009.pdf

Ux Consulting Company LLC. (2009, August 1). *Nuclear Fuel Cycle.* Retrieved June 2, 2010, from World Uranium Production: http://www.uxc.com/fuelcycle/uranium/uxc_UProdTable.aspx

Ux Consulting Company, LLC. (2011, August 29). *UxC Nuclear Fuel Price Indicators (Delayed).* Retrieved August 31, 2011, from Weekly Spot Ux U3O8 Price as of August 29, 2011: http://www.uxc.com/review/uxc_Prices.aspx

Wald, M. L. (2010, April 5). *U.S. Sued Over Nuclear Waste Fees.* Retrieved September 6, 2011, from The New York Times: http://www.nytimes.com/2010/04/06/business/energy-environment/06nuke.html?scp=1&sq=Utilities%20Sue%20U.S.%20to%20Halt%20Nuclear%20Waste%20Fees.&st=cse

Westinghouse Electric Company. (2011, February). *Nuclear Fuel/Fuel Engineering.* Retrieved September 2, 2011, from BWR Control Rod CR 82M-1: http://www.westinghousenuclear.com/Products_&_Services/docs/flysheets/NF-FE-0002.pdf

World Health Organization. (2011). *Ionizing Radiation.* Retrieved September 6, 2011, from What is Ionizing Radiation?: http://www.who.int/ionizing_radiation/about/what_is_ir/en/index.html

World Nuclear Association. (2006, June). *What is uranium? How does it work?* Retrieved May 19, 2010, from http://www.world-nuclear.org/education/uran.htm

World Nuclear Association. (2010, September). *Physics of Uranium and Nuclear Energy*. Retrieved December 9, 2011, from Control of Fission: http://world-nuclear.org/education/phys.htm

World Nuclear Association. (2011, October). *Uranium Enrichment*. Retrieved December 9, 2011, from Conversion, In Detail: http://world-nuclear.org/info/inf28.html

World Nuclear Association. (2011b, July). *Radioactive Waste Management*. Retrieved December 15, 2011, from Managing HLW from used fuel: http://world-nuclear.org/info/inf04.html

Corn Cobs and Wood Chips Ready for Combustion,
Credit/Courtesy of USDA FSA, Farm Service Agency

Chapter 6
Biomass

Recently Living Matter

Biomass is the oldest energy resource harnessed by humanity. The "bio" part of the term tells us that organic material—plant matter—is its essential component. The most obvious source is wood, but this is just one of many energy resources derived from plants. Other biomass examples include vegetable oils, grasses, corn, oil-rich algae, manure, municipal sewage, landfill gas, and municipal solid waste (MSW). Think of MSW as combustible garbage consisting of paper, wood, food, leather, textiles and yard trimmings. Capturing the energy contained in these resources requires different techniques. Immediate combustion is the most straightforward method, where material is simply burned. Other techniques are more complex, such as the extensive processing required to transform various types of plants into ethanol, which can then be used as a vehicle fuel. Decay processes are also important because biomass that decomposes in an oxygen-deficient (anaerobic) environment releases methane (CH_4), which as we know from Chapter 4 is also an important source of energy.

Figure 6-1, Algae is a Biomass Energy Resource, *Credit/Courtesy of Peggy Greb, United States Department of Agriculture*

Figure 6-2, Corn is a Biomass Energy Resource, *Credit/Courtesy of Keith Weller, United States Department of Agriculture*

Biomass energy sources are not fossil fuels. Recall that fossil fuels are energy resources that come from the remains of plants and animals that lived millions of years ago. While biomass energy resources also come from plants and animals, this flora and fauna lived recently—as recently as right now, in fact, while you are reading this sentence. There is no deep burial and thus no naturally-occurring heat-and-pressure-induced transformation inside the earth's crust. There is nothing "fossil" about biomass. The living, green corn you drive by today, for example, could become fuel in your car's tank later this year. In the same respect, the garbage you throw out today could decompose and release methane, which could then be burned to generate electricity that powers your television. Both the corn and the garbage are also quickly replenished—the corn with a new crop season and the garbage with the next complete landfill pit. **Biomass is a renewable energy resource because it is replenished in a short time span.**

In 2010 renewable energy sources constituted about 8% of the total energy used in the United States. This equates to 8 quadrillion Btu, or more simply, 8 quads. **Biomass is the largest renewable energy source, supplying 4.24 quads, or 53% of all the renewable energy in the United States in 2010** (United States Energy Information Administration, 2011). Biomass energy resources are broadly divided into three categories: 1) wood, 2) waste, and 3) alcohol fuels.

Figure 6-3, Municipal Solid Waste is a Biomass Energy Resource

Figure 6-4, Wood is a Biomass Energy Resource, *Credit/Courtesy of David Nance, United States Department of Agriculture*

Figure 6-5, Sugarcane is a Biomass Energy Resource, *Credit/Courtesy of Ryan Viator, United States Department of Agriculture*

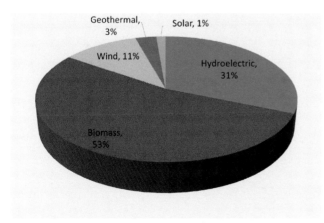

Figure 6-6, The Important Role of Biomass as a Renewable Energy Source

Wood

Wood has always been, and continues to be, the most important component of biomass energy. Its role as a readily-available energy resource is possible because logs, branches, bark, woodchips and other woody waste can be burned to generate heat and light. Indoor cooking with wood remains a vital part of life for tens of millions of people throughout the developing world. This traditional energy use, however, is fraught with problems because it exposes women and children to hazardous carbon monoxide (CO) levels and unhealthy particulate matter concentrations as ash becomes airborne. Respiratory illnesses and low birth weight are both associated with using wood as a fuel source, especially in enclosed areas. This problem is so prevalent that the World Health Organization (WHO) lists health problems caused by indoor smoke as contributing to 2 million deaths annually, with the majority occurring in low-income countries (World Health Organization, 2009, p. 11).

Figure 6-7, Indoor Cooking with Biomass is a Health Threat in the Developing World, *Credit/Courtesy of The Shell Foundation*

Figure 6-8, The 21 MW Tracy Biomass Power Plant Near San Francisco Burns Wood Residues, *Credit/ Courtesy of Andrew Carlin, National Renewable Energy Laboratory*

Figure 6-9, Truck Unloading Wood Chips for the Tracy Biomass Plant in Tracy, California, *Credit/ Courtesy of Andrew Carlin, National Renewable Energy Laboratory*

In the developed world, by contrast, wood is rarely used as a cooking and heating fuel. Instead, it and any derived fuels (such as black liquor) are burned in large-scale industrial facilities to generate electricity. In 2009, for example, 36,243,438 thousand kWh of electricity was generated in the United States from wood and its derived fuels (United States Energy Information Administration, 2011b). Wood-based electrical generation facilities capitalize on particular natural and/or industrial locations to burn wood and paper waste, often in the same power plant. This requires long-term logistical planning and cooperation to ensure financing, an uninterrupted fuel supply, and a committed buyer of any generated electricity (Greer, 2007). Compared to burning biomass in the developing world, this developed world, industrial-scale use of biomass is safer and delivers more reliable energy (in the form of electricity) to consumers. Not only is electricity more useful than biomass, much of the particulate matter is captured at the power plant and either recycled or disposed of in landfills.

Pulping liquor, usually called "black liquor", is more important as a wood-based energy source than wood itself. It is a combustible waste product that results from producing wood pulp—a precursor to making paper products—and is one of the "derived fuels" mentioned above. **Black liquor is the largest source of energy derived from wood** (United States Department of Energy, 2008). Wood pulp is produced by

Figure 6-10, Processing Woodchips into Biofuel, *Credit/Courtesy of National Database of State and Local Wildfire Hazard Mitigation Programs*

soaking wood chips in a "digester", a pressurized container holding a hot water and chemical solution. After several hours the chips are sufficiently broken down and the pulp fiber is removed. The solution remaining in the digester is then processed and concentrated to become energy-rich black liquor. Black liquor has significant energy content—about 5,800 to 6,600 Btu per pound of dried solids (United States Environmental Protection Agency, 2001, p. 3). This dark, viscous, energy-dense liquid is sprayed into a recovery boiler where it burns. **A recovery boiler is a large furnace that burns black liquor to 1) recover pulping chemicals and 2) boil water to produce steam.** Pulp producers use recovery boilers to lower their input costs and reduce or even eliminate their dependency on the grid. For example, the Östrand pulp mill, in Timrå, Sweden (owned by SCA Forest Products, a European paper manufacturer) annually generates 500 GWh of electricity with its recovery boiler, thereby making the facility "self-sufficient in both electricity and heat" (SCA Forest Products, Sundsvall, 2011).

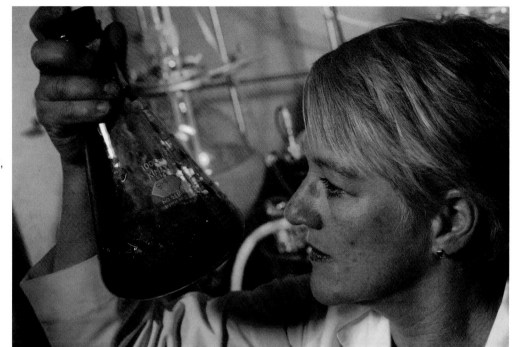

Figure 6-11, Black Liquor, *Credit/Courtesy of Keith Weller, USDA/ARS*

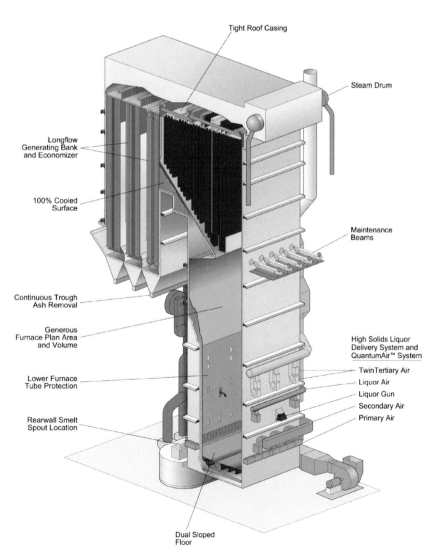

Figure 6-12, A Chemical and Heat Recovery Boiler for the Pulp and Paper Industry, *Credit/Courtesy of Babcock & Wilcox*

Figure 6-13, A Recovery Boiler, *Credit/Courtesy of Wikipedia*

Figure 6-14, Östrand Pulp Mill, in Timrå, Sweden, *Credit/Courtesy of SCA Forest Products*

Waste to Energy (WTE)

Most of the things we throw away—call it trash or garbage—can be burned, and as a society we throw away a lot of stuff. In 2009, for example, Americans generated a total of 243 million tons of municipal solid waste (MSW), which equates to 4.34 pounds of waste per person per day (United States Environmental Protection Agency, 2011). Much of this material, however, was not bound for landfills. After sorting and recycling, about 12% (29 million tons) of the remaining trash generated in 2009 was burned to generate electricity (United States Environmental Protection Agency, 2010, p. 2). **Waste-to-energy facilities, often called "incinerators" or "combustion plants," are steam-driven electrical generation plants that use garbage as fuel.** In 2009, 8,342,265 thousand kWh of electricity was generated in the United States from MSW, which includes paper, wood, food, leather, textiles and yard trimmings (United States Energy Information Administration, 2011b).

Ideal waste-generation behavior of course begins with reducing the amount of waste produced per person per year. This is followed by appropriate compost-

Figure 6-15, Municipal Solid Waste (MSW)

Figure 6-16, A Waste-to-Energy (WTE) Power Plant

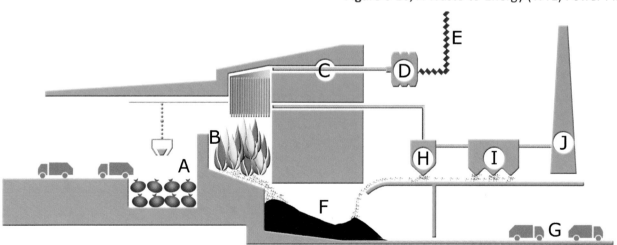

A   Waste deposited into enclosed pit
B   Waste transferred to combustion chamber
C   Heat recovered to generate steam
D   Steam turbine generates electricity
E   Electricity delivered to grid
F   Ash collected from combustion chamber, scrubber & electrostatic precipitator
G   Ash recycled or landfilled
H   Combustion gas passes through scrubber
I   Combustion gas passes though electrostatic precipitator
J   Cleaned gas emitted into atmosphere up stack

ing and recycling. In communities without a waste-to-energy (WTE) facility, anything not composted or recycled is bound for a landfill. Communities with a WTE facility, by contrast, might see some or all of their garbage headed for the combustion chamber. In 2010, 86 WTE plants existed in 24 states across America, with Florida leading in both number of plants (11) and generating capacity (530.4 MW) (plants vary in size from 10 to 77 MW) (Michaels, 2010, pp. 15-16). About 80% of the WTE plants in the United States do not recover recyclable materials before combustion, they simply "mass burn" the waste "as-is" (United States Environmental Protection Agency, 2011b). A preferable method is to recover recyclables and non-combustibles before combustion. **Refuse derived fuel (RDF) is combustible garbage—often processed into standard, pelletized form—after the recyclables and non-combustibles are recovered.**

Burning most things produces ash and gases, and in this respect garbage is no different. Specifically, burning MSW emits particulate matter (PM), metals (mostly as solids, except for mercury which is emitted as a vapor), acid gases such as hydrochloric acid (HCl) and sulfur dioxide (SO_2), carbon monoxide (CO), nitrogen oxides (NO_x), and various dioxin chemicals that exist as a gas or solid precipitate that condenses out on the PM (United States Environmental Protection Agency, 1996, pp. 12-14). These emissions are pollutants that can pose a health threat to humans and the environment.

As with coal-fired power plants, emission control features exist to reduce pollution volume from biomass-fired power plants. Controls include electrostatic precipitators and filters to capture the PM, scrubbing technologies that introduce lime into a combustion chamber or flue to capture acid gases, and combustion controls to ensure an incinerator operates at temperatures that destroy certain compounds and produce as little carbon monoxide as possible (United States Environmental Protection Agency, 1996, pp. 14-19). **Emission controls exist because the United States Clean Air Act (CAA), last amended in 1990, limits emissions of hazardous air pollutants (HAPs) according to their source category.** Under the CAA, the United States Environmental Protection Agency is charged with establishing stringent emission standards. The agency does this by requiring firms to install the maximum achievable control technology (MACT) available for each emitted source category of HAP (United States Environmental Protection Agency, 2007).

Figure 6-17, Map of Operating WTE Plants in the United States, 2010

States with Operating Plants

Calculating SO₂ & NOₓ Emissions from a WTE Power Plant

Calculation 6-1,
Calculating the SO$_2$ & NO$_x$ Emissions from a WTE Power Plant

A) Miami-Dade County Resource Recovery Facility has a 77 MW capacity
B) WTE facilities typically have a 90% capacity
C) 77 MW x .9 = 69.3 MW
D) 69.3 MW x 24 hours = 1,663.2 MWh generated in 1 day
E) 1,663.2 x 365 = 607,068 MWh generated in 1 year

So...

F) 1.2 pounds of SO$_2$ are produced per MWh from a WTE power plant
G) 1.2 pounds x 607,068 MWh = 728,482 pounds of SO$_2$ emitted in 1 year

And...

H) 6.7 pounds of NO$_x$ are produced per MWh from a WTE power plant
I) 6.7 pounds x 607,068 MWh = 4,067,356 pounds of NOx emitted in 1 year

For a straightforward example of WTE emissions, consider the following: The Environmental Protection Agency estimates that the average emission rate from a WTE facility includes 1.2 pounds of SO$_2$ per MWh and 6.7 pounds of NO$_x$ per MWh (United States Environmental Protection Agency, 2010b). Using an example from Florida, the Miami-Dade County Resource Recovery Facility has a 77 MW capacity that uses RDF (Miami-Dade County, 2011). Operating with a 90 capacity factor (this is normal for WTE plants), this facility annually produces 607,068 MWh. Thus in one year, 728,482 pounds of SO$_2$ are produced, along with 4,067,356 pounds of NO$_x$. Under the United States CAA, these gases cannot be released into the atmosphere and are thus scrubbed out of the plant's emissions using the techniques described above. On-site continuous emission monitors (CEMS) check the facility's emissions to ensure it maintains current standards established and enforced by the EPA.

Landfill Gas to Energy (LFGTE)

Of the 243 million tons of MSW generated in the United States in 2009, about 54% (132 million tons) ended up in landfills (these sites are often simply called "dumps") (United States Environmental Protection Agency, 2010, p. 2). At the landfill this material is buried in a series of clay-lined pits called "cells". Over the working life of a landfill (usually measured in decades) new pits are created as "active" cells fill with garbage and are then covered with clay. Eventually a large hill is created which is finally capped with clay and topsoil, and then seeded with grasses, shrubs and trees. The site may become a recreational location, a native ecosystem habitat, or remain off-limits to the public,

depending on local municipal ordinances and circumstances. Whatever the case, beneath the manicured hill is a mountain of garbage, steadily rotting away and releasing many tons of carbon dioxide (CO$_2$) and methane (CH$_4$) over the next 10 to 60 years.

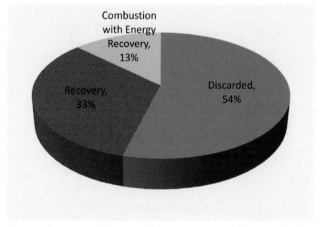

Figure 6-18, Pie-Chart of Management of MSW in the United States, 2008

CO$_2$ and CH$_4$ are naturally released by decomposition. **About 50% of the gas released from a landfill is CO$_2$, and about 50% is CH$_4$** (United States Environmental Protection Agency, 2011c). Both of these gases are odorless, so the unpleasant "rotten egg" smell associated with landfills comes from trace amounts of other gases. Hydrogen sulfide (H$_2$S) is often the culprit, detectable by many people at the very miniscule level of .00047 parts per million (ppm) (Iowa State University, 2004, p. 2). These and other gases are called "landfill gas" and often simply referred to as "LFG". For our purposes, whether or not a landfill smells bad is irrelevant—we are interested in the odorless CH$_4$.

When garbage decomposes it is first consumed by "aerobic" bacteria—these thrive in the presence of oxygen. Eventually the aerobic bacteria use most of the waste's oxygen and thereby make it available for "anaerobic" bacteria—which thrive in oxygen-deficient environments. The anaerobic bacteria break the waste into cellulose, amino acids and sugars, which are then further broken down through fermentation into gases and short-chained organic compounds. At this stage the fermenting garbage creates an ideal habitat for another type of microorganism—methanogens—which, as their name implies, create large quantities of methane gas (United States Environmental Protection Agency, 2010c, p. 2).

We know from Chapter 4 that methane is an important energy source and normally considered a non-renewable, fossil fuel. When generated from a landfill, however, this gas is appropriately considered a renewable energy source because it is continually replenished, so long as humans continue to create garbage. **In 2008, landfills in the United States emitted 6,016 Gg of methane** (a "Gg" is 1 Gigagram, or 1 billion grams) (United States Environmental Protection Agency, 2010c, p. 2). In landfills equipped with LFG collection and control systems, this methane is captured and either burned to generate electricity or

put to direct use. The United States has about 2,400 MSW landfills either operating or recently closed, and 520 (about 22%) of them have operating LFG utilization projects (United States Environmental Protection Agency, 2011d). In 2009, 7,351,052 thousand kWh of electricity was generated in the United States from LFG (United States Energy Information Administration, 2011b).

Collecting LFG requires building an interconnected network of vacuum-driven gas wells and mainlines throughout a landfill, preferably as each cell is filled with garbage. **A LFG well is a perforated plastic pipe, surrounded by gravel and placed vertically or horizontally within the garbage.** The pipe's perforations allow LFG to flow from the waste and into the pipe. LFG migrates to the well because low pressure is created by an electric vacuum at the end of the line. The vacuum literally draws LFG from the waste into the wells and then into a mainline. **A LFG mainline is a non-perforated plastic pipe that transports gas from a landfill to an adjacent treatment station, flare station or blower.** Landfill managers must continually monitor the quality and quantity of LFG and adjust the network's vacuum pressure accordingly in order to

Figure 6-19, Map of Operational and Candidate LFG Utilization Projects, 2010

Figure 6-20, Establishing a Horizontal LFG Well (Note Perforations in Pipe)

maximize LFG collection without drawing atmospheric air into the landfill. Recall that in order to create methane, anaerobic bacteria must be able to thrive. If the vacuum is too strong, atmospheric oxygen will be drawn into the landfill's rotting garbage, thereby killing the very bacteria needed to create the methane.

For an example of LFGTE, consider the following: Countryside Landfill in Grayslake, Illinois, holds 6,500,000 tons of MSW and is equipped with a LFG collection and control system of over 120 wells that supply 3,000 ft³ of LFG per minute. Across the street from the landfill, Biogas Energy Solutions operates a 7.8 MW power plant using several modified diesel engines that burn methane (United States Environmental Protection Agency, 2011e). Operating at a 90 capacity factor, this facility produces 61,320 MWh in one year. In 2009 the average Illinois household used 728 kWh per month (United States Energy Information Administration, 2011c), so over the course of a year 8,736 kWh are used. This is more simply expressed as 8.7 MWh. A division step reveals that this power plant satisfied the annual electrical needs of about 7,048 homes. Another way to state this is that in 2009 over 7,000 homes in Illinois got all of their electricity from garbage at Countryside Landfill.

Figure 6-21, LFG Wellhead

Figure 6-22, LFG Mainline (note non-perforated pipe)

Calculation 6-2, Calculating Electricity Production from the 7.8 MW Biogas Energy Solutions LFGTE Power Plant, Near Countryside Landfill, & the Number of Homes it Satisfies

Calculating Electricity Production from the 7.8 MW Biogas Energy Solutions LFGTE Power Plant, Near Countryside Landfill, & the Number of Homes it Satisfies

A) Biogas Energy Solutions has a 7.8 MW capacity
B) WTE facilities typically have a 90% capacity
C) 7.8 MW x .9 = 7 MW
D) 7 MW x 24 hours = 168 MWh generated in 1 day
E) 168 MWh x 365 = 61,320 MWh generated in 1 year

So...
A) Average Illinois household uses 728 kWh per month
B) 728 kWh x 12 = 8,736 kWh used in 1 year
C) 8,736 kWh = 8.7 MWh
D) 61,320 MWh / 8.7 = 7,048 households

Ethanol

Ethanol (C$_2$H$_5$OH) is a type of alcohol—the same kind that is in vodka, gin, beer or any other alcoholic beverage. It has been mixed into gasoline in the United States since 1988, beginning in Denver, Colorado, in order to reduce carbon monoxide (CO) emissions from vehicles' internal combustion engines (United States Energy Information Administration, 2008b). When vehicles use gasoline, not all the carbon in the fuel is completely burned because all engines have some degree of inefficiency. One result of this incomplete fuel-burn is that some of the leftover carbon chemically combines with oxygen, and is then emitted as CO from tailpipes. Carbon monoxide is poisonous to humans and emissions are regulated by the United States Environmental Protection Agency. Adding ethanol (an "oxegenate") to gasoline reduces carbon monoxide tailpipe emissions because the oxygen in ethanol increases the combustibility of gasoline (United States Department of Energy, 2003). Adding anything to gasoline "reformulates" it—that is, makes it into something other than 100% gasoline—and thus the term "reformulated gasoline" (abbreviated "RFG") is often used when ethanol or other oxygenates are blended into gasoline. **Reformulated gasoline (RFG) contains additives that result in a cleaner-burning fuel and thus less air pollution. Ethanol is an oxegenate bacuse it contains oxygen, and when blended with gasoline, makes gasoline burn more efficiently.**

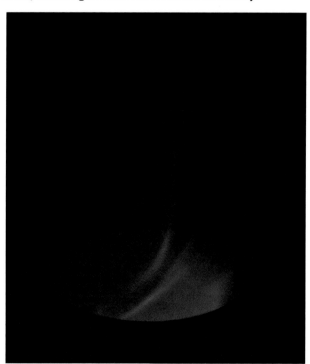

Figure 6-23, Combusting a Small Puddle of Ethanol, *Credit/Courtesy of Wikipedia*

Three different feedstocks are used to make ethanol: 1) sugar feedstocks, 2) starchy feedstocks, and 3) cellulosic feedstocks (United States Department of Energy, 2010). Sugar feedstocks are crops such as sugarcane (especially important in Brazil) and sugar beets. Starchy feedstocks are grains such as corn (especially important in the United States). Cellulosic feedstocks are crops such as switchgrass, miscanthus, willow, poplar, and plant wastes such as corn stover (the stalk and leaves of corn) and bagasse (the crushed stalks of sugarcane after the juice is extracted). Notice that all three feedstocks are plants. Through photosynthesis, plants have the amazing ability to create their own food—glucose (C$_6$H$_{12}$O$_6$), a simple form of sugar, with just three ingredients, CO$_2$, H$_2$O & sunlight. **All three ethanol feedstocks contain sugar, the essential ingredient for making alcohol.**

The sugar contained in sugar feedstocks is the easiest and cheapest to extract. For example, one need only squeeze sugarcane and sugar beets to obtain their juice—a liquid rich in dissolved sugar. The sugar contained in starchy feedstocks is more difficult to obtain because the starch (which is simply a chain of chemically combined glucose molecules) must first be broken apart. The sugar contained in cellulosic feedstocks is the most difficult to obtain because the cellulose (also chain of chemically combined glucose molecules, although arranged differently than starch) must also be broken apart, and it exists within a tough, interconnected matrix of other molecules that form the cell walls of plants (United States Department of Energy, 2006, pp. 44-47). Breaking apart starch and cellulose is accomplished by combining the feedstocks with water and then soaking the mixture in "slurry" and "liquefaction" tanks while applying digestive enzymes. **Digestive enzymes in slurry and liquefaction tanks break complex feedstocks into simpler component parts.** Cellulosic feedstock is pretreated with heat, pressure or acid to first break down the feedstock, thereby increasing the surface area available for later enzymatic breakdown.

After sugar is obtained, the juice, grain slurry, or cellulosic slurry is pumped into a fermentation tank holding several thousand gallons of liquid where specific strains of yeast are added. Yeasts are a type of fungus that eat sugar (they also make bread rise). **During fermentation, yeasts consume sugar and produce carbon dioxide (CO$_2$) and ethanol (C$_2$H$_5$OH) as waste products.** The carbon dioxide can either be vented into the atmosphere or chilled and sold as an industrial byproduct of the fermentation process. In the case of grain slurry, about 50 hours must pass before enough

Figure 6-24, Train Loaded with Sugarcane, a Sugar Feedstock for Ethanol Production, *Credit/ Courtesy of Scott Bauer, United States Department of Agriculture*

Figure 6-25, Corn, a Starchy Feedstock for Ethanol Production, *Credit/ Courtesy of Doug Wilson, United States Department of Agriculture*

Figure 6-26, Switchgrass, a Cellulosic Feedstock for Ethanol Production, *Credit/Courtesy of United States Department of Agriculture*

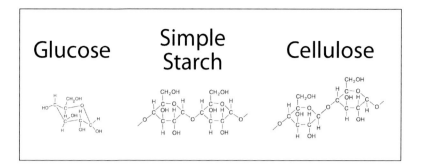

Figure 6-27, The Chemical Structures of Glucose, Starch and Cellulose Molecules

of the slurry is ethanol (Renewable Fuels Association, 2011). Percentages vary, but when approximately 15% of the slurry is ethanol, it is time to pump the mixture into a distillation column (ICM, Inc., 2009). 15% may seem like a small amount, but it is the maximum amount of ethanol that the yeast can produce before they die in an environment made toxic by their own waste.

Ethanol distillation is not unlike petroleum distillation described in Chapter 3. In the case of grain, the slurry & ethanol mixture is boiled at the bottom of a distillation column which causes the ethanol to evaporate and rise. The vapor rises up the column, cools and then condenses at the top, after which the liquid is routed to another distillation column where the pro-

cess is repeated. After several distillations, the condensed ethanol is 190-proof (Renewable Fuels Association, 2011). **Proof is a number twice as large as the percentage of alcohol in a solution.** 190-proof means that the freshly distilled ethanol is 95% alcohol by volume (190/2 = 95). Compare this to a bottle of vodka purchased at your local liquor store—it typically has a proof of 80, which means that ethanol comprises only 40% of the bottle's contents. After removing the remaining 5% of water, the now 200-proof ethanol is denatured and stored, ready for transport. **Denaturing ethanol renders it unfit for human consumption.**

Figure 6-28, An Ethanol Plant in West Burlington, Iowa, *Credit/Courtesy of Steven Vaughn, United States Department of Agriculture*

Transporting ethanol to gasoline blending terminals normally occurs via train (the primary method), truck or barge (United States Department of Agriculture, 2007). While pipelines may seem like a better option, especially considering the network of existing hydrocarbon pipelines that exist in the United States, they are not appropriate for transporting ethanol because of three reasons. First, ethanol absorbs water, and since water can collect in low-points of hydrocarbon pipeline systems, it may be contaminated while in transit. Second, most ethanol is manufactured in the Midwest, while most pipeline staging areas exist in the south along the Gulf coast. Thus, any pipeline transport would first require shipping ethanol south to new, dedicated ethanol staging tanks. Finally, the low relative volume of ethanol compared to other products normally shipped via pipeline would make it difficult for the market to capture any volume discounts (Whims, 2002).

Figure 6-29, Many Tons of Grain Stored in an Ethanol Plant in West Burlington, Iowa, *Credit/Courtesy of Steven Vaughn, United States Department of Agriculture*

The United States produces the most fuel ethanol in the world (United States Energy Information Administration, 2011d). In 2010, 13,231 million gallons (13,321,000,000 gallons) of denatured ethanol was produced in the United States for fuel (United States Energy Information Administration, 2011e). This amount is remarkable when considering that in 2006 the United States produced only 4,884 million gallons. The 171% increase in 4 years is due to state and federal policy initiatives designed to boost ethanol production from the massive corn supplies in the American Midwest (United States Department of Energy, 2010b). Most ethanol manufacturing facilities are thus logically located where corn is grown— Minnesota, Iowa, Nebraska and several other Midwest states. As of September 2011, 209 biorefineries existed throughout the United States, heavily concentrated in the American Midwest (Renewable Fuels Association, 2011b).

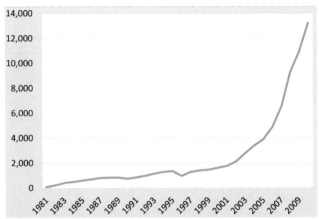

Figure 6-30, Graph of United States Ethanol Consumption, 1981-2010 (Millions of Gallons)

All passenger vehicles in the United States that run on gasoline can also run on E-10. **E-10 is a blend of 10% ethanol and 90% gasoline.** This is classified as a low-level ethanol blend and is "substantially similar" to gasoline (and therefore legal to sell), as defined under The Clean Air Act, Title 42, Chapter 85 (United States House of Representatives, 2009). Next time you are at a gasoline station, look at the stickers on the pump. If you are filling up in a large metropolitan area known for smog problems, there is a good chance that one of the stickers may prominently display "E-10" and state that the gasoline you are buying is "reformulated" to contain 10% ethanol. In many places RFG is mandated by law and implemented on a county-by-county basis. Places such as southern California, parts of Texas, the metropolitan areas of St. Louis and Chicago, and much of the eastern United States from Boston through New Jersey, are sometimes known as "RFG Areas" (United States Environmental Protection Agency, 2011f). Intermediate-level ethanol blends include E-15 and E-20. Such blends can be used in Flexible Fuel Vehicles (abbreviated FFV), but not yet in standard passenger vehicles in the United States (United States Department of Energy, 2010c). For a county-by-county list of RFG areas throughout the United States, check out this web site hosted by the Environmental Protection Agency:

RFG Areas: http://www.epa.gov/otaq/fuels/gasoline-fuels/rfg/areas.htm

A "high-level" ethanol blend is E-85. **E-85 is a blend of 85% ethanol and 15% gasoline.** The only passenger vehicles in the United States that can legally use E-85 are Flexible Fuel Vehicles. **Flexible Fuel Vehicles can run on E-10, E-85, or any combination of both.** FFVs are often designated with a small "Flexible Fuel" decal somewhere on the rear of the vehicle. There are approximately 8 million FFVs on the road today (United

Figure 6-31, A typical E-10 Sticker on a Gasoline Pump

CONTAINS UP TO 10% ETHANOL

States Department of Energy, 2011), which is just 6% of the 137 million registered passenger automobiles in the United States (United States Department of Transportation, 2009). The number of gasoline stations selling E-85 is also quite small (but growing fast), with the total number being 2,454 as of September 2011 (United States Department of Energy, 2011b). For some perspective on these numbers, consider that the United States currently has about 119,000 retail gasoline stations (United States Bureau of the Census, 2007). So, 6% of the passenger vehicles in the United States can use E-85, but only 2% of the nation's gasoline stations provide this type of fuel. This mismatch illustrates the infrastructural challenges to widespread adoption of alternative vehicle fuels.

Finding a station selling E-85 can be challenging, so the United States Department of Energy established the Alternative Fueling Station Locator, an interactive web-based program asking users to select an alternative fuel (there are 7 from which to choose) and then enter an address or zip code. After clicking "Get Results" a map appears, along with text, showing users the location, address, phone number and distance to the desired station. This is a fun way to see how much

market penetration E-85 and other alternative fuels have made in your area.

Alternative Fueling Station Locator: http://www.afdc.energy.gov/afdc/locator/stations/

In the United States, the energy content in 1 gallon of conventional gasoline varies from summer to winter in some locations because of air pollution regulations prescribed by the Environmental Protection Agency. **A gallon of conventional gasoline purchased in the summer has an average energy content of 114,500 Btu, while a gallon conventional gasoline purchased in the winter has an average energy content of 112,500 Btu** (United States Environmental Protection Agency, 2007b). Thus "summer-blend" gasoline has about 1.7% more Btu than "winter-blend". This difference in energy content is due to the Clean Air Act mandating oxygenates—often ethanol—be mixed into gasoline during cold winter months. Adding oxygenates reduces a gallon's Btu content, but also reduces carbon monoxide tailpipe emissions, which increase during cold weather (United States Environmental Protection Agency, 2011g). You may already know that your vehicle averages fewer miles per gallon during the winter than during the summer. This is thus partially explained by the fact that there is literally less energy in each gallon of gasoline you buy between October 1 and March 31.

Adding ethanol to gasoline reduces the gasoline's energy content. The effect of this is that a car will literally travel fewer miles with an ethanol/gasoline mixture compared to straight gasoline. Why does the United States promote this less-efficient product? It is partly due to environmental reasons, as when ethanol is used to oxygenate gasoline. It is partly due to geopolitical reasons, as a means to promote domestic energy supplies and thereby reduce our dependency on foreign energy sources. Finally, it is partly due to internal political reasons, as a means to increase rural employment and thereby promote stability and prosperity across the American Midwest.

Figure 6-32, A Gasoline Pump Selling E-85, Credit/Courtesy of Indiana.gov

Fuel	Btu/gallon
Summer Blend	114,500
Winter Blend	112,500
E-10	111,836
Pure Ethanol	76,100

Table 6-1, Average Btu Content of Summer Blend, Winter Blend, E-10 and Pure Ethanol

Biomass's Virtues and Vices

Recall that the major biomass energy sources are wood, waste and alcohol fuels. While all three contain carbon, none of this carbon is derived from ancient sources. Instead, plants pull *today's* carbon (in the form of CO_2) from the air during photosynthesis and release *that same carbon* during combustion. Since no additional carbon is added to the air, biomass is said to have a net-zero carbon impact on our atmosphere. Advocates of using biomass energy thus argue that it is "carbon neutral" (National Renewable Energy Laboratory, 2011). **Carbon neutral energy sources absorb and release the same amount of carbon, and thus do not increase the volume of atmospheric carbon dioxide (CO_2).** Since the combustion of fossil fuels releases—thus adds—geologically-trapped carbon into the air, they are said to have a "net-positive" impact on our atmosphere.

Is biomass really carbon neutral? Take a look at Figures 6-3, 6-9, 6-24 and 6-29. Notice in each of these pictures that a diesel-powered vehicle is processing or transporting the biomass. Even in figures where no vehicle is pictured, a fossil fuel powered vehicle and/or fossil fuel derived product is probably responsible for some aspect of production and processing. Given these variables, a more thoughtful and careful claim is that biomass *may* be a carbon neutral energy source, depending on "how much energy was used to grow, harvest, and process the fuel" (National Renewable Energy Laboratory, 2010).

Carbon neutrality is an important issue because political and economic entities around the world are concerned that carbon emissions are responsible for trapping heat and warming our planet. The Intergovernmental Panel on Climate Change (IPCC), a scientific body within the United Nations, states in its 2007 assessment report (called "AR4" because it is the fourth Assessment Report on this issue) that there is now unequivocal evidence of global warming from GHG emissions, and that these emissions are "very likely" due to anthropogenic (human) activities (Intergovernmental Panel on Climate Change, 2007, p. 72). The same report states that the widespread warming observed on our planet over the past 50 years would be "extremely unlikely" without anthropogenic activities, and that any temperature changes due to natural events, such as solar and/or volcanic activities, over this same period would have "likely" produced a cooling effect instead of the observed warming effect (Intergovernmental Panel on Climate Change, 2007, p. 39).

Recall that about half of the gas emitted from landfills is methane (CH_4). This is a significant GHG because, like CO_2, methane traps heat and thus warms the planet. CH_4 is of particular concern, however, because even though the volume emitted during any year is much less than CO_2, it is 25 times more effective at trapping heat than CO_2 (United States Energy Information Administration, 2011f). Consider this straightforward example: just 1 kilogram of CH_4 traps as much heat as 25 kilograms of CO_2 over a 100 year period. The number "25" is thus methane's global warming potential (GWP). **Global warming potential (GWP) is the ratio of atmospheric heating that results from the emission of 1 kilogram of a GHG compared to 1 kilogram of carbon dioxide over a 100 year time span.** Both WTE and LFGTE facilities help mitigate the release of methane by either burning the garbage before it rots, or burning the methane that is produced as the garbage rots. In both cases, less methane is emitted into the atmosphere.

Because carbon dioxide is the most significant GHG by volume of emissions, other GHG emissions are often recorded in comparison to CO_2. This comparison technique is called carbon dioxide equivalent (CO_2e). **Carbon dioxide equivalent (CO_2e) is calculated by multiplying the weight of the GHG gas by its GWP.** In 2008, for example, landfills in the United States emitted methane in the amount of 184.3 million metric tons of carbon dioxide equivalent, also written as 184.3 $MMTCO_2e$. This amount compares to 260.2 $MMTCO_2e$ emitted in 1990 (United States Energy Information Administration, 2009). This 29% reduction over 18 years is due to federal laws requiring landfills to burn LFG, as well as tax incentives for LFGTE operations.

Using biomass as a source of energy eliminates the need to mine, drill or import some fossil fuels to generate electricity. For example, burning 1 metric tonne (1 metric tonne = 1,000 kilograms = 2,204.6 pounds) of MSW generates 600 kWh of electricity and thereby avoids the mining of 551 pounds of high-quality coal (Psomopoulos, Bourka, & Themelis, 2009, p. 1719). Since it has a domestic origin, advocates also argue that biomass increases a country's energy security because fewer resources need to be imported, especially from unstable and non-democratic regimes. This seems ostensibly true, but it is wise to proceed carefully when such claims are heard. There is no doubt that burning MSW and LFG generates electricity, but this fact has no bearing on the United States' dependency upon foreign oil because nearly our entire vehicle fleet runs on petroleum products, not electricity.

Using biomass as an energy source has the advantage of spurring economic growth, both inside and outside rural areas. A single commercial ethanol manufacturing plant, for example, employs between 50-75 people, mostly from the surrounding rural community (Southern Illinois University Edwardsville, 2011). The 209 biorefineries operating in the United States as of September 2011 therefore, employ between 10,450 and 15,675 people. The economic ripple effect of biomass energy through the United States economy is also significant. In 2007, for example, ethanol production generated 238,000 jobs in all sectors of the United States economy, thereby increasing household income by $12.3 billion and generating billions in federal and state tax revenue (United States Department of Energy, 2010d).

When the marketplace demands energy crops, land is taken out of food production and placed into energy production. With less land devoted to food, upward pricing pressure on crops such as corn, wheat, soybeans and vegetables is likely. Meat and dairy prices also rise as crops head to ethanol production facilities instead of food troughs for animal. **Increasing food prices spark arguments about what kind of crops ought to be grown in a country—this is known as the "food versus fuel" debate.** The United States Congressional Budget Office (CBO) estimates that between .5% and .8% of the over 5% increase in food prices in the United States between 2007 and 2008 was due to land being taken out of food production and put into ethanol production (Congress of the United States Congressional Budget Office, 2009, p. 6).

In a developed country such as the United States where, both at and away from home, only 9.4% of disposable personal income is spent on food (United States Department of Agriculture, 2011), an increase due to ethanol production is somewhat noticeable but not crippling. Citizens of developing countries, however, are often hit hard by even modest increases in food prices because a comparatively large percentage of household expenditures are devoted to buying food (Food and Agriculture Organization of the United Nations, 2008, pp. 6-7). For example, consider these four countries and the percent of household expenditures spent on food at home: Mexico = 22.7%, India = 27.7%, Indonesia = 32.2%, Kenya = 41.7% (United States Department of Agriculture, 2011b). The food versus fuel debate may soften as food prices rise less steeply (or even decline) in conjunction with advances made in cellulosic ethanol research. Switchgrass, for example, is a cellulosic ethanol feedstock that thrives on marginal lands ill-suited for food production. **Switchgrass can be cultivated for energy without impacting food prices.**

Planting, cultivating, harvesting and transporting energy crops are obviously not energy-free activities. Since tractors and trucks need fuel, fertilizers are often petro-chemical based, and ethanol conversion plants use energy, a pertinent question to ask is whether or not planting energy crops yields a positive net energy balance. **The term "energy balance" is the difference between 1) the total Btu in 1 gallon of ethanol (76,100 Btu) and 2) the amount of fossil energy required to produce that gallon.** The preferred scenario is to produce a greater number of Btu (a "positive" balance) than are expended.

The United States Department of Agriculture calculates a positive energy balance for corn-based ethanol production in the United States, ranging from 1.4 to 2.3 Btu gained for every 1 Btu of fossil energy used in production (United States Department of Agriculture, 2010, p. 5). Considering Btu/gallon and net energy balance, most current studies calculate a positive energy yield of between 15,000 and 40,000 Btu per gallon (United States Department of Energy, 2007). The variations in these findings, along with the fact that several studies find a negative energy balance, illustrate the challenge accurately calculating the correct input variables used in these projects.

You may have heard that corn is not an ideal crop from which to produce ethanol. One reason for this is that other crops yield a greater energy balance. Recall, for example, that sugarcane is a sugar feedstock. Few inputs are needed to extract its sugar because it is not locked up as starch or cellulose. Less energy is therefore required to make ethanol from sugarcane because there is no need for digestive enzymes or slurry tanks. Sugarcane's energy balance is 8.3 (United States Agency for International Development, 2011). Compare this to corn's 2.3 energy balance. For another example, consider the energy balance of various cellulosic feedstocks. Low energy balance estimates are around 13.5 (Schmer, Vogel, Mitchell, & Perrin, 2007), while high estimates are around 23 (Farrell, Plevin, Turner, Jones, O'Hare, & Kammen, 2006). Dropping enzyme costs, coupled with research into various fast-growing cellulosic energy crops, could make cellulosic ethanol the most important biomass fuel in the near future.

Another reason you may have head that corn is not an ideal crop from which to produce ethanol is tied to the Conservation Reserve Program (CRP) in the United States. The CRP was established to protect topsoil from erosion, reduce water runoff and sedimentation, and protect ground and surface water resources (United States Department of Agriculture, 2012). Farmers participating in the CRP are paid rent by the United States government to plant and maintain resource-conserving vegetative covers, such as native grasses, windbreaks and shade trees, instead of commercial crops. So long as the commercial crops are not excessively profitable to grow, millions of acres of farmland remain within the CRP. Farmers make economic decisions like anyone else, however, and high corn prices are now compelling some to remove their land from the CRP in order to plant corn (Marcotty, 2011). As of March, 2012, corn is priced at $6.72 per bushel (CME Group, 2012), a highly inflated price compared to most recent history and due in large part because corn-based ethanol producers are bidding up the price. Economic modeling of this issue indicates that as the price of corn rises, more acreage is removed from the CRP. The effect is that nitrogen and phosphorous are introduced into local waterways, soil erosion increases and carbon-sequestering acreage shrinks (Secchi & Babcock, 2007).

Feedstock	Energy Balance
Corn	1.4 - 2.3
Sugarcane	8.3
Cellulosic	13.5 - 23

Table 6-2, Energy Balance for Corn, Sugarcane and Cellulosic Ethanol

Important Ideas:

- Biomass is the oldest energy resource harnessed by humanity.

- Biomass energy resources are not fossil fuels.

- Biomass is a renewable energy resource because it is replenished in a short time span.

- Biomass is the largest renewable energy source, supplying 3.87 quads, or 53% of all the renewable energy in the United States in 2008.

- Black liquor is the largest source of energy derived from wood.

- A recovery boiler is a large furnace that burns black liquor to 1) recover pulping chemicals and 2) boil water to produce steam.

- Waste-to-energy facilities, often called "incinerators" or "combustion plants," are steam-driven electrical generation plants that use garbage as fuel.

- Refuse derived fuel (RDF) is combustible garbage—often processed into standard, pelletized form—after the recyclables and non-combustibles are recovered.

- Emission controls exist because the United States Clean Air Act (CAA), last amended in 1990, limits emissions of hazardous air pollutants (HAPs) according to their source category.

- About 50% of the gas released from a landfill is CO_2, and about 50% is CH_4.

- In 2008, landfills in the United States emitted 6,016 Gg of methane.

- A LFG well is a perforated plastic pipe, surrounded by gravel and placed vertically or horizontally within the garbage.

- A LFG mainline is a non-perforated plastic pipe that transports gas from a landfill to an adjacent treatment station, flare station or blower.

- Ethanol (C_2H_5OH) is a type of alcohol—the same kind that is in vodka, gin, beer or any other alcoholic beverage.

- Reformulated gasoline (RFG) contains additives that result in a cleaner-burning fuel and thus less air pollution.

- Ethanol is an oxegenate because it contains oxygen, and when blended with gasoline, makes gasoline burn more efficiently.

- Three different feedstocks are used to make ethanol: 1) sugar feedstocks, 2) starchy feedstocks, and 3) cellulosic feedstocks.

- All three ethanol feedstocks contain sugar, the essential ingredient for making alcohol.

- Digestive enzymes in slurry and liquefaction tanks break complex feedstocks into simpler component parts.

- During fermentation, yeasts consume sugar and produce carbon dioxide (CO_2) and ethanol (C_2H_5OH) as waste products.

- Proof is a number twice as large as the percentage of alcohol in a solution.

- Denaturing ethanol renders it unfit for human consumption.

- The United States produces the most fuel ethanol in the world.

- E-10 is a blend of 10% ethanol and 90% gasoline.

- E-85 is a blend of 85% ethanol and 15% gasoline.

- Flexible Fuel Vehicles can run on E-10, E-85, or any combination of both.

- A gallon of conventional gasoline purchased in the summer has an average energy content of 114,500 Btu, while a gallon conventional gasoline purchased in the winter has an average energy content of 112,500 Btu.

- Adding ethanol to gasoline reduces the gasoline's energy content.

- Carbon neutral energy sources absorb and release the same amount of carbon, and thus do not increase the volume of atmospheric carbon dioxide (CO_2).

- Global warming potential (GWP) is the ratio of atmospheric heating that results from the emission of 1 kilogram of a GHG compared to 1 kilogram of carbon dioxide over a 100 year time span.

- Carbon dioxide equivalent (CO_2e) is calculated by multiplying the weight of the GHG gas by its GWP.

- When the marketplace demands energy crops, land is taken out of food production and placed into energy production.

- Increasing food prices spark arguments about what kind of crops ought to be grown in a country—this is known as the "food versus fuel" debate.

- Switchgrass can be cultivated for energy without impacting food prices.

- The term "energy balance" is the difference between 1) the total Btu in 1 gallon of ethanol (76,100 Btu) and 2) the amount of fossil energy required to produce that gallon.

- The United States Department of Agriculture calculates a positive energy balance for corn-based ethanol production in the United States, ranging from 1.4 to 2.3 Btu gained for every 1 Btu of fossil energy used in production.

Works Cited

CME Group. (2012, March 14). *Agricultural Products*. Retrieved March 14, 2012, from Corn Futures: http://www.cmegroup.com/trading/agricultural/grain-and-oilseed/corn.html

Congress of the United States Congressional Budget Office. (2009, April). *The Impact of Ethanol Use on Food Prices and Greenhouse-Gas Emissions*. Retrieved October 21, 2011, from Ethanol Production and Food Prices: http://www.cbo.gov/ftpdocs/100xx/doc10057/04-08-Ethanol.pdf

Farrell, A. E., Plevin, R. J., Turner, B. T., Jones, A. D., O'Hare, M., & Kammen, D. M. (2006, January 27). Ethanol Can Contribute to Energy and Environmental Goals. *Science*, 506-508.

Food and Agriculture Organization of the United Nations. (2008, June 3-5). *Bioenergy Policy, Markets and Trade and Food Security (HLC/08/BAK/7)*. Retrieved October 21, 2011, from Technical Background Document from the Expert Consultation Held On 18 to 20 February 2008: ftp://ftp.fao.org/docrep/fao/meeting/013/ai788e.pdf

Greer, D. (2007, July). Financing Wood-Fired Electricity Generation. *BioCycle, 48*(7), 66-70.

ICM, Inc. (2009). *Ethanol Production Process*. Retrieved October 20, 2011, from Simultaneous Saccharification Fermentation: http://www.icminc.com/ethanol/production_process/

Intergovernmental Panel on Climate Change. (2007). *Climate Change 2007: Synthesis Report. Contribution of Working Groups I, II and III to the Fourth Assessment Report of the Intergovernmental Panel on Climate Change*. (R. K. Pachauri, & A. Reisinger, Editors) Retrieved October 21, 2011, from 6.1 Observed changes in climate and their effects, and their causes: http://www.ipcc.ch/pdf/assessment-report/ar4/syr/ar4_syr.pdf

Iowa State University. (2004, May). *The Science of Smell Part 1: Odor Perception and Physiological Response*. Retrieved September 7, 2010, from http://www.extension.iastate.edu/Publications/PM1963A.pdf

Marcotty, J. (2011, November 28). *High crop prices a threat to nature?* Retrieved March 14, 2012, from StarTribune: http://www.startribune.com/local/134566683.html

Miami-Dade County. (2011). *Public Works and Waste Management*. Retrieved October 19, 2011, from Resources Recovery Facility: http://www.miamidade.gov/dswm/factsheet_resources_recovery_facility.asp

Michaels, T. (2010, December). *The 2010 ERC Directory of Waste-to-Energy Plants*. Retrieved October 19, 2011, from Energy Recoverty Council: http://www.energyrecoverycouncil.org/userfiles/file/ERC_2010_Directory.pdf

National Renewable Energy Laboratory. (2010, February 9). *Learning About Renewable Energy*. Retrieved March 13, 2012, from Biomass Energy Basics, Benefits of Using Biomass: http://www.nrel.gov/learning/re_biomass.html

National Renewable Energy Laboratory. (2011, August 19). *Climate Neutral Research Campuses*. Retrieved March 13, 2012, from Biomass Energy: http://www.nrel.gov/applying_technologies/climate_neutral/biomass.html

Psomopoulos, C. S., Bourka, A., & Themelis, N. J. (2009). Waste-to-energy: A review of the status and benefits in USA. *Waste Management, 29*, 1718–1724.

Renewable Fuels Association. (2011). *How Ethanol is Made*. Retrieved October 20, 2011, from The Ethanol Production Process - Dry Milling: http://www.ethanolrfa.org/pages/how-ethanol-is-made

Renewable Fuels Association. (2011b, September 26). *Biorefinery Locations*. Retrieved October 20, 2011, from Operational and Under Construction: http://www.ethanolrfa.org/bio-refinery-locations/

SCA Forest Products, Sundsvall. (2011, March 22). *Östrand pulp mill*. Retrieved October 19, 2011, from http://www.sca.com/en/Sundsvall/SCA-Sundsvall/Ostrand-pulp-mill/

Schmer, M. R., Vogel, K. P., Mitchell, R., & Perrin, R. K. (2007). Energy balance of switchgrass grown for cellulosic ethanol in the Northern Great Plains, USA. *23rd American Chemical Society Annual Meeting*. Chicago, IL: American Chemical Society.

Secchi, S., & Babcock, B. A. (2007). *Impact of High Corn Prices on Conservation Reserve Program Acreage*. Retrieved March 14, 2012, from Center for Agricultural and Rural Development, Iowa State University: http://www.card.iastate.edu/iowa_ag_review/spring_07/article2.aspx

Southern Illinois University Edwardsville. (2011). *National Corn-to-Ethanol Research Center (NCERC)*. Retrieved October 21, 2011, from How many jobs does a typical ethanol plant provide?: http://www.siue.edu/ethanolresearch/faq.shtml

United States Agency for International Development. (2011, March 9). *Energy Team, Office of Infrastructure & Engineering, Bureau for Economic Growth, Agriculture, and Trade*. Retrieved January 24, 2012, from An Introductory Guide for Assessing the Potential of Biofuels in Developing Countries: http://pdf.usaid.gov/pdf_docs/PNADO644.pdf

United States Bureau of the Census. (2007). *2007 Economic Census*. Retrieved October 21, 2011, from Gasoline stations, NAICS: 447, Table 1. Selected Industry Statistics for the U.S. and States: 2007: http://factfinder.census.gov/servlet/IQRTable?_bm=y&-ds_name=EC0700A1&-NAICS2007=447&-_lang=en

United States Department of Agriculture. (2007, September). *Ethanol Transportation Backgrounder*. Retrieved October 20, 2011, from Expansion of U.S. Corn-based Ethanol from the Agricultural Transportation Perspective: http://www.ams.usda.gov/AMSv1.0/getfile?dDocName=STELPRDC5063605&acct=atpub

United States Department of Agriculture. (2010, June). *2008 Energy Balance for the Corn-Ethanol Industry*. Retrieved October 21, 2011, from Energy Consumption by Corn Producers: http://www.usda.gov/oce/reports/energy/2008Ethanol_June_final.pdf

United States Department of Agriculture. (2011, July 13). *Economic Research Service*. Retrieved October 21, 2011, from Food CPI and Expenditures: Table 7, Food expenditures by families and individuals as a share of disposable personal income: http://www.ers.usda.gov/Briefing/CPIFoodAndExpenditures/Data/Expenditures_tables/table7.htm

United States Department of Agriculture. (2011b, July 13). *Food CPI and Expenditures: Food Expenditure Tables*. Retrieved October 21, 2011, from Table 97. Expenditures spent on food and alcoholic beverages that were consumed at home, by selected countries, 2010: http://www.ers.usda.gov/Briefing/CPIFoodAndExpenditures/Data/Table_97/table97_2010.xls

United States Department of Agriculture. (2012, March 7). *Farm Service Agency*. Retrieved March 14, 2012, from Conservation Reserve Program: http://www.fsa.usda.gov/FSA/webapp?area=home&subject=copr&topic=crp

United States Department of Energy. (2003, August). *Office of Energy Efficiency and Renewable Energy* . Retrieved October 20, 2011, from FreedomCAR & Vehicle Technologies Program, Just the Basics: Ethanol: http://www1.eere.energy.gov/vehiclesandfuels/pdfs/basics/jtb_ethanol.pdf

United States Department of Energy. (2006, June). *Breaking the Biological Barriers to Cellulosic Ethanol: A Joint Research Agenda*. Retrieved October 20, 2011, from DOE/SC-0095: http://genomicscience.energy.gov/biofuels/2005workshop/b2blowres63006.pdf

United States Department of Energy. (2007, March). *Office of Energy Efficiency and Renewable Energy*. Retrieved October 21, 2011, from Ethanol, the Complete Energy Lifecycle Picture, Figure 4: The majority of corn ethanol/fossil energy studies (especially more recent studies) show that corn ethanol has a positive net fossil energy value: http://www1.eere.energy.gov/vehiclesandfuels/pdfs/program/ethanol_brochure_color.pdf

United States Department of Energy. (2008, October). *United States of America, Third National Report for the Joint Convention on the Safety of Spent Fuel Management and on the Safety of Radioactive Waste Management*. Retrieved September 7, 2011, from http://www.em.doe.gov/pdfs/3rd%20US%20Rpt%20on%20SNF%20JC--%20COMPLETE%20REPORT%20-%2010%2013%2008.pdf

United States Department of Energy. (2010, September 24). *Alternative Fuels and Advanced Vehicles Data Center (AFDC)*. Retrieved October 20, 2011, from Ethanol Feedstocks: http://www.afdc.energy.gov/afdc/ethanol/feedstocks.html

United States Department of Energy. (2010b, September 16). *Alternative Fuels and Advanced Vehicles Data Center (AFDC)*. Retrieved October 20, 2011, from Ethanol Incentives and Laws: http://www.afdc.energy.gov/afdc/ethanol/incentives_laws.html

United States Department of Energy. (2010c, December 3). *Alternative Fuels and Advanced Vehicles Data Center*. Retrieved October 21, 2011, from E15-E20: Intermediate Ethanol Blends: http://www.afdc.energy.gov/afdc/ethanol/blends_e15_e20.html

United States Department of Energy. (2010d, December 27). *Office of Energy Efficiency and Renewable Energy*. Retrieved October 21, 2011, from Biomass Program, Economic Growth, Employment: http://www1.eere.energy.gov/biomass/economic_growth.html

United States Department of Energy. (2011, October 20). *Alternative Fuels and Advanced Vehicles Data Center*. Retrieved October 21, 2011, from Flexible Fuel Vehicles: http://www.afdc.energy.gov/afdc/vehicles/flexible_fuel.html

United States Department of Transportation. (2009, December). *Federal Highway Administration*. Retrieved October 21, 2011, from Highway Statistics 2008, State Motor-Vehicle Registrations - 2008: http://www.fhwa.dot.gov/policyinformation/statistics/2008/mv1.cfm

United States Energy Information Administration. (2008, April). *Independent Statistics and Analysis*. Retrieved October 19, 2011, from Biomass: http://www.eia.doe.gov/cneaf/solar.renewables/page/biomass/biomass.html

United States Energy Information Administration. (2008b, June). *Energy Timelines*. Retrieved October 20, 2011, from Ethanol: http://tonto.eia.doe.gov/kids/energy.cfm?page=tl_ethanol

United States Energy Information Administration. (2009, December 8). *Independent Statistics and Analysis*. Retrieved October 20, 2011, from Emissions of Greenhouse Gases Report, Methane Emissions, Waste Management, Table 19: U.S. Methane Emissions from Waste Management, 1990-2008: http://www.eia.doe.gov/oiaf/1605/ggrpt/methane.html

United States Energy Information Administration. (2011, July 14). *Independent Statistics and Analysis*. Retrieved October 19, 2011, from What Role Does Renewable Energy Play in the United States?: http://www.eia.gov/energyexplained/index.cfm?page=renewable_home

United States Energy Information Administration. (2011b). *Independent Statistics and Analysis*. Retrieved October 19, 2011, from Electricity Net Generation From Renewable Energy by Energy Use Sector and Energy Source, 2005 - 2009: http://www.eia.gov/cneaf/solar.renewables/page/table3.html

United States Energy Information Administration. (2011c, June 1). *Frequently Asked Questions – Electricity, Question: How much electricity does an American home use?* Retrieved October 19, 2011, from Average monthly residential electricity consumption, prices, and bills by state, Excel Spreadsheet: http://www.eia.gov/tools/faqs/faq.cfm?id=97&t=3

United States Energy Information Administration. (2011d). *Independent Statistics and Analysis*. Retrieved October 20, 2011, from International Energy Statistics, Fuel Ethanol Production (Thousand Barrels Per Day), 2009: http://www.eia.gov/cfapps/ipdbproject/iedindex3.cfm?tid=79&pid=80&aid=1&cid=regions&syid= 2009&eyid=2009&unit=TBPD

United States Energy Information Administration. (2011e, October 19). *Independent Statistics and Analysis*. Retrieved October 20, 2011, from Annual Energy Review 2010, Table 10.3 Fuel Ethanol Overview, 1981-2010: http://www.eia.gov/totalenergy/data/annual/showtext.cfm?t=ptb1003

United States Energy Information Administration. (2011f). *Independent Statistics and Analysis*. Retrieved October 21, 2011, from Voluntary Reporting of Greenhouse Gases Program, Greenhouse Gases and Global Warming Potentials (GWP): http://www.eia.gov/oiaf/1605/gwp_tbl.html

United States Environmental Protection Agency. (2010, July 19). *RadTown USA*. Retrieved September 6, 2011, from What is Radiation?: http://www.epa.gov/radtown/basic.html#what

United States Environmental Protection Agency. (1996, October). *AP 42, Fifth Edition, Volume I, Chapter 2: Solid Waste Disposal*. Retrieved October 21, 2011, from 2.1 Refuse Combustion, Final Section - Supplement B: http://www.epa.gov/ttn/chief/ap42/ch02/final/c02s01.pdf

United States Environmental Protection Agency. (2001, September). *MACT II Pulp and Paper Combustion Sources National Emission Standards for Hazardous Air Pollutants (NESHAP): A Plain English Description: EPA-456/R-01-003. Chapter 2*. Retrieved August 24, 2010, from National Service Center for Environmental Publications (NSCEP): http://www.epa.gov/ttn/atw/pulp/pulppg.html

United States Environmental Protection Agency. (2007, June 6). *Technology Transfer Network, Air Toxics Web Site*. Retrieved October 19, 2011, from CAA Section 112(d)-Emission Standards: http://www.epa.gov/ttn/ atw/112dpg.html

United States Environmental Protection Agency. (2007b, August 14). *Fuel Economy Impact Analysis of RFG*. Retrieved October 21, 2011, from Conventional Gasoline, Average Energy Content (btu per gallon): http:// www.epa.gov/otaq/datafiles/rfgecon.htm

United States Environmental Protection Agency. (2010, December). *Office of Solid Waste (5306P)*. Retrieved October 19, 2011, from Municipal Solid Waste Generationl in the United States:2009 Facts and Figures, EPA530-R-10-012: http://www.epa.gov/osw/nonhaz/municipal/pubs/msw2009rpt.pdf

United States Environmental Protection Agency. (2010b, March 17). *Municipal Solid Waste, Environmental Impacts*. Retrieved October 19, 2011, from Air Emissions Impacts: http://www.epa.gov/cleanenergy/ energy-and-you/affect/municipal-sw.html

United States Environmental Protection Agency. (2010c, April 15). *2010 U.S. Greenhouse Gas Inventory Report, U.S. EPA # 430-R-10-006*. Retrieved October 19, 2011, from Inventory of U.S. Greenhouse Gas Emissions and Sinks: 1990-2008, Chapter 8: Waste: http://www.epa.gov/climate/climatechange/emissions/down-loads10/508_Complete_GHG_1990_2008.pdf

United States Environmental Protection Agency. (2011, July 26). *Text Version of Municipal Solid Waste Charts*. Retrieved October 19, 2011, from MSW Generation Rates, 1960-2009: http://www.epa.gov/epawaste/ facts-text.htm#chart1

United States Environmental Protection Agency. (2011b, July 27). *Wastes - Non-Hazardous Waste - Municipal Solid Waste*. Retrieved October 19, 2011, from Combustion: http://www.epa.gov/epawaste/nonhaz/mu-nicipal/combustion.htm

United States Environmental Protection Agency. (2011c, July 25). *Landfill Methane Outreach Program*. Retrieved October 19, 2011, from Methane Emissions From Landfills: http://www.epa.gov/lmop/basic-info/index. html

United States Environmental Protection Agency. (2011d, August 10). *Landfill Methane Outreach Program*. Retrieved October 19, 2011, from How can landfill gas be used for energy?: http://www.epa.gov/lmop/faq/lfg.html

United States Environmental Protection Agency. (2011e, August 3). *Landfill Methane Outreach Program*. Retrieved October 19, 2011, from Operational Projects--Electricity Excel Spreadsheet: http://www.epa.gov/lmop/projects-candidates/operational.html

United States Environmental Protection Agency. (2011f, September 16). *Reformulated Gas*. Retrieved October 21, 2011, from Map of Current RFG Areas: http://www.epa.gov/otaq/fuels/gasolinefuels/rfg/areas.htm

United States Environmental Protection Agency. (2011g, September 16). *State Winter Oxygenated Fuel Program*. Retrieved October 21, 2011, from http://www.epa.gov/otaq/fuels/gasolinefuels/winterprograms/index.htm

United States House of Representatives. (2009, January 5). *Downloadable United States Code*. Retrieved October 21, 2011, from Chapter 85 - Air Pollution Prevention and Control: http://uscode.house.gov/download/pls/42C85.txt

Whims, J. (2002, August). *Agricultural Marketing Resource Center*. Retrieved January 23, 2012, from Pipeline Considerations for Ethanol: http://www.agmrc.org/media/cms/ksupipelineethl_8BA5CDF1FD179.pdf

World Health Organization. (2009). *Global health risks: mortality and burden of disease attributable to selected major risks*. Retrieved November 7, 2011, from Table 1: Ranking of selected risk factors: 10 leading risk factor causes of death by income group, 2004: http://www.who.int/healthinfo/global_burden_disease/GlobalHealthRisks_report_full.pdf

A Kaplan Turbine, *Credit/Courtesy of Voith Hydro*

Chapter 7
Water

Capturing Water's Kinetic Energy

Water is the second most important renewable energy source in the United States. In 2010 it accounted for 31% of all the renewable energy sources consumed in the country, which adds up to 2.48 quadrillion Btu, or more simply, 2.48 quads (United States Energy Information Administration, 2011). Nearly all of these Btu were generated by capturing the kinetic energy of moving water and using it to spin hydroelectric turbines. Recall from Chapter 1 that kinetic energy is motion. Under gravity's influence water naturally moves from high elevations to low elevations. We normally experience this when seeing a running river and intuitively knowing that water flows downhill. Since water has significant density (about 1,000 kg/m^3—or roughly 1 short ton per cubic yard), and is relatively plentiful in many parts of the world, it is an especially convenient substance to capture and put to use.

Water is a renewable energy source because it is constantly replenished. This perpetual energy supply continues as long as rain and snow keep falling, tributaries keep feeding rivers, currents keep moving, tides keep rising and falling, and waves continue crashing onto shorelines. These kinetic energy sources are all part of the hydrologic cycle. The hydrologic cycle describes the unending circulation of water as it transitions from liquid to vapor (evaporation) and back to liquid (condensation and/or precipitation). Water thus precipitates from clouds and flows downhill to rivers, lakes and oceans. Along the way we harness its energy—consider any large dam, a "conventional" hydroelectric power plant. Even after water reaches the ocean, the earth's spin and the moon's gravitational pull continue inducing movement, and this too can be exploited. Thus, gigantic hydroelectric turbines with nameplate capacities of hundreds of MW are only part of water's energy narrative. Much smaller hydroelectric operations also exist, with .1 to 5 MW nameplate capacities, and the number of these operations is growing fast.

Water is the leading renewable energy source used to generate electricity in the United States. This is abundantly clear when comparing the electricity production of conventional hydroelectric power plants to the output of other renewable energy sources, including biomass, geothermal, solar/photovoltaic and wind. For example, in 2010, conventional hydroelectric plants in the United States generated 257 million MWh, which accounts for a whopping 60% of the electrical generation of all renewable energy sources during that year (United States Energy Information

Administration, 2011b). The next closest renewable energy source in terms of electrical generation was wind, with 94.6 million MWh (United States Energy Information Administration, 2011c).

Water is the leading renewable energy source used to generate electricity in the world. For example, in 2009 China led the world with about 549 million MWh generated with conventional hydroelectric facilities. Electrical generation from all other renewable energy sources combined totaled only about 28 million MWh that same year. A similar pattern is observed in Brazil, with 387 million MWh generated with conventional hydroelectric facilities in 2009, and only 20 million MWh generated from all other renewable energy sources that same year. Most other countries follow this same pattern, with some notable exceptions that provide a glimpse into policy decisions and incentive programs within various governments. Spain is one example where in 2009 conventional hydroelectric facilities generated about 26 million MWh while wind generated 35 million MWh (United States Energy Information Administration, 2011d).

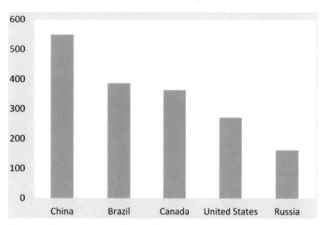

Figure 7-1, Conventional Hydroelectric Generation, Leading Countries, 2009 (million MWh)

Discussing hydroelectric power requires familiarizing ourselves with some of the most common vocabulary terms used in the industry. We'll use conventional hydroelectric plants because these are the biggest and best known facilities, and some readers have undoubtedly toured one or more of these power plants. In conventional hydroelectric plants a large concrete dam impounds river water (the term "impoundment facility" is fairly common) and stores it in a man-made lake called a "reservoir" that is situated behind the dam. Reservoir water is a dam's "headwater". After water passes through a dam it continues flowing along its existing river channel, and here it is called "tailwater". Reservoir depths vary depending on local geology and water flow, which is a critical issue because the height of the headwater compared to the tailwater partially determines the capacity of a hydroelectric facility. **"Head" is the vertical difference in elevation between the reservoir's surface (the "headwater") and the river's surface downstream from the dam (the "tailwater").**

Head tells us how much the water will drop and is measured in feet or meters. Notice in the above paragraph that head *partially* determines a hydroelectric facility's capacity. The famous Hoover Dam, for example (located at the Nevada and Arizona border southeast of Las Vegas), operates with an average head of about 520 feet and a capacity of 2,080 MW (United States Department of Interior, 2009). In contrast, the Grand Coulee Dam (located in central Washington State) has a significantly lower head—about 330 feet—but a much a higher capacity of 6,809 MW (United States Department of Interior, 2010). Also notice that head is not the same thing as the height of a dam. The Hoover dam, for example, measures 726 feet high and is much taller than the Grand Coulee, which measures 550 feet high (United States Department of Interior, 2010b, p. 2).

The amount of available water is also a critical issue for a hydroelectric facility because, like head, water volume is instrumental in determining the generating capacity of a dam. **"Flow" is water volume, measured in cubic feet or cubic meters, passing a point within one second.** Flow is most often reported in cubic feet per second (ft³/s) or cubic meters per second (m³/s). To generate electricity, headwater must drop through a dam's turbines. To do this, the first step is to allow headwater to flow into a penstock. **A penstock is a pipe that routes headwater to a hydroelectric turbine.** A single, large dam may have over a dozen turbines—the Hoover Dam, for example, has 17. From the penstock, flow is adjusted and guided to the turbine's blades—called the runner—which is the heart of any hydroelectric facility. **A turbine's runner is the series of blades or vanes that spin by capturing the kinetic energy of falling water.** Exiting the runner, flow is routed downstream to the tailwater.

Intuition correctly tells us that large hydroelectric facilities produce more power than small facilities. We can quantify this intuition once a river's head and flow are determined, and then use these values to calculate gross available power in Watts, kilowatts or megawatts. There are numerous calculation methods, and 5 of the most common methods are presented below along with hypothetical head and flow rates for a large river. Note that the equations are all very similar, with differences primarily existing between English and metric measurements.

Figure 7-2,
A View from the Crest of the Hoover Dam (Height = 726 feet, Head = 520 feet, Capacity = 2,080 MW), *Credit/Courtesy of U.S. Department of Interior, Bureau of Reclamation, Lower Colorado Region*

Figure 7-3,
The Grand Coulee Dam (Height = 550 feet, Head = 350 feet, Capacity = 6,809 MW), *Credit/ Courtesy of Bonneville Power Administration*

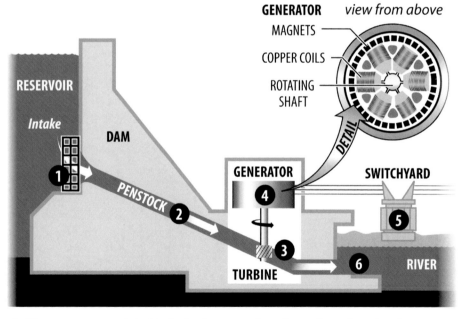

Figure 7-4, A Conventional Hydroelectric Facility Model, *Credit/Courtesy of National Energy Education Development Project*

1. Water in a reservoir behind a hydropower dam flows through an intake screen, which filters out large debris, but allows fish to pass through.

2. The water travels through a large pipe, called a penstock.

3. The force of the water spins a turbine at a low speed, allowing fish to pass through unharmed.

4. Inside the generator, the shaft spins coils of copper wire inside a ring of magnets. This creates an electric field, producing electricity.

5. Electricity is sent to a switchyard, where a transformer increases the voltage, allowing it to travel through the electric grid.

6. Water flows out of the penstock into the downstream river.

Calculating Hypothetical Gross Available Power in a River

Head = 500 ft (152.40 m)
Flow = 15,000 ft^3/s (or 424.75 m^3/s or 6,732,420 gallons/minute)
Efficiency = .9

Equation #1
Power calculated in kilowatts:
kW = H x F x E x .0846
- H= head (ft)
- F = flow (ft^3/s)
- E = efficiency, stated between 0 and 1
- 0.0846 = a constant value

kW = 500 x 15,000 x .9 x .0846
kW = 571,050 or <u>571 MW</u>

Equation #2
Power calculated in Watts
W = (H x F x E) / 5.3
- H = head (ft)
- F = flow (gallons/minute)
- E = efficiency, stated between 0 and 1
- 5.3 = a constant value

W = (500 x 6,732,420 x .9) / 5.3
W = 3,029,589,000 / 5.3
W = 571,620,566
kW = 571,620 or <u>571 MW</u>

Equation #3
Power calculated in kilowatts
kW = H x F x E / 11.8
- H = head (ft)
- F = flow (ft^3/s)
- E = efficiency, stated between 0 and 1
- 11.8 = a constant value

kW = 500 x 15,000 x .9 / 11.8
kW = 6,750,000 / 11.8
kW = 572,033 or <u>572 MW</u>

Equation #4
Power calculated in horsepower (HP) & converted to Watts
HP = (H x F x E)/8.8
- H = head (ft)
- F = flow (ft^3/s)
- E = stated between 0 and 1
- 8.8 = a constant value
- 1 HP = 746 watts, so multiply HP by 746 to obtain Watts

HP = (500 x 15,000 x .9)/8.8
HP = 6,750,000 / 8.8
HP = 767,045
W= HP x 756, so...
W = 767,045 x 746 = 572,215,570
kW = 572,215 kW or <u>572 MW</u>

Equation #5
Power calculated in kilowatts
kW = H x F x E x 9.8
- H = height in meters
- F = flow (m^3/s)
- E = efficiency, stated between 0 and 1
- 9.8 = a constant value

kW = 152.40 x 424.75 x .9 x 9.8
kW = 570,935 or <u>570 MW</u>

Calculation 7-1, Calculating Hypothetical Gross Available Power in a River

Notice that in Calculation 7-1 an efficiency of 90% (.9) is assumed. Recall from chapter 1 that average, age-ing, coal-fired power plants in the United States are only 35% efficient because a lot of energy is lost when water is converted from liquid to steam and then back again. Hydropower facilities have no such state-change requirement and thus have no comparative energy-losses. Nearly all (between 85-90%) of the kinetic energy of the moving water is captured and used to generate electricity (United States Department of the Interior, Bureau of Land Management, 2004, p. 4).

Hydropower turbines are not 100% efficient because of three main issues—turbulence, friction and cavitation. Turbulence is random change in water's flow patterns, pressures and velocities as it moves through a penstock, wicket gates (these are valves specific to particular kinds of turbines), and the runner itself. These same physical constructs also cause friction—or resistance to flow—as the water moves over these surfaces. Cavitation is the formation of bubbles created as a turbine's spinning runner creates zones of low pressure within the water flow.

While the efficiency of hydropower turbines is impressive, this does not mean that conventional hydroelectric facilities generate 85-90% of their nameplate capacity. Recall from Chapter 1 that capacity factor is the percent of time a generating system operates at maximum capacity. One might reasonably expect, given that hydroelectric turbines operate at 90% efficiency, that the facilities that house them would have similarly high capacity factors. This is not the case, however, as illustrated between the years 1998 and 2009, where the average capacity factor of conven-tional hydroelectric facilities in the United States was only 39.8% (United States Energy Information Administration, 2011e). For an especially vivid example, consider the Hoover Dam and its nameplate capacity of 2,080 MW. With this capacity, the Hoover Dam could, in a perfect world, theoretically generate 18,220,800 MWh of electricity in one year if it operated at maximum capacity 100% of the time. From 1947 through 2008, however, the average annual net generation was 4,200,000 MWh (United States Department of Interior, 2009). Thus, the Hoover Dam operates at only 23% of its capacity.

Conventional hydroelectric facilities operate at a fraction of their capacity because they are used to meet peak electricity demand rather than serving as baseload power suppliers. Thus a conventional hydroelectric facility may have several turbines sitting idle during evening hours when electricity demand is low and rates are cheap. When demand increases during the daytime, however, these idle turbines may be brought online, thereby allowing the power producer to capture relatively high electricity prices. In addition, as investments grow in renewable sources of electricity such as wind and sunlight, hydroelectric facilities serve to mitigate the intermittent character of these energy sources and thus stabilize the electricity supply serving the grid. These two advantages are possible because unlike coal-fired power plants and nuclear reactors, hydroelectric turbines are quickly turned on and off without compromising their structural integrity or efficiency.

Calculation 7-2, Calculating the Capacity Factor of the Hoover Dam

Calculating the Capacity Factor of the Hoover Dam

A) The Hoover Dam has a 2,080 MW capacity

In a perfect world...
B) 2,080 MW x 24 hours x 365 days = 18,220,800 MWh generated in 1 year

In the real world...
C) Net annual generation averages only 4,200,000 MWh

So...
D) 4,200,000 MWh / 18,220,800 MWh = .23 x 100 =23%

Types of Hydropower Turbines

Old-fashioned waterwheels often come to mind when thinking about capturing the power of falling water. Situated alongside rivers, these large wooden structures once transferred water's kinetic energy, via a series of gears and axes, into a millhouse where it was used to grind grains, spin saw blades, or power other useful tools. In the late nineteenth century, however, this strictly mechanical energy transfer was steadily being replaced by electricity, which was often generated in the very same millhouses. Electrical generators gradually replaced gear assemblies, and machines that once ran on mechanical energy were modified or rebuilt to operate strictly on electricity. Electricity's efficiency and transferability proved revolutionary, and by the start of the twentieth century riverside locations that once merely ground wheat or cut logs began appealing to a new type of entrepreneur—the power producer.

Figure 7-5, Waterwheel, *Credit/Courtesy of Bureau of Land Management/Oregon*

The twentieth century witnessed the birth of large-scale hydroelectric projects with their accompanying massive turbines. Gone were the wooden waterwheels, replaced by colossal steel runners weighing many tons and able to operate at very high speeds under massive pressure loads. As strong and impressive as these modern turbines are, however, they still do essentially the same thing as their earlier prototypes—spin. Modern hydroelectric turbines are classified into two categories—impulse and reaction (United States Department of Energy, 2011). **Impulse turbines spin in an aerated space at atmospheric pressure as high-velocity water hits each bucket or blade on a runner.** The most common impulse hydroelectric turbine design is the Pelton, which has obvious similarities to a waterwheel. **Reaction turbines spin while completely submerged in water as high-velocity and high-pressure water streams past each vane or blade on a runner.** Two common reaction hydroelectric turbine designs are the Francis and the Kaplan. While the Francis looks rather uncommon, the Kaplan has obvious similarities to propellers used to power boats.

Both impulse and reaction turbines exist because each hydroelectric facility has a site-specific head and flow. Over the course of the twentieth century, hydroelectric engineers discovered that some turbine designs operate most efficiently at high-head sites, while others are better suited for low-head sites. The same is true regarding flow. Thus we have the principle that impulse turbines are generally suited for high-head & low-flow sites, while reaction turbines are generally suited for low-head & high-flow sites. These generalizations have significant exceptions and overlap, however, so it is perhaps most productive to approach the understanding of hydroelectric turbines by keeping in mind the major differences among the three types.

The Pelton turbine spins under the force of a high pressure water jet (or jets) carefully aimed at the center of a double-cupped bucket (United States Department of Energy, 2011b). Large Pelton buckets can measure 60 centimeters wide by 1 meter long, and a single runner often contains 18 individual buckets. Large Pelton runners can have diameters over 4 meters. The water jet splits as it hits each bucket—half flows into to the right cup, half into the left cup. The force of the spray causes the runner to spin, thereby offering up another double-cupped bucket to the water jet. A single Pelton turbine will have one, two or up to six water jets aimed at the runner (Voith Hydro, 2009, p. 6). Under ideal circumstances, water exits a Pelton bucket with as little momentum as possible, with about 90% of its kinetic energy having been captured by the turbine.

Figure 7-6, The Most Common Impulse Hydroelectric Turbine Design—The Pelton, *Credit/Courtesy of Voith Hydro*

Figure 7-7, A Common Reaction Hydroelectric Turbine Design—The Francis, *Credit/Courtesy of United States Army Corps of Engineers*

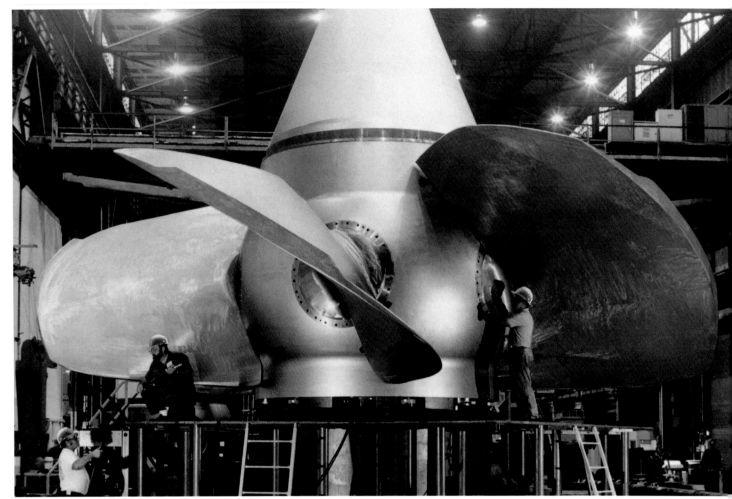

Figure 7-8, A Common Reaction Hydroelectric Turbine Design—The Kaplan, *Credit/Courtesy of Voith Hydro*

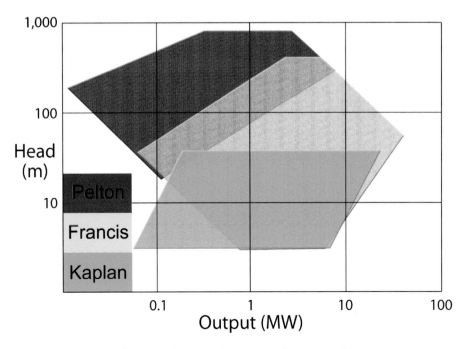

Figure 7-9, Graph of General Optimal Heads and Outputs for Various Turbine Designs

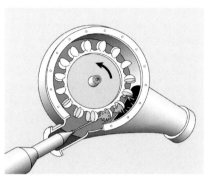

Figure 7-10, A Pelton Turbine Model, *Credit/Courtesy of United States Department of Energy*

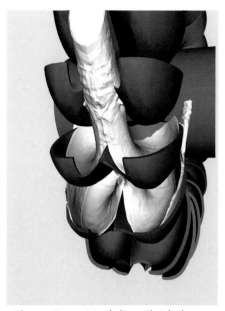

Figure 7-11, Modeling Fluid Flow Over a Pelton Runner, *Credit/ Courtesy of Voith Hydro*

Figure 7-12, The Spray Nozzle and Double-Cupped Buckets of a Pelton Turbine, *Credit/Courtesy of Voith Hydro*

Figure 7-13, A Pelton Turbine Spray Nozzle, *Credit/ Courtesy of Voith Hydro*

Figure 7-14, Worker Standing Next to a Pelton Turbine, *Credit/Courtesy of Voith Hydro*

Figure 7-15, A Newly Manufactured Pelton Turbine, *Credit/Courtesy of Voith Hydro*

The Francis turbine spins while completely submerged in a stream of high pressure water. Water is routed from the penstock, into a scroll case, and finally to the turbine. **A scroll case is a spiral-shaped pipe that routes high-pressure water toward a reaction turbine.** Water flow is controlled with a series of wicket gates that surround the turbine's housing. **Wicket gates are adjustable steel flaps that open and close, thereby controlling the flow of water into a reaction turbine.** With fully opened wicket gates, high-pressure water flows into all of the runner's vanes at once, causing the turbine's spin. After passing through the vanes, lower-pressure water exits through the runner's center and into another pipe, called a "draft tube", and then falls towards the tailwater. Some of the largest Francis turbines measure over 6 meters in diameter (Voith Hydro, 2009b, p. 6) and have as many as 12 individual vanes, with each vane measuring over three meters high.

The Kaplan turbine, like the Francis, also spins while completely submerged in a stream of high pressure water. Like the Francis, water is routed to the turbine via a penstock and scroll case, and here also wicket gates control the flow of water to the runner. Recall that Kaplan turbines look like a boat's propeller, and when the wicket gates are open, high pressure water

Figure 7-16, Worker Standing Next to a Francis Turbine, *Credit/Courtesy of Voith Hydro*

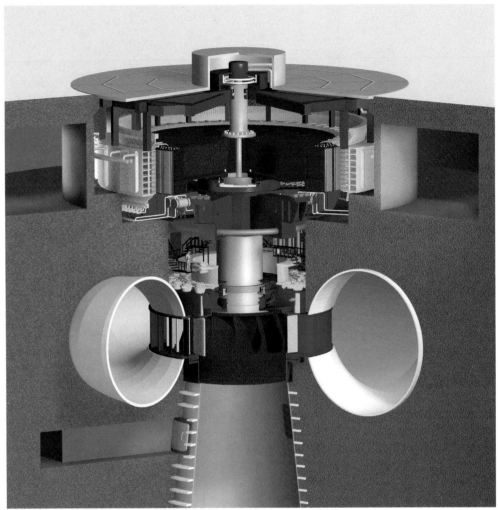

Figure 7-17, A Francis Turbine Model for the Xingo Hydro Power Plant in Brazil,
Credit/Courtesy of Voith Hydro

Figure 7-18, A Scroll Case at the Grand Coulee Dam,
Credit/Courtesy of United States Bureau of Reclamation

Wicket Gates (yellow)

Closed (no flow)   Opened (high flow)

Figure 7-19, Closed & Opened Wicket Gates that Control Water Flow into a Reaction Turbine's Runner, *Credit/Courtesy of Wikipedia (adapted)*

Figure 7-20, A Francis Turbine Model Test (Notice Wicket Gates 6-13 Above Runner), *Credit/Courtesy of Voith Hydro*

Figure 7-21, Lowering a Francis Turbine Runner into the Itaipu Dam, in Brazil & Paraguay, *Credit/Courtesy of Voith Hydro*

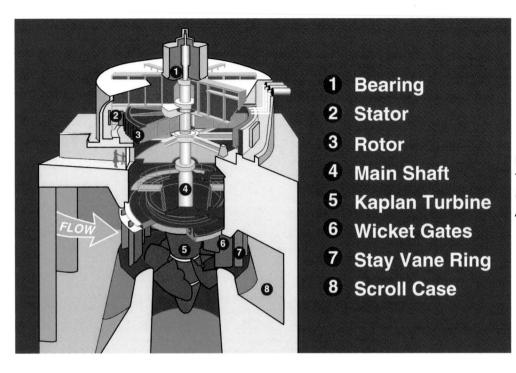

Figure 7-22, A Kaplan Turbine Model, *Credit/ Courtesy of United States Army Corps of Engineers*

1. Bearing
2. Stator
3. Rotor
4. Main Shaft
5. Kaplan Turbine
6. Wicket Gates
7. Stay Vane Ring
8. Scroll Case

flows over all of the runner's blades at once, thereby causing the turbine's spin (think about a boat's propeller spinning while completely submerged). After passing through the blades, lower-pressure water exits toward the tailwater through a draft tube. Kaplan turbines may have 5 to 7 blades, with some of the largest turbines measuring over 9 meters in diameter (Voith Hydro, 2009c, p. 7). An important advantage unique to Kaplan turbines is the ability of some designs to adjust the pitch of the blades. Adjustable blades give hydropower operators the ability to compensate for seasonal water flow variations.

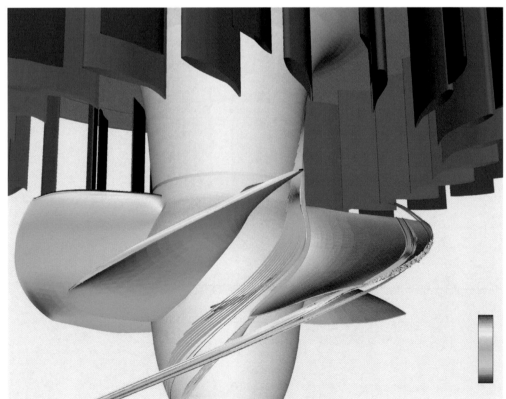

Figure 7-23, Modeling Fluid Flow Over a Kaplan Runner, *Credit/Courtesy of Voith Hydro*

Figure 7-24, The Kaplan Turbines Inside Bonneville Dam's Powerhouse, Oregon, *Credit/Courtesy of Bob Heims, United States Army Corps of Engineers*

Figure 7-25, A Kaplan Turbine with Adjustable Pitch Blades, *Credit/ Courtesy of Voith Hydro*

Figure 7-26, Workers Next to a Kaplan Turbine, *Credit/Courtesy of Laurie Driver, United States Army Corps of Engineers*

Conventional Hydroelectric Plants & Sediment

Conventional hydroelectric power plants are the largest capacity power plants on earth. Recall that these kinds of hydroelectric facilities trap—impound—water in a reservoir behind a dam. Water is released to accommodate varying electricity demand and to control reservoir depth. Dams are typically classified as gravity, arch, or a combination of the two. **Gravity dams impound water with sheer bulk and mass, while arch dams are thinner, typically built from concrete, and impound water with an arch that curves upstream.**

In addition to providing the power to spin turbines, reservoirs are often created to prevent flooding, provide drinking and irrigation water, and create recreational areas. This means that impoundment facilities often shoulder several responsibilities, all of which may not be mutually compatible. For example, if a drought strikes, many of a dam's penstocks might be closed in order to conserve water, meaning of course that the dam generates fewer kilowatt-hours. On the flip side, if an electric grid is overly dependent upon a hydroelectric facility—essentially rendering it a base load power plant—irrigation needs may be left unfulfilled as reservoir water is used for electrical generation rather than food production. Imagine the potential conundrum then, when hydroelectric engineers and policy makers meet to discuss how best to combat drought in a region with a rising population and thus rising food, water and electricity demands. Hopefully the means to resolve such questions will have already been tackled during the reservoir's planning stages where optimization models are used to best address these often competing demands.

In addition to the tug-of-war regarding water use, hydroelectric engineers must also mitigate a reservoir's inevitable sediment build-up (this is usually called "sedimentation"). Plainly stated, dams are sediment traps. All streams and rivers carry sediment. **Sediment is the eroded and transported mineral and rock debris within a streamflow.** Sediment varies in size, is classified by average diameter, and includes (from smallest to largest dimension): clay, silt, sand, gravel, cobbles and boulders. You have undoubtedly noticed that some rivers flow with clear water, while others flow with an opaque, muddy color. In the United States, this contrast is especially visible at the confluence of the Mississippi and Missouri Rivers, where the high silt content of the Missouri (nicknamed the "Big Muddy") meets the much clearer water of the Mississippi.

Various-sized sediment naturally travels in a streamflow until the flow velocity reaches zero (this is called a "base level"), at which time the sediment settles out of the water column. In uninterrupted rivers, ultimate zero-velocity occurs at an ocean, so sea level is aptly named "ultimate base-level". Reservoirs create artificial base-levels where streamflow stops and sediment settles to the reservoir's bottom. The same process happens in lakes and ponds, but in a reservoir's case, the natural bottom scouring and flushing that occurs in uncontrolled high-water events is absent (dams, after all, *prevent* uncontrolled high-water events). Thus sediment accumulates, raises a reservoir's bottom and reduces storage capacity. Many other challenges (both upstream and downstream) arise from sedimentation, including the formation of undesirable deltas, navigation interruption, turbine runner abrasion, and ecological damage (Morris & Fan, 1997, pp. 2.8-2.14).

Preventing sedimentation is in everyone's interest. The challenge lies in implementation, which is often site-specific and subject to distinct variables such as the unique geology of the reservoir and its sediment, the design of the dam, seasonal variations in water flow and sediment load, and riverbank ecology. The best solution, of course, is to avoid sedimentation in the first place. This requires upstream watershed management to reduce streambank erosion. Upstream anti-

Table 7-1, World's Largest Capacity Conventional Hydroelectric Power Plants

Facility & Country	Nameplate Capacity (MW)
Three Gorges (China)	22,400
Itaipu (Brazil & Paraguay)	14,000
Guri Simón Bolívar (Venezuela)	10,200
Tucurui (Brazil)	8,370
Grand Coulee (United States)	6,809

Class	Grain Diameter Range (mm)
Boulders	256-4,096
Cobbles	64-256
Gravel	2-64
Sand	.062-2
Silt	.004-.062
Clay	.00024-.004

Table 7-2, Sediment Classes

sedimentation practices include structural measures such as terracing, vegetation measures such as reforestation, and tillage measures such as contour farming (United States Department of Interior, 2006, p. 6.17).

Hydroelectric facility operators must often allow sediment-laden "density currents" to quickly flow through a reservoir by opening a dam's undersluice gate. **A density current is a unique flow of sediment-laden water that remains separate from clearer reservoir water due to its higher relative density and forward momentum.** Density currents are naturally-occurring events that happen as a result of temperature differences within a streamflow, and/or sudden sedimentation somewhere upstream. Since a density current is by definition more dense than standing reservoir water, it flows across a reservoir's bottom toward the dam. Opening an undersluice gate the moment before

such a current arrives allows nearly all of it to pass to the tailwater, thereby avoiding any significant settling within the reservoir. This practice is often called "venting" a density current. (Durgunoglu & Singh, 1993, p. 8).

Taking advantage of seasonal peak discharges (i.e., floods, springtime snowmelts, or rainy seasons) is another way to mitigate reservoir sedimentation. Higher-volume, higher-flow water literally picks up and transports more, and larger, sediment than lower-volume, lower-flow water. Thus riverbeds are naturally scoured during normal, high-water events when more water flows through an existing river channel. These "events" may rise, peak and diminish over the course of months. When high-volume water enters a reservoir, engineers can steadily open undersluice gates and thus use the water's own growing power

Figure 7-27,
The Silt-Laden Knife River flowing into the Missouri River, *Credit/Courtesy of United States Environmental Protection Agency*

Figure 7-28, Contour Farming Reduces Erosion, *Credit/Courtesy of Tim McCabe, Natural Resources Conservation Service*

to naturally scour the reservoir's bottom. These same outlets are steadily closed after peak discharge when river flow returns to pre-flood stages. The practice of opening and closing large-capacity undersluice gates during flood season is often called "routing bed material" (Morris & Fan, 1997, p. 15.2.2).

Recovering lost reservoir capacity due to sedimentation is sometimes accomplished with siphoning, flushing or dredging. Siphoning, as the name suggests, involves using the impoundment's head to route sediment-laden water up and over a dam to the tailwater. For rivers with especially high sediment loads, a siphon may be permanently established to mitigate deposition. **Flushing a reservoir drains it, creates a temporary streamflow across a reservoir's exposed bed, and thereby opens a scoured channel in the deposited sediment.** Flushing releases a large amount of sediment in a short period—over the course of many days or weeks—and is often correlated with seasonal flooding to take advantage of the increased power and scouring ability of the flowing water. Flushing is no panacea to sedimentation, however, as it only scours a deep, narrow channel in the sediment while most of the rest of the reservoir bed remains undisturbed (Morris & Fan, 1997, p. 15.1).

Dredging removes reservoir sediment by excavating it while the standing, impounded water remains present. This can be a controversial practice because the sediment may be toxic, a result of upstream pollution that may become concentrated within a reservoir's sediment (Durgunoglu & Singh, 1993, p. 16). There also may be disagreements about where to place the sediment after it is removed. In some cases it is appropriate to truck it away from the dam site, dump the material, and then stabilize it with vegetation. Under other circumstances—if the sediment is clean and the correct size—it may be beneficial to reintroduce the sediment to the river by gradually dumping it into the tailwater, in which case it would be naturally carried downstream. If this is the preferred solution, dam operators in the United States need to acquire permits through the Environmental Protection Agency (EPA) under Section 404 of the Clean Water Act. This law states that timely notice and public hearings must take place before the issuance of any permit to move sediment. The EPA may deny permits if the Secretary determines that sediment relocated will adversely influence water supplies, ecosystems or recreational areas (United States Environmental Protection Agency, 2009).

Figure 7-29, Sluice Gate at the Three Gorges Dam, *Credit/Courtesy of Thomas Froats*

Figure 7-30, Sluice Gates at the Three Gorges Dam, *Credit/Courtesy of Thomas Froats*

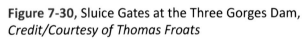

Figure 7-31, A Bucket Loader Dredging Sediment, *Credit/Courtesy of Ricky Garcia, US Army Corps of Engineers*

Exemplary Hydroelectric Plants

The world's largest capacity conventional hydro-electric facility—and largest power plant—is China's Three Gorges Dam, containing 32 Francis turbines (each rated at 700 MW) and a nameplate capacity of 22, 400 MW, (32 x 700 = 22,400 MW) (China Three Gorges Corporation, 2010). This dam impounds the Yangtze River at Sandouping, in the province of Hubei. It creates a winding, linear reservoir that expands river navigation over 400 miles southwest to the major

city of Chongqing in the direct-controlled municipality of the same name. The reservoir displaced a record 1,130,000 people as the Yangtze waters rose and flooded nearly 250 miles^2 of riverside farms and communities (China Three Gorges Corporation, 2010b). Vessels regularly transit the dam via a 5-step, 2-lane lock system located on the north side of the river, with each lock measuring 1,000 feet long and 110 feet wide. Several large vessels can therefore fit into a single lock as they transit up or downstream and common sights are cargo carriers, container vessels and especially coal barges.

The world's largest capacity pumped storage hydro-electric facility is the Bath County Pumped Storage Station in Virginia, containing 6 Francis turbines (each rated 462 MW) and a nameplate capacity of 2,772 MW, (6 x 462 = 2,772 MW) (Dominion, 2011). This facility pumps water into its upper reservoir when electricity rates are inexpensive, and releases water into its lower reservoir when electricity rates are expensive. The upper reservoir water level is about 1,000 feet higher than the lower reservoir, but this elevation difference fluctuates by about 100 feet as water is either pumped up to, or drawn down from, the upper reservoir. The naturally mountainous terrain of Virginia's Allegheny Mountains makes this design possible. Similar designs exist is countries all over the world and all of them are premised on the same idea—gather and store water when electricity

Figure 7-32, China's Three Gorges Dam, *Credit/Courtesy of Wikimedia*

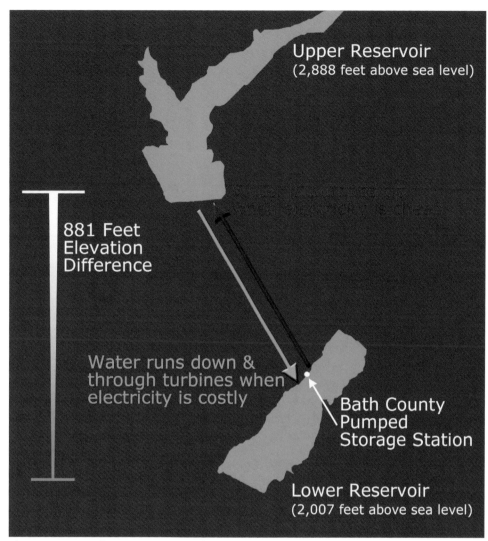

Figure 7-33, Bath County Pumped Storage Station, Aerial Model

is cheap, release water and generate electricity when it is expensive. **In pumped storage hydroelectric facilities, the same turbines that *generate* electricity during peak electricity use hours, *consume* electricity during off-peak hours as they are spun backwards to pump water uphill.**

The world's largest capacity run-of-river hydroelectric facility is the Chief Joseph Dam in Washington, containing 27 Francis turbines (16 rated at 64 MW, 11 rated at 95 MW) and a nameplate capacity of 2,069 MW (16 x 64 = 1,024 MW, and, 11 x 95 = 1,045 MW, so, 1,024 + 1,045 = 2,069 MW) (University of Washing-

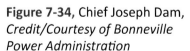

Figure 7-34, Chief Joseph Dam, *Credit/Courtesy of Bonneville Power Administration*

ton, School of Aquatic & Fishery Sciences, 2008). This facility is located downstream from the Grand Coulee Dam. Unlike the Grand Coulee, however, the Chief Joseph Dam does not create a massive impoundment reservoir and thus cannot store comparatively large quantities of water (United States Army Corps of Engineers, 2011). While Rufus Woods Lake (a reservoir) is indeed created by the Chief Joseph Dam, it contains a fraction of the amount of water in Franklin D. Roosevelt Lake (the reservoir created by the Grand Coulee Dam). To illustrate, remember that both lakes are narrow, winding bodies of water located on the Columbia River: Franklin D. Roosevelt Lake is 151 miles long (United States Department of Interior, 2011), while Rufus Woods Lake is only 51 miles long (United States Army Corps of Engineers, 2011b). Lacking a comparatively large impoundment area, Chief Joseph Dam and all run-of-river hydroelectric facilities are much more dependent upon consistent water flow provided by nature than conventional hydroelectric facilities. **Run-of-River hydroelectric facilities depend on the continuing flow of river water, rather than massive quantities of impounded water, to spin their turbines.** This fact makes them more vulnerable to drought because a significant reduction in flow volume could, in a worst-case scenario, render the turbines inoperative.

The world's largest capacity tidal hydroelectric facility is the Rance Tidal Power Station in Brittany, France, containing 24 specially-designed Kaplan variety turbines (each rated at 10 MW) and a nameplate capacity of 240 MW (24 x 10 = 240 MW) (Électricité de France, 2010). The station consists of 300 meter long tidal barrage that crosses the Rance River about two miles upstream from the Atlantic Ocean. This area is an estuary, a coastal body of water that experiences both fresh and salt-water flows due to shifting tides. **A**

Figure 7-35, Rance Tidal Power Station, *Credit/Courtesy of Wikimedia*

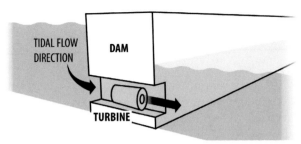

Figure 7-36, Model of Tidal Barrage, *Credit/Courtesy of National Energy Education Development Project*

tidal barrage hydroelectric facility consists of a large barrier stretching across a coastal area that exploits the natural rise and fall of tidal water to generate electricity. Think of a barrage as a kind of dam—rather than impounding river water, however, a barrage traps inflowing, high-tide water and releases it during out-flowing, ebb tides (this is the simplest of several possible designs) (World Energy Council, 2011). Unlike some other tidal facilities, the Rance Tidal Power Station has completely immersed, bulb-shaped "two-way" or "reversible" turbines, and therefore captures energy from rising as well as falling tidal flows (Buigues, Zamora, Mazón, Valverde, & Pérez, 2006, p. 2). The tides documented at Saint-Malo (a small port near the power station) illustrate why the Rance estuary is a good site for a tidal power station. While tidal ranges vary every day, an 8 to 10 meter (26 to 32 feet!) difference between high and low tides is common.

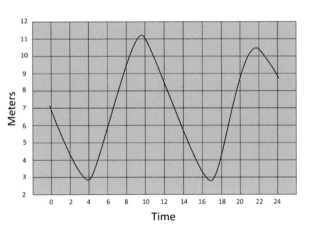

Figure 7-37, Tidal Chart for Saint-Malo, France

Water's Virtues and Vices

A balanced examination of water as an energy source reveals that it has advantages and disadvantages. At first glance this may appear strange given that water is a natural substance that falls from our skies and flows across earth's surface. Many residents living near large hydroelectric facilities, however, regularly encounter controversial environmental, social and economic issues connected with using water for utility-scale elec-

---

Calculating CO_2 Emission Savings from Hydroelectric Dams Compared to Coal-Fired Power Plants

A) Average annual net generation of Hoover Dam is 4,200,000 MWh

B) 4,200,000 MWh = 4,200,000,000 kWh

C) 2.08 lbs. of CO_2 is emitted for every kWh generated with coal

D) 4,200,000,000 kWh x 2.08 lbs. = 8,736,000,000 lbs. of CO_2

E) 8,736,000,000 lbs. /2,000 = 4,368,000 short tons of CO_2

F) The Hoover Dam annually spares the atmosphere 4.4 million short tons of CO_2

And...

G) In 2010, hydroelectric plants in the United States generated 257 million MWh

H) 257 million MWh = 257,000,000,000,000 Watt hours = 257,000,000,000 kWh

I) 257,000,000,000 kWh = 2.57×10^{11} kWh

J) $(2.57 \times 10^{11}$ kWh) $(2.08 \times 10^0$ lbs.) = 5.35×10^{11} lbs. of CO_2

K) 5.35×10^{11} lbs. / 2×10^3 = 2.7×10^8 tons of CO^2

L) In 2010, hydroelectric plants in the United States spared the atmosphere 2.7×10^8 short tons (270 million short tons) of CO^2

Calculation 7-3, Calculating CO_2 Emission Savings from Hydroelectric Dams Compared to Coal-Fired Power Plants

257 million MWh (2.57×10^{11} kWh). Using the same technique as above, we can state that hydroelectric plants in the United States spared the atmosphere from 270,000,000 short tons of CO_2 in 2010.

tricity generation. Communities need to understand the strengths and weaknesses of harnessing water's power before any hydroelectric projects are planned. Only then can sustainable practices be implemented to maximize water's benefits while effectively mitigating its drawbacks (Sustainable Hydropower, 2011).

The most prominent environmental virtue of using water as an energy source is the generation of emission-free electricity. Spinning a hydroelectric turbine does not require fossil fuel combustion or nuclear fission—both of which create undesirable waste products. Mother Nature readily provides the two most important ingredients for hydroelectricity—water and gravity. Every MWh of electricity produced by water does not have to be generated with an emission-producing energy source. To illustrate, recall that the average annual net generation of the Hoover Dam is 4,200,000 MWh (4,200,000,000 kWh). Since an average of 2.08 lbs of CO_2 is emitted for every kWh generated with coal (United States Energy Information Administration, 2011f), the Hoover Dam alone spares the atmosphere from 4,368,000 short tons of CO_2 every year. To further ramp up this emission-free virtue of hydroelectric power, recall that in 2010, conventional hydroelectric plants in the United States generated

The most prominent environmental vice of using water as an energy source is the deliberate flooding of communities and natural habitats behind a dam as a reservoir fills. We came across this issue once before with the forced relocation of over 1 million Chinese as a result of the Three Gorges Dam project. Entirely new towns were constructed as homes, farms, historic sites from China's dynastic past, as well as riparian (riverbank) ecosystems were inundated with rising Yangtze River water. Because dams alter a river's natural flow, up and downstream plant and animal biodiversity is compromised. Ecologists now openly lament the environmental losses due to the Three Gorges Dam, and doubt that the conservation efforts of the China Three Gorges Corporation will sufficiently mitigate these impacts (López-Pujol & Ming-Xun, 2009).

The United States also struggles with environmental consequences of large dams. This issue is well illus-

Figure 7-38, Glen Canyon Dam, *Credit/ Courtesy of U.S. Department of Interior, Bureau of Reclamation, Lower Colorado Region*

trated with the Glen Canyon Dam, located near Page, Arizona, upstream from Grand Canyon National Park and just south of the Utah border. The dam creates Lake Powell reservoir, the largest of four water impoundments on the Colorado River that comprise the Colorado River Storage Project (United States Department of Interior, 2010c). Controversy began in 1956, before construction, when the Sierra Club, the oldest, largest and most influential environmental conservation group in the United States, dropped its opposition to the dam during congressional negotiations about two other water impoundment projects further upstream. Then executive director David Brower bitterly lamented this decision, citing the Sierra Club's own policy of "no major scenic resource should be sacrificed for a power project" (Brower, 2008). The Glen Canyon Institute currently leads opposition to the dam, citing the loss of scenic natural beauty as well as up and downstream riparian habitat. The institute advocates (as Brower did) decommissioning the dam while leaving it in place due to the expense of tearing it down. The Institute advocates restoring Glen Canyon by lowering Lake Powell reservoir and drilling

bypass tunnels around the base of the dam (Glen Canyon Institute, 2011).

Flood protection is an important social and economic virtue of large hydroelectric facilities. This is a significant argument made by China when extolling the virtures of the Three Gorges Dam. While the Yangtze River is an essential waterway for millions, its floods have killed over 327,000 people within the past century (the fatalities are staggering: 1931 = 145,000, 1935 = 142,000, 1949 = 5,699, 1954 = 33,169, and 1998 = 1,526) (China Three Gorges Corporation, 2010c). These same floodwaters inundated over 10 million hectares (>38,000 square miles) of farmland, destroying crops and adding financial hardship to families already devastated by the loss of loved ones.

The United States, too, benefits from the flood control capacity of various hydroelectric projects. For example, consider the Tennessee Valley Authority (TVA). TVA is the largest public power company in the United States. In addition to coal-fired and nuclear power plants, TVA operates 29 hydroelectric dams and 1

Figure 7-39, Lake Powell, *Credit/Courtesy of U.S. Department of Interior, Bureau of Reclamation, Lower Colorado Region*

pumped-storage facility (Tennessee Valley Authority, 2011). These facilities currently prevent about $240 million in flood damage in an average year, and have averted over $5.4 billion in losses since the mid-1930s when construction began (Tennessee Valley Authority, 2011b).

Large hydroelectric facilities may damage fisheries by trapping silt and thereby making fish habitats nutrient-deficient. This issue is well illustrated in Egypt where the Aswan High Dam impounds the Nile River. The dam generates electricity and prevents annual flooding, but also traps silt. Nutrient-rich silt (containing organic matter, nitrogen and phosphorous) once flowed freely down the Nile River and into the Mediterranean Sea. At the Delta, this silt formed the foundation of an ecosystem that supported large varieties of fish, including jacks, mullets, redfish, bass and herring, as well as shrimp. Lake Nasser, the reservoir behind the dam, began filling and trapping silt in 1964. Some ecologists suspect that the subsequent 15

Figure 7-40, The Aswan High Dam Power Station, *Credit/Courtesy of Wikimedia*

year decline in Egyptian Mediterranean fish yield was caused by the dam trapping the silt (this is conjecture, however, as no silt-specific nutrient data was collected at that time). Interestingly, since the 1980s Egypt's Mediterranean fishery has rebounded many times over, leading still others to hypothesize that anthropogenic nutrient sources (fertilizer and sewage run-off) are fostering a silt-free but artificially-nutrient-rich ecosystem in the Mediterranean Sea off Egypt's Nile River Delta (Nixon, 2004, p. 165).

Large hydroelectric facilities can lower stream-levels and act as barriers for migrating fish, and turbines can injure or kill fish. The National Oceanic and Atmospheric Administration (NOAA) states that hydroelectric dams often account for "the largest single impact on Endangered Species Act-listed fish within a specific river basin" (National Oceanic and Atmospheric Administration, 2011). This is well-illustrated in the Columbia and Snake Rivers of the United States and Canada where local, state, federal and tribal agencies have worked for decades to mitigate this problem. Columbia and Snake River salmon and steelhead in particular are threatened or endangered due to the hydroelectric facilities built across their migratory paths. These fish are born in fresh water and then migrate to the ocean where they mature. Adults return to the same hatchery years later to spawn. Migrating fish must pass through or around many dams on both journeys.

Assisting juvenile fish (called "smolt" at this stage in life) on their Columbia and Snake River downstream passage includes the following methods: directing fish towards spillways, diversions or slides rather than allowing them to pass through penstocks and turbines; augmenting river flow by releasing more water from upstream dams (thereby mimicking high-volume and

Figure 7-41, Bonneville Dam, *Credit/Courtesy of Paul Stahmann*

Figure 7-42, Salmon Smolts at Bonneville Lock and Dam, *Credit/Courtesy of United States Army Corps of Engineers*

Figure 7-43, Bonneville Dam and Fish Ladder, *Credit/Courtesy of Paul Stahmann*

Figure 7-44, Bonneville Dam Fish Ladder, *Credit/Courtesy of Paul Stahmann*

Figure 7-45, Little Goose Lock and Dam, *Credit/Courtesy of United States Army Corps of Engineers. 2010 Fish Passage Plan*

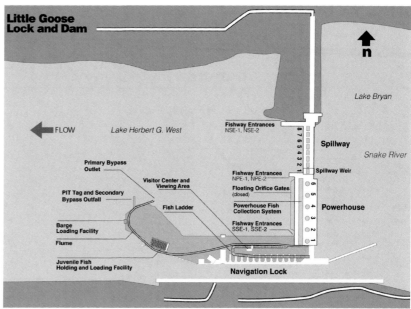

Figure 7-46, Spiraling Juvenile Fish Flume Over Adult Fish Ladder, Little Goose Lock and Dam, *Credit/Courtesy of United States Army Corps of Engineers*

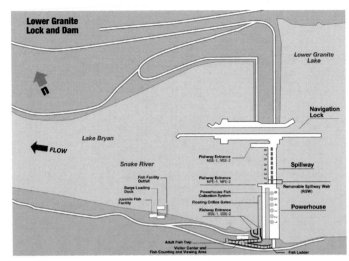

Figure 7-47, Lower Granite Lock and Dam, *Credit/Courtesy of United States Army Corps of Engineers. 2010 Fish Passage Plan*

Figure 7-48, Fish Barge on the Columbia River, *Credit/Courtesy of United States Army Corps of Engineers*

thus faster-flowing spring run-off); and, somewhat surprisingly, barge transport (Bonneville Power Administration, 2006, p. 1). Barge transport is managed and operated at four federal dams (McNary, Lower Monumental, Little Goose and Lower Granite) by the United States Army Corps of Engineers (United States Army Corps of Engineers, 2010, pp. 132, 196, 230, 265). These same dams (as well as many others) have fish ladders for returning adults to swim upstream in order to return to their spawning habitats.

Important Ideas:

- Water is the second most important renewable energy source in the United States.

- Water is a renewable energy source because it is constantly replenished.

- Water is the leading renewable energy source used to generate electricity in the United States.

- Water is the leading renewable energy source used to generate electricity in the world.

- "Head" is the vertical difference in elevation between the reservoir's surface (the "headwater") and the river's surface downstream from the dam (the "tailwater").

- "Flow" is water volume, measured in cubic feet or cubic meters, passing a point within one second.

- A penstock is a pipe that routes headwater to a hydroelectric turbine.

- A turbine's runner is the series of blades or vanes that spin by capturing the kinetic energy of falling water.

- Hydropower turbines are not 100% efficient because of three main issues—turbulence, friction and cavitation.

- Conventional hydroelectric facilities operate at a fraction of their capacity because they are used to meet peak electricity demand rather than serving as baseload power suppliers.

- Impulse turbines spin in an aerated space at atmospheric pressure as high-velocity water hits each bucket or blade on a runner.

- Reaction turbines spin while completely submerged in water as high-velocity and high-pressure water streams past each vane or blade on a runner.

- A scroll case is a spiral-shaped pipe that routes high-pressure water toward a reaction turbine.

- Wicket gates are adjustable steel flaps that open and close, thereby controlling the flow of water into a reaction turbine.

- Conventional hydroelectric power plants are the largest capacity power plants on earth.

- Gravity dams impound water with sheer bulk and mass, while arch dams are thinner, typically built from concrete, and impound water with an arch that curves upstream.

- Sediment is the eroded and transported mineral and rock debris within a streamflow.

- A density current is a unique flow of sediment-laden water that remains separate from clearer reservoir water due to its higher relative density and forward momentum.

- Flushing a reservoir drains it, creates a temporary streamflow across a reservoir's exposed bed, and thereby opens a scoured channel in the deposited sediment.

- Dredging removes reservoir sediment by excavating it while the standing, impounded water remains present.

- The world's largest capacity conventional hydroelectric facility—and largest power plant—is China's Three Gorges Dam, containing 32 Francis turbines (each rated at 700 MW) and a nameplate capacity of 22, 400 MW.

- The world's largest capacity pumped storage hydroelectric facility is the Bath County Pumped Storage Station in Virginia, containing 6 Francis turbines (each rated 462 MW) and a nameplate capacity of 2,772 MW.

- In pumped storage hydroelectric facilities, the same turbines that *generate* electricity during peak electricity use hours, *consume* electricity during off-peak hours as they are spun backwards to pump water uphill.

- The world's largest capacity run-of-river hydroelectric facility is the Chief Joseph Dam in Washington, containing 27 Francis turbines (16 rated at 64 MW, 11 rated at 95 MW) and a nameplate capacity of 2,,069 MW.

- Run-of-River hydroelectric facilities depend on the continuing flow of river water, rather than massive quantities of impounded water, to spin their turbines.

- The world's largest capacity tidal hydroelectric facility is the Rance Tidal Power Station in Brittany, France, containing 24 specially-designed Kaplan variety turbines (each rated at 10 MW) and a nameplate capacity of 240 MW.

- A tidal barrage hydroelectric facility consists of a large barrier stretching across a coastal area that exploits the natural rise and fall of tidal water to generate electricity.

- The most prominent environmental virtue of using water as an energy source is the generation of emission-free electricity.

- The most prominent environmental vice of using water as an energy source is the deliberate flooding of communities and natural habitats behind a dam as a reservoir fills.

- Flood protection is an important social and economic virtue of large hydroelectric facilities.

- Large hydroelectric facilities may damage fisheries by trapping silt and thereby making fish habitats nutrient-deficient.

Works Cited

Bonneville Power Administration. (2006, December). *Managing the Columbia River System to Help Fish*. Retrieved October 27, 2011, from http://www.bpa.gov/corporate/pubs/backgrounder/06/bg120106.pdf

Brower, D. R. (2008). *Sierra Magazine, Let the River Run Through It*. Retrieved October 27, 2011, from More than forty years ago David Brower made a mistake. Now he says it's time to bring Glen Canyon back to life: http://www.sierraclub.org/sierra/199703/brower.asp

Buigues, G., Zamora, I., Mazón, A., Valverde, V., & Pérez, F. (2006, April 5,6,7). *Sea Energy Conversion: Problems and Possibilities*. Retrieved October 27, 2011, from International Conference on Renewable Energies and Power Quality, ICREPQ'06: http://www.icrepq.com/icrepq06/242-buigues.pdf

China Three Gorges Corporation. (2010). *Three Gorges Project*. Retrieved October 27, 2011, from Remarkable Power Generation Benefit: http://www.ctgpc.com.cn/en/benefifs/benefifs_a_2.php

China Three Gorges Corporation. (2010b). *Three Gorges Project*. Retrieved October 27, 2011, from The Relocation Population and Difficulty of Resettlement: http://www.ctgpc.com.cn/en/benefifs/benefifs_a_7.php

China Three Gorges Corporation. (2010c). *Biggest Flood Control Benefit in the World*. Retrieved October 27, 2011, from Floods Records of Yangtze River (1931, 1935, 1949, 1954, and 1998): http://www.ctgpc.com.cn/en/benefifs/benefifs_a.php

Dominion. (2011). *Hydro Power*. Retrieved October 27, 2011, from Bath County Pumped Storage Station: http://www.dom.com/about/stations/hydro/bath-county-pumped-storage-station.jsp

Durgunoglu, A., & Singh, K. P. (1993, May). *The Economics of Using Sediment-Entrapment Reduction Measures in Lake and Reservoir Design*. Retrieved October 27, 2011, from Illinois State Water Survey: http://www.isws.illinois.edu/pubdoc/CR/ISWSCR-552.pdf

Électricité de France. (2010). *Marine Energies, How it Works*. Retrieved October 27, 2011, from Tidal Power: http://businesses.edf.com/generation/hydropower-and-renewable-energy/marine-energies/how-it-works-43777.html

Glen Canyon Institute. (2011). *Frequently Asked Questions About Restoring Glen Canyon*. Retrieved October 27, 2011, from Restoration of Glen Canyon: http://www.glencanyon.org/aboutgci/faq.php

López-Pujol, J., & Ming-Xun, R. (2009). Biodiversity and the Three Gorges Reservoir: a Troubled Marriage. *Journal of Natural History, 43*, 2765-2786.

Morris, G. L., & Fan, J. (1997). *Reservoir Sedimentation Handbook*. New York: McGraw-Hill Professional.

National Oceanic and Atmospheric Administration. (2011, September 6). *Northwest Regional Office*. Retrieved October 27, 2011, from NOAA's National Marine Fisheries Service, Salmon & Hydropower: http://www.nwr.noaa.gov/Salmon-Hydropower/index.cfm

Nixon, S. W. (2004, March/April). The Artificial Nile: The Aswan High Dam Blocked and Diverted Nutrients and Destroyed a Mediterranean Fishery, But Human Activities May Have Revived It. *American Scientist, 92*, 158-165.

Sustainable Hydropower. (2011). *About Sustainability in the Hydropower Industry*. Retrieved October 27, 2011, from Sustainability Challenges: http://www.sustainablehydropower.org/site/info/aboutsustainability.html

Tennessee Valley Authority. (2011). *Frequently Asked Questions About TVA*. Retrieved October 27, 2011, from How Does TVA Generate Electricity?: http://www.tva.com/abouttva/keyfacts.htm#elecgen

Tennessee Valley Authority. (2011b). *Flood Damage Reduction*. Retrieved October 27, 2011, from http://www.tva.com/river/flood/index.htm

United States Army Corps of Engineers. (2010, March). *Fish Passage Plan*. Retrieved October 27, 2011, from Corps of Engineers Projects, CENWD-PDW-R: http://www.nwd-wc.usace.army.mil/tmt/documents/fpp/2010/final/FPP_2010_Complete_Doc.pdf

United States Army Corps of Engineers. (2011, July 16). *Seattle District, Chief Joseph Dam*. Retrieved October 27, 2011, from A Run of the River Dam: http://www.nws.usace.army.mil/PublicMenu/Menu.cfm?sitename=cjdam&pagename=hydropower

United States Army Corps of Engineers. (2011b, June 8). *Seattle District, Chief Joseph Dam*. Retrieved October 27, 2011, from Recreation: http://www.nws.usace.army.mil/PublicMenu/Menu.cfm?sitename=cjdam&pagename=recreation

United States Department of Energy. (2011, October 11). *Wind and Water Power Program*. Retrieved October 26, 2011, from Types of Hydropower Turbines: http://www1.eere.energy.gov/water/hydro_turbine_types.html

United States Department of Energy. (2011b, February 9). *Energy Savers*. Retrieved October 26, 2011, from Microhydropower System Turbines, Pumps, and Waterwheels: http://www.energysavers.gov/your_home/electricity/index.cfm/mytopic=11120

United States Department of Interior. (2006, November). *Bureau of Reclamation*. Retrieved October 27, 2011, from Erosion and Sedimentation Manual: http://www.usbr.gov/pmts/sediment/kb/ErosionAndSedimentation/Contents.pdf

United States Department of Interior. (2009, February). *Bureau of Reclamation: Lower Colorado Region*. Retrieved October 26, 2010, from Hydropower at Hoover Dam: http://www.usbr.gov/lc/hooverdam/faqs/powerfaq.html

United States Department of Interior. (2010, December). *Bureau of Reclamation*. Retrieved October 26, 2011, from Grand Coulee Powerplant, Rated Head: http://www.usbr.gov/projects/Powerplant.jsp?fac_Name=Grand+Coulee+Powerplant

United States Department of Interior. (2010b, May). *Bureau of Reclamation*. Retrieved October 26, 2011, from Grand Coulee Dam Statistics and Facts: http://www.usbr.gov/pn/grandcoulee/pubs/factsheet.pdf

United States Department of Interior. (2010c, May 4). *Bureau of Reclamation*. Retrieved October 27, 2011, from Colorado River Storage Project, Glen Canyon Unit: http://www.usbr.gov/projects/Project.jsp?proj_Name=Colorado%20River%20Storage%20Project

United States Department of Interior. (2011, February 15). *Bureau of Reclamation, Columbia Basin Project*. Retrieved October 27, 2011, from Benefits, Recreation: http://www.usbr.gov/projects/Project.jsp?proj_Name=Columbia%20Basin%20Project

United States Department of the Interior, Bureau of Land Management. (2004). *American Energy for America's Future*. Retrieved October 26, 2011, from The Role of the U.S. Department of the Interior: http://www.blm.gov/pgdata/etc/medialib/blm/wo/MINERALS__REALTY__AND_RESOURCE_PROTECTION_.Par.61407.File.dat/EnergyBro.pdf

United States Energy Information Administration. (2011, July 14). *Independent Statistics and Analysis*. Retrieved October 25, 2011, from What Role Does Renewable Energy Play in the United States?: http://www.eia.gov/energyexplained/index.cfm?page=renewable_home

United States Energy Information Administration. (2011b, September 15). *Independent Statistics and Analysis*. Retrieved October 25, 2011, from Net Generation by Energy Source: Total (All Sectors): http://www.eia.gov/electricity/data.cfm#generation

United States Energy Information Administration. (2011c, October 19). *Independent Statistics and Analysis*. Retrieved October 25, 2011, from Electricity Net Generation: Total (All Sectors), 1949-2010: http://www.eia.gov/totalenergy/data/annual/showtext.cfm?t=ptb0802a

United States Energy Information Administration. (2011d). *Independent Statistics and Analysis*. Retrieved October 26, 2011, from International Energy Statistics, Renewables, Electricity Generation: http://www.eia.gov/cfapps/ipdbproject/iedindex3.cfm?tid=6&pid=alltypes&aid=12&cid=&syid=2005&eyid=2009&unit=BKWH

United States Energy Information Administration. (2011e, April). *Independent Statistics and Analysis*. Retrieved October 26, 2011, from Average Capacity Factors by Energy Source, Electric Power Annual With Data for 2009, Average Capacity Factors by Energy Source, 1998 through 2009 (percent): http://www.eia.doe.gov/cneaf/electricity/epa/epat5p2.html

United States Energy Information Administration. (2011f, April 28). *Independent Statistics and Analysis*. Retrieved October 27, 2011, from How much carbon dioxide (CO2) is produced per kilowatt-hour when generating electricity with fossil fuels?: http://www.eia.gov/tools/faqs/faq.cfm?id=74&t=11

United States Environmental Protection Agency. (2009, December 15). *Policy & Guidance, Wetlands*. Retrieved October 27, 2011, from Clean Water Act, Section 404: http://water.epa.gov/lawsregs/guidance/wetlands/sec404.cfm

University of Washington, School of Aquatic & Fishery Sciences. (2008, December 22). *Columbia Basin Research, Hydroelectric Information for Columbia and Snake River Projects*. Retrieved October 27, 2011, from Chief Joseph Dam - General Information: http://www.cbr.washington.edu/crisp/hydro/chj.html

Voith Hydro. (2009, October). *Pelton Turbines*. Retrieved October 26, 2011, from Voith Hydro: http://www.voithhydro.com/media/VSHP090041_Pelton_t3341e_72dpi.pdf

Voith Hydro. (2009b, May). *Francis Turbines*. Retrieved October 26, 2011, from Voith Hydro: http://www.voithhydro.com/media/t3339e_Francis_72dpi.pdf

Voith Hydro. (2009c, May). *Kaplan Turbines*. Retrieved October 26, 2011, from Voith Hydro: http://www.voithhydro.com/media/t3340e_Kaplan_72dpi.pdf

World Energy Council. (2011). *Survey of Energy Resources 2007, Tidal Energy*. Retrieved October 27, 2011, from Harnessing the Energy in the Tides, Tidal Barrage Methods: http://www.worldenergy.org/publications/survey_of_energy_resources_2007/tidal_energy/755.asp

A Wind Turbine's Nacelle, *Credit/Courtesy of Patrick Corkery, National Renewable Energy Laboratory*

Chapter 8
Wind

Wind's Capacity Growth

Generating electricity by capturing the kinetic energy of wind is experiencing tremendous growth. If you happen to live in a windy area, you may intuitively know this as massive turbines sprout up across the landscape. The installed capacity of wind power across the United States, as of June 30, 2011, is 42,432 MW (United States Department of Energy, 2011). To compare, consider that the year-end installed capacity of wind power in 1999 was just 2,472 MW. This means that 1,616% growth occurred in 12 years. No other energy source (non-renewable or renewable) experienced anything close to this growth over the same time period. While this growth is impressive, recall that all renewable energy sources accounted for only 8% of the United States energy consumption by energy source in 2010. Wind specifically contributed to 11% of that total, or .88 quadrillion Btu (8 x. 11 = .88) (United States Energy Information Administration, 2011).

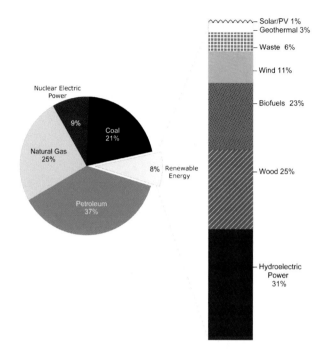

Figure 8-2, Wind's 11% Contribution to United States Renewable Energy Consumption, 2010, *Credit/Courtesy of United States Energy Information Administration*

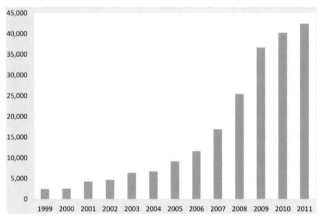

Figure 8-1, United States Installed Wind Power Capacity (MW), 1999-2011

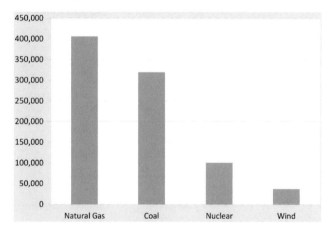

Figure 8-3, Wind's Installed Capacity (MW) Compared to Non-Renewables, 2010

Wind's phenomenal growth should be tempered with the realization that its installed capacity pales next to the installed capacity of non-renewable energy sources in the United States. Consider the 2010 capacities of the following energy resources: natural gas = 406,100 MW, coal = 319,400 MW, nuclear = 101,000 MW (United States Energy Information Administration, 2011b). Thus, while its growth is impressive, wind's total contribution to the energy profile of the United States is relatively small. Clearly, however, something significant is driving the electricity-generation industry to adopt wind quickly. This is evident across the country where private industries, municipalities and public institutions are tackling issues associated with this energy source. It is ironic that wind has been "discovered" in the past decade, while it has been used around the world for millennia to sail ships and pump water.

The United States is just one player in this current wind-frenzy. China leads the world with 44,733 MW of installed capacity (World Wind Energy Association, 2011). This is impressive growth given that in 2008 China had only 12,170 MW of installed capacity. Other leading countries include Germany, Spain, and India (United States Energy Information Administration, 2011c).

Wind is growing as an energy source partly because of the fact that it generates carbon-free electricity. Recall from chapter one that most electricity is generated by making a rotor spin. Coal-fired power plants

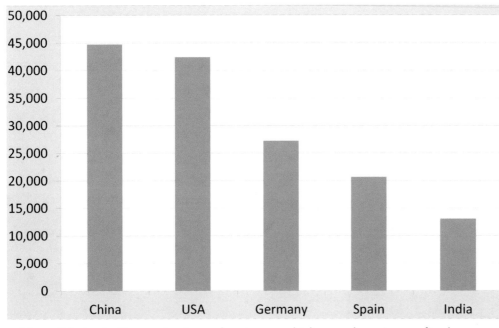

Figure 8-4, Installed Wind Capacity (MW), Leading Countries, 2010

achieve this by boiling water to make steam, which then spins a turbine, which in turn spins the rotor in a generator. Natural gas plants skip the steam part by simply burning the CH_4 in a gas turbine, which then spins the rotor in a generator. Also recall that both coal and natural gas emit carbon dioxide as an unavoidable byproduct of combustion.

Wind turbines require no combustion to induce rotor spin. The force of air molecules against a turbine's blades accomplishes what steam does in coal-fired power plants, and what high-pressure combustion does in gas-fired power plants. As environmentally conscious companies increasingly look for ways to "green" their energy consumption by using carbon-free electricity, wind power is increasingly becoming a popular option. For example, Kohl's Department Stores, Motorola, TD Bank and Whole Foods Market were all honored at the 2010 Renewable Energy Markets Conference as leading green power purchasers (United States Environmental Protection Agency, 2011). TD bank in particular purchased over 240 mil-

Figure 8-5, Commercial Wind Turbines, *Credit/ Courtesy of National Renewable Energy Laboratory*

lion kWh of wind-derived electricity via renewable energy certificates, thereby satisfying 100% of its electricity needs in 2010.

Wind is growing as an energy source partly because of state tax incentives. Consider the following examples: In Nebraska, wind energy development projects are 100% exempt from paying state sales tax on the gross receipts from property purchased, leased or rented for community-based energy development projects (Database of State Incentives for Renewables and Efficiency (DSIRE), 2011). In Maine, community wind generation projects—defined by Maine's legislature as 10 MW or less—can claim a 100% sales tax reimbursement for any property physically incorporated into the project (Database of State Incentives for Renewables and Efficiency (DSIRE), 2011b). In 2007, Illinois taxation of wind energy devices was standardized and simplified across all counties, thereby easing many tax burdens and project complications. Today in Illinois any wind turbine larger than 500 kW that is used to generate commercial electricity is practically valued at $360,000/MW (Database of State Incentives for Renewables and Efficiency (DSIRE), 2011c). Numerous other tax incentives exist across many other states and are often combined in laws encompassing other renewable energy sources such as solar and geothermal.

Wind is growing as an energy source partly because of renewable energy certificates (RECs). Keep in mind that on any electric grid, all the electricity is identical. Electricity delivered to location A is indistinguishable from electricity delivered to location B, even if 45% of was generated in a coal-fired power plant, 10% by wind turbines and 45% by nuclear fission. A kilowatt-hour is a kilowatt-hour, regardless of its origin. How then can an individual, organization or company claim, as TD Bank does, that 100% of its electricity is wind-derived? They do so by buying RECs. **Renewable energy certificates (RECs) represent the environmental and social attributes of renewable energy sources, and are bought and sold apart from the physical commodity of electricity** (United States Environmental Protection Agency, 2011b)**.**

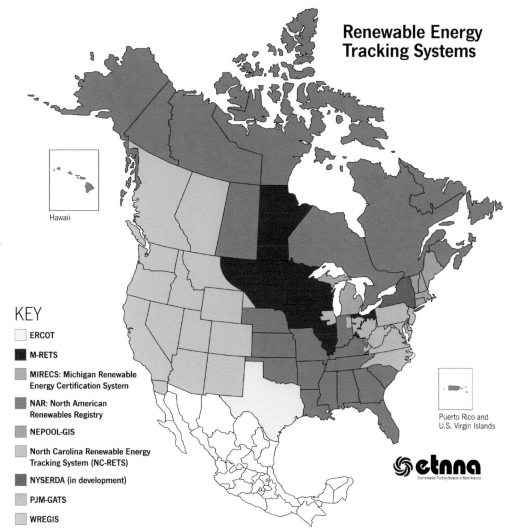

Figure 8-6, Map of the Nine REC Regional Tracking Systems, *Credit/ Courtesy of Environmental Tracking Network of North America*

Renewable Energy Tracking Systems

KEY

☐ ERCOT
■ M-RETS
▨ MIRECS: Michigan Renewable Energy Certification System
▨ NAR: North American Renewables Registry
☐ NEPOOL-GIS
☐ North Carolina Renewable Energy Tracking System (NC-RETS)
■ NYSERDA (in development)
☐ PJM-GATS
☐ WREGIS

A single REC is created for every MWh of electricity generated via renewable energy sources, including biomass, moving water, wind, geothermal and sunlight. Nine regional tracking systems (some currently in development) register RECs as they are generated, assign each one a unique number, and match each one to its current owner (United States Environmental Protection Agency, 2008). RECs accumulate in participants' accounts, much like money in a bank account. REC owners are free to do whatever they like with their RECs, including transferring, retiring, or exporting them to another regional tracking system. RECs are retied—and therefore rendered no longer useable—when environmental or renewable energy claims are made specific to the tracking numbers associated with individual RECs. Registration fees, annual fees, as well as volumetric fees are often collected from each participant in order to finance the accounting within each of the nine regional tracking systems.

Buying and selling RECs takes place in REC exchanges, which are different entities from the regional tracking systems. The United States Department of Energy currently lists 24 different exchanges (United States Department of Energy, 2011b). Two kinds of buyers purchase RECs—those wishing to satisfy renewable energy compliance mandates, such as renewable energy portfolio standards (which are explained below), and voluntary buyers wishing to purchase renewable energy for progressive environmental, social or

marketing reasons (Environmental Tracking Network of North America, 2011, p. 4). The latter is the case with TD Bank—in addition to paying its normal electricity bill, the company also purchased and retired RECs equal to the total number of MWh it consumed in 2010 (the amount TD Bank spent purchasing these RECs is proprietary information and therefore unavailable). Thus, REC buyers receive proof of compliance and/or environmental stewardship when they retire their purchased RECs, and REC generators receive money, which then spurs construction of more renewable energy projects and offsets risk associated with a nascent industry.

Wind is growing as an energy source partly because of state legislatures adopting renewable energy portfolio standards (RPS). Just as environmentally conscious companies are increasingly looking for ways to "green" their energy consumption, so too are states. One way to do this is by creating market demand for electricity created from renewable energy sources. A renewable energy portfolio standard (RPS) is a law requiring utilities to supply a specific amount of electricity from renewable energy sources. States adopt their own unique RPS depending on their diverse interests, and often have gradual, phase-in schedules, thereby allowing utilities several years to either build

Figure 8-7, Map of States with Renewable Energy Portfolio Standards, 2011, *Credit/Courtesy of Database of State Incentives for Renewables and Efficiency (DSIRE)*

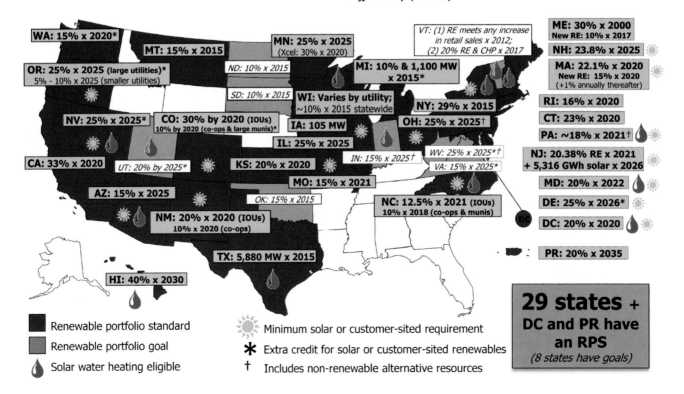

renewable capacity and/or finance REC purchasing plans (purchasing RECs is one way a utility can satisfy its responsibilities under a RPS). There are currently 33 states, as well as the District of Columbia, with RPS laws. Many states encourage growth in all renewable energy, while others target specific sectors (United States Environmental Protection Agency, 2010).

For examples of the diversity amongst states with RPS, as well as how wind power is uniquely encouraged, consider Virginia, New Mexico and Illinois. Unlike New Mexico and Illinois, Virginia's RPS is voluntary. The state's legislature encourages public utilities to supply electricity via renewable energy sources, and provides an increased rate of return as an incentive to reach specific RPS goals (Database of State Incentives for Renewables and Efficiency (DSIRE), 2011d). Public utilities achieve RPS goals by accumulating credits, and triple credits are available to any utility supplying energy derived from offshore wind (State of Virginia, 2007).

By contrast, investor-owned utilities in New Mexico must generate 20% of retail electricity sales via a diversified portfolio of renewable energy sources by 2020 (Database of State Incentives for Renewables and Efficiency (DSIRE), 2011e). The state legislature's definition of a "fully diversified renewable energy portfolio" states that no less than 20% of the RPS requirement must be met using wind energy (State of New Mexico, 2007).

Finally, the state of Illinois specifies that 25% of retail electricity sales come from renewable energy sources by 2025 (Database of State Incentives for Renewables and Efficiency (DSIRE), 2011f). While numerous renewable energy sources are considered in the state's RPS, wind-generated electricity is clearly preferred. By 2025, investor-owned electric utilities in Illinois must supply a full 75% of the abovementioned retail electricity sales with wind-derived generation (State of Illinois, 2007).

Air Pressure, Wind & Power

Just 4 different gases comprise over 99% of our atmosphere: nitrogen (N_2), oxygen (O_2), argon (Ar) & carbon dioxide (CO_2) (National Weather Service, 2010). It may surprise you to know that most of the air we breathe is not oxygen but nitrogen. Consider this: each time you inhale, about 77% of what fills your lungs is nitrogen, a gas that does little more than take up space. Nitrogen, along with all the other atmospheric gases, presses down, up and side-to-side on every square

inch of everything. The average amount of pressing is 14.7 pounds for every square inch of every surface (recall from chapter 3 that this average pressure is called "1 standard atmosphere" and is abbreviated "1 atm"). Another way to consider this is to realize that air has mass, approximately 1.225 kg/m^3 at 59 °F (15 °C).

We rarely think about air pressure because we have evolved into this atmosphere and as a result our bodies are perfectly suited to the force these gases exert upon us. In addition, we seldom think about changes in air pressure unless we happen to be riding a fast elevator or flying in an airplane and experience the unpleasant sensation of having our ears pop. Yet, because the atmosphere acts like a fluid, always churning and mixing in an effort to become homogenized, air pressure variations are a constant reality.

Many interrelated variables heat up and cool down the molecules in our atmosphere. These include a location's latitude, surface topography, nearby bodies of water, prevailing winds, weather patterns, earth's 24 hour rotation, as well as the 23.5° tilt of our planet's axis. Increasing temperature makes a parcel of air less dense than the surrounding molecules, and it thus rises (similar to a hot air balloon). Comparatively, decreasing temperature renders that same parcel denser than the surrounding molecules, and it thus sinks.

When a cubic meter of air rises it expands because there are literally fewer air molecules surrounding it at this higher elevation. Molecules that were once confined to that cubic meter are now free to leave. Thus, within that cubic meter, air pressure drops. This is illustrated on weather maps with a large letter "L", meaning low relative pressure. Conversely, when a cubic meter of air falls, there are literally more molecules

Figure 8-8, Rising & Sinking Air Columns and Resulting Low and High Pressures

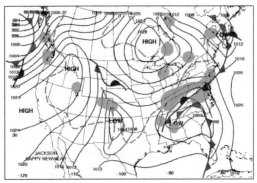

Figure 8-9, Weather Map Illustrating Low and High Pressure, *Credit/Courtesy National Oceanic and Atmospheric Administration*

surrounding it at this lower elevation. Molecules that were once outside the cubic meter are now squeezed into this space. Thus, within this cubic meter, air pressure rises. This too is illustrated on weather maps, this time with a large letter "H", meaning high relative pressure.

The difference between a high and low pressure center is often described as a gradient. We normally think of a gradient as a hill, where, if we happen to be riding a bicycle, a "steep" gradient means a fast ride down a precipitous slope. Similar reasoning is used in meteorology to describe the pressure difference between highs and lows and the resulting air movement. A steep pressure gradient means a lot of pressure difference exists between two adjacent pressure centers and, like our bicycle, air will quickly flow down that slope, from high to low pressure.

Air moving down a pressure gradient is called wind. Since it has mass, its kinetic energy can be captured and transformed into electricity. **Wind-power density is how much energy is available in any square meter of wind and is calculated in Watts/m².** Calculating wind-power density requires keeping two important caveats in mind. First, the power calculations initially assume a constant, perfect wind velocity, which is

---

Wind-Power Density, Calculated for "Perfect" Wind (measured in Watts/m²)

$$W/m^2 = .5 \times 1.225 \times V^3$$

- .5 = a constant value
- 1.225 = air's density at 59 °F (15 °C)
- V = wind's velocity, expressed in meters/second (m/s), cubed

Power in "perfect" wind at 1 m/s
$$W/m^2 = .5 \times 1.225 \times 1^3$$

$$W/m^2 = .5 \times 1.225 \times 1^3$$
$$W/m^2 = .5 \times 1.225 \times 1$$
$$= .61 \ Watt/m^2$$

Power in "perfect" wind at 3 m/s
$$W/m^2 = .5 \times 1.225 \times 3^3$$

$$W/m^2 = .5 \times 1.225 \times 3^3$$
$$W/m^2 = .5 \times 1.225 \times 27$$
$$= 17 \ Watts/m^2$$

Power in "perfect" wind at 5 m/s
$$W/m^2 = .5 \times 1.225 \times 5^3$$

$$W/m^2 = .5 \times 1.225 \times 5^3$$
$$W/m^2 = .5 \times 1.225 \times 125$$
$$= 77 \ Watts/m^2$$

Power in "perfect" wind at 7 m/s
$$W/m^2 = .5 \times 1.225 \times 7^3$$

$$W/m^2 = .5 \times 1.225 \times 7^3$$
$$W/m^2 = .5 \times 1.225 \times 343$$
$$= 210 \ Watts/m^2$$

Power in "perfect" wind at 9 m/s
$$W/m^2 = .5 \times 1.225 \times 9^3$$

$$W/m^2 = .5 \times 1.225 \times 9^3$$
$$W/m^2 = .5 \times 1.225 \times 729$$
$$= 447 \ Watts/m^2$$

Calculation 8-1, Wind-Power Density, Calculated for "Perfect" Wind (measured in Watts/m²)

never the case—half the time it is faster and half the time it is slower than the stated average velocity. Second, a non-linear relationship exists between wind speed and power, which simply means that *doubling wind's velocity does not merely double its power.* For example, 3 m/s wind contains 17 Watts/m^2 of power, while 6 m/s wind contains 139 Watts/m^2—over 8 times as much power. Thus the math in Calculation 8-1 only applies to a "perfect" world and is useful only in demonstrating the exponential rise in Watts/m^2 with increasing wind velocity.

Each place on earth has "real" wind, with velocities that gust, taper off, and change according to dozens of possible variables. We need to couple this knowledge with the non-linear relationship between wind velocity and wind-power density explored in the above calculations, while remembering that a little bit more wind velocity yields a lot more power. At any site then,

an average wind velocity occurs, but during the 50% of time that the wind is blowing faster than the average, many, many more Watts are available for capture. The likelihood of capturing these Watts can be calculated with each increase in any unit of wind speed, but the probability statistics are beyond our immediate scope. **Wind-power density in "real" wind is nearly double that of "perfect" wind.** Notice that the equations in Calculation 8-2, simply take the formula for calculating energy in "perfect" wind and add one additional multiplication step in order to calculate the available Watts in "real" wind.

Way back in 1986 the United States Department of Energy created the Wind Energy Resource Atlas of the United States (United States National Renewable Energy Laboratory, 1986). The effort mapped the average annual winds across the country at 10 and 50 meters above the ground (50 meters being the most common

Calculation 8-2, Wind-Power Density, Calculated for "Real" Wind (measured in Watts/m^2)

Wind-Power Density, Calculated for "Real" Wind (measured in Watts/m^2)

$$W/m^2 = 1.91 \times .5 \times 1.225 \times V^3$$

- 1.91 = a constant value
- .5 = a constant value
- 1.225 = air's density at 59^0 F (15^0 C)
- V = wind's velocity, expressed in meters/second (m/s), cubed

Power in "real" wind at 1 m/s
$$W/m^2 = 1.91 \times .5 \times 1.225 \times 1^3$$

$W/m^2 = 1.91 \times .5 \times 1.225 \times 1^3$
$W/m^2 = 1.91 \times .5 \times 1.225 \times 1$
$= 1$ Watt/m^2

Power in "real" wind at 3 m/s
$$W/m^2 = 1.91 \times .5 \times 1.225 \times 3^3$$

$W/m^2 = 1.91 \times .5 \times 1.225 \times 3^3$
$W/m^2 = 1.91 \times .5 \times 1.225 \times 27$
$= 32$ Watts/m^2

Power in "real" wind at 5 m/s
$$W/m^2 = 1.91 \times .5 \times 1.225 \times 5^3$$

$W/m^2 = 1.91 \times .5 \times 1.225 \times 5^3$
$W/m^2 = 1.91 \times .5 \times 1.225 \times 125$
$= 146$ Watts/m^2

Power in "real" wind at 7 m/s
$$W/m^2 = 1.91 \times .5 \times 1.225 \times 7^3$$

$W/m^2 = 1.91 \times .5 \times 1.225 \times 7^3$
$W/m^2 = 1.91 \times .5 \times 1.225 \times 343$
$= 401$ Watts/m^2

Power in "real" wind at 9 m/s
$$W/m^2 = 1.91 \times .5 \times 1.225 \times 9^3$$

$W/m^2 = 1.91 \times .5 \times 1.225 \times 9^3$
$W/m^2 = 1.91 \times .5 \times 1.225 \times 729$
$= 853$ Watts/m^2

Average Wind Speed (m/s)	Average Wind Speed (mph)	Wind-Power Density (W/m²)
1	2.24	1
2	4.47	9
3	6.71	32
4	8.95	75
5	11.18	146
6	13.42	253
7	15.66	401
8	17.90	599
9	20.13	853
10	22.37	1,170
11	24.61	1,557
12	26.84	2,022

Table 8-1, Power in "Real" Wind, Measured in Watts/meter² (W/m²)

height of wind turbines at that time). Today the effort continues, with the United States National Renewable Energy Lab (NREL) mapping winds at 80 meters above the ground. 80 meters is a common height for contemporary, commercial-scale wind turbines. Producing wind maps requires a wind-classification scheme, thus cartographers (map-makers) created the now commonly-used "Wind Power Classes". **Wind Power Classes range from 1 to 7, with class 1 winds having poor resource potential, and class 7 winds having superb resource potential.** Generally, sites with class 3 winds and higher are suitable for commercial-scale wind turbine applications (United States National Renewable Energy Laboratory, 2011).

Wind Class	Wind-Power Density (W/m²)	Average Wind Speed (m/s)	Average Wind Speed (mph)
1	0-200	0 - 5.6	0 - 12.5
2	200-300	5.6 - 6.4	12.5 - 14.3
3	300-400	6.4 - 7	14.3 - 15.7
4	400-500	7 - 7.5	15.7 - 16.8
5	500-600	7.5 - 8	16.8 – 17.9
6	600-800	8 - 8.8	17.9 – 19.7
7	800-2,000	8.8 - 11.9	19.7 – 26.6

Table 8-2, Wind Power Classes, Classified at 50 Meters Above Ground

Figure 8-10, Wind Farm near Spanish Fork, Utah *Credit/ Courtesy of Paul Stahmann*

Figure 8-11, Wind Resource Map of the United States at 80 Meters, *Credit/Courtesy National Renewable Energy Laboratory*

United States - Annual Average Wind Speed at 80 m

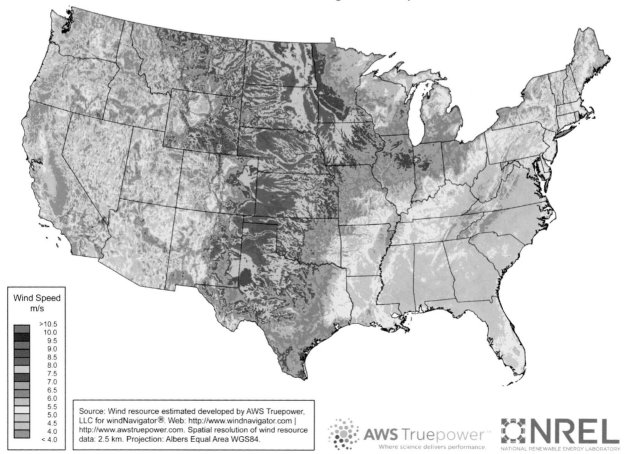

Wind Speed m/s

>10.5
10.0
9.5
9.0
8.5
8.0
7.5
7.0
6.5
6.0
5.5
5.0
4.5
4.0
< 4.0

Source: Wind resource estimated developed by AWS Truepower, LLC for windNavigator®. Web: http://www.windnavigator.com | http://www.awstruepower.com. Spatial resolution of wind resource data: 2.5 km. Projection: Albers Equal Area WGS84.

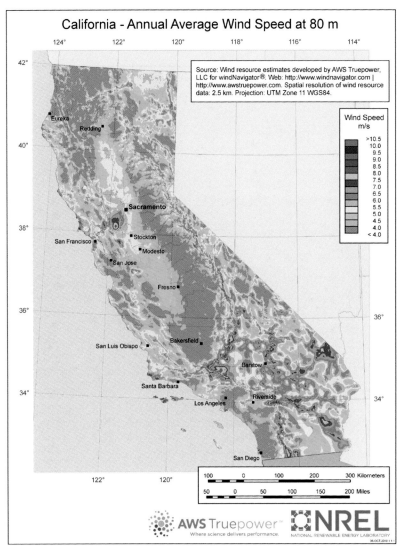

Figure 8-12, Wind Resource Map of California at 80 Meters, *Credit/Courtesy National Renewable Energy Laboratory*

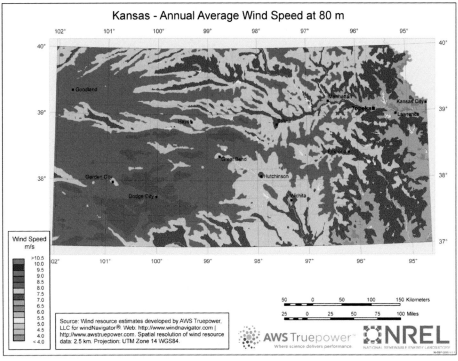

Figure 8-13, Wind Resource Map of Kansas at 80 Meters, *Credit/ Courtesy National Renewable Energy Laboratory*

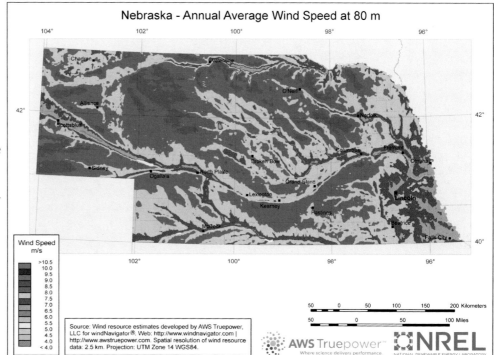

Figure 8-14, Wind Resource Map of Nebraska at 80 Meters, *Credit/Courtesy National Renewable Energy Laboratory*

Figure 8-15, Wind Farm in Indiana, *Credit/Courtesy of Paul Stahmann*

Bigger is Better

A cursory examination of wind power over the past decade quickly reveals that commercial-scale wind turbines are trending ever-larger. Energy companies are interested in capturing greater economies of scale by taking advantage of the non-linear relationship between wind speed and power, so increasingly higher towers and longer rotor blades are the norm. **Wind turbine towers are commonly made of tubular steel, with a base measuring several meters in diameter, and gently tapering at about 80 to 100 meters above the ground.** Situated at the top of the tower is a nacelle. **A nacelle is the large, enclosed structure atop a wind turbine that houses the gears, generator, controller and brake.** Attached to the nacelle via a steel shaft is the rotor. **A rotor includes the turbine's hub and blades.** Blades are constructed in one seamless piece of fiberglass and epoxy, and have the ability to change pitch and therefore feather into and out of the wind. As we'll see in the calculations below, the length of the blades—more specifically the diameter of the rotor—is a critical variable in determining the power output from any wind turbine.

Wind turbine blades are tapered similar to an airplane wing. The blades move when the wind blows, but since they are attached at the hub, the result is a spin. The shaft holding the rotor spins slowly, about 18 revolutions per minute (rpm), which is far too slow to generate significant amounts of electricity (United States Department of Energy, 2011c). A set of gears inside the nacelle, therefore increases the spin of the shaft to about 1,800 rpm, an appropriate amount for electrical generation. Since wind shifts direction, and most commercial-scale turbines are "upwind turbines" (meaning they face into the wind), the rotor must continually adjust its bearing, the direction it is facing, in order to capture the best wind. This is done with a yaw drive, which rotates the nacelle so it always faces into the wind. Finally, wind turbines do not spin in very light or very fast winds. Exceedingly light winds do not have enough wind-power density to move the blades, or they simply do not produce enough power to make up of for the unavoidable efficiency losses as kinetic energy is transformed into electrical energy. Very fast winds, on the other hand, pose a hazard to

Figure 8-16, Transporting Wind Turbine Rotor Blades, *Credit/Courtesy of Siemens*

the blades' structural integrity. **Cut-in wind speed is the speed at which a turbine's blades begin to spin, and cut-out wind speed is the speed at which a turbine's blades are deliberately stopped.**

For a wind turbine example, consider one of Siemens Energy's average size turbines, the SWT-2.3-101. The rather cryptic name is deciphered as follows: "SWT" = Siemens Wind Turbine, "2.3" = 2.3 MW capacity, and "101" = the rotor's diameter, measured in meters. Each of the 3 blades is 49 meters long (160 feet). They are situated atop a tower that stands 80 meters high (262 feet). For comparison, consider that the height of the Statue of Liberty, from its heel to the top of its head, is 111 feet, 6 inches (The Statue of Liberty-Ellis Island Foundation, Inc., 2011). Thus, the tower is taller, and each blade is longer, than the Statue of Liberty.

Depending on winds, the SWT-2.3-101 rotor spins between 6 and 16 rpm. This rotation is increased with gears in the nacelle that have a ratio of 1:91. Thus for every 1 spin of the rotor, the shaft that holds the rotor spins 91 times. In low winds the rotor spins at 6 rpm while the shaft spins 546 rpm (6 x 91 = 546 rpm), and in high winds the rotor spins at 16 rpm while the shaft

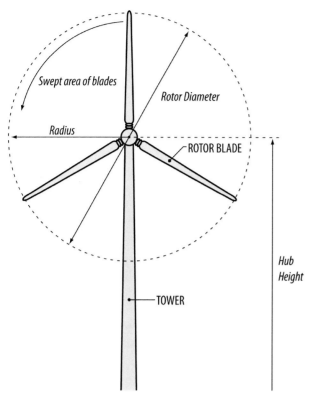

Figure 8-17, Model of a Wind Turbine's Geometry, *Credit/Courtesy of National Energy Education Development Project*

Figure 8-18, A Wind Turbine's Nacelle, *Credit/ Courtesy of Siemens*

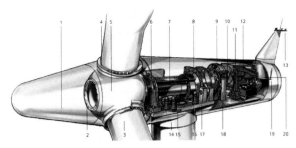

1. Spinner
2. Spinner bracket
3. Blade
4. Pitch Bearing
5. Rotor Hub
6. Main Bearing
7. Main Shaft
8. Gearbox
9. Brake disc
10. Coupling
11. Generator
12. Service Crane
13. Meteorological sensors
14. Tower
15. Yaw Ring
16. Yaw Gear
17. Nacelle bedplate
18. Oil Filter
19. Canopy
20. Generator fan

Figure 8-19, SWT-2.3-101, *Credit/Courtesy of Siemens*

spins at 1,456 rpm (16 x 91 = 1,456 rpm). The turbine's cut-in speed is 4 m/s, while its cut-out speed is 25 m/s (Siemens Energy, 2010).

A very common land-based commercial-scale wind turbine is rated at 1.5 MW. 1.5 MW is the turbine's nameplate capacity, a term we've come across before when examining other energy resources and respective power plants. Recall from Chapter 7, for example, that the Grand Coulee Dam has a nameplate capacity of 6,809 MW, or from Chapter 1 that the Palo Verde Nuclear Generating Station has a nameplate capacity of 3,942 MW. Just as with these generating methods, the nameplate capacity of wind turbines must be understood in the context of their capacity factor. Recall that capacity factor is the percent of time a generating system operates at maximum capacity. **Wind turbines typically have capacity factors between 25-30%** (United States Department of Energy, 2010, p. 49). Thus,

under impossible "perfect" circumstances, a 1.5 MW turbine would generate 13,140 MWh of electricity in 1 year (1.5 MW x 24 hours x 365 days = 13,140 MWh/year). Since the real world, however, has 1) variable winds, 2) turbines with efficiencies designed for specific wind velocities, and 3) an average capacity factor of about 30%, a 1.5 MW turbine will never generate anywhere near 13,140 MWh/year.

No matter how well engineers craft the next generation of wind turbines, they will never exceed 59.3% efficiency—this is known as Betz's Law after the German physicist that first calculated the limit in the early twentieth century. To better illustrate Betz's Law, consider that a turbine spins because wind moves into *and also out of* the turbine. If a turbine captured 100% of the wind's kinetic energy, wind would exit the turbine at 0 m/s—that is, not exit at all. With no wind exiting the turbine, it would not spin because no wind could enter it (Danish Wind Industry Association, 2011). Betz's Law states that the maximum amount of the wind's kinetic energy that any turbine can capture is 59.3%.

Presented in Calculation 8-3 is an equation that can be used to determine the power output of a typical three-blade commercial wind turbine. The math in Calculation 8-3 illustrates the importance of rotor diameter (D) because all other variables are held constant. Note two issues with this calculation method. First, it does not explicitly use a turbine's nameplate capacity. It does, however, us D, which stands for a turbine's rotor diameter. This is an important detail because turbines of the same nameplate capacity often have slightly longer or shorter blade lengths. Second, this calculation method uses turbine efficiency and states it as a

Turbine	Nameplate Capacity	Hub Height (meters)	Rotor Diameter (meters)	Area Swept by Rotor (meters^2)
Vestas V52-850 kW	850 kW	44-74 (depending on model)	52	2,124
General Electric TC2-1.5xle	1.5 MW	80	82.5	5,346
Enercon E-82/3 MW	3 MW	78-138 (depending on model)	82	5,281
RePower 5 MW	5 MW	117 (onshore model)	126	12,469

Table 8-3, Several Examples of Wind Turbines and Respective Sizes

Figure 8-20, A 1.5 MW Wind Turbine Base

constant 30%. In practice, however, turbine efficiency is increasing and also specific to individual turbine designs from different manufacturers.

In Calculation 8-3, 4 increasingly large wind turbines from Enercon (a German wind turbine manufacturing company) (Enercon, 2011) are theoretically placed in the same location with 7 m/s wind. The only variable that changes is the rotor diameter, from 53, to 82, to 101 and finally to 126 meters. This exercise gives us insight into why the wind-power industry prefers ever-larger rotors (and thus necessitates ever-taller turbine towers). Notice how a small percentage increase in rotor diameter yields a much greater relative percentage increase in power output.

The math in Calculation 8-3 also reveals additional insight into wind turbines and their capacity factors. Recall that a 25-30% capacity factor is the norm (the average in the United States reached 34% in 2008 before returning to 30% the next year). When examining our calculations with this in mind, then, we ought to get close to this capacity factor range. The first three turbines from Enercon yield capacity factors of 33%, 32% and 32% respectively, and thus we are can be relatively comfortable with these calculations. The last turbine, however, yields a capacity factor of only 20%, and yet it is the largest and most powerful turbine made by Enercon.

Why then such a dismal capacity factor? Simply this, the Enercon E-126 is not designed for winds at 7 m/s. No doubt an energy firm could erect one in such a

Figure 8-21, SWT-3.0-101, *Credit/Courtesy of Siemens*

place, but the economics of the project would prove disastrous because decades would pass before any capital expenditures were recouped, and by then the turbine would have exceeded its expected lifetime. Thus, Enercon's largest turbine only makes sense when situated in places where winds blow faster than 7 m/s, and thus has a wind-power density greater than 401 Watts/m^2. If, for example, we placed this turbine at a site with 9 m/s winds (Class 7, superb winds) with 852 W/m^2 wind-power density, the resulting calculations reveal a much more impressive 42% capacity factor.

Wind Farms

Individual wind turbines have very small nameplate capacities compared to individual fossil fuel, nuclear, and conventional hydroelectric power plants. For perspective, consider that the largest proposed wind turbine in the world is 10 MW. Patented by Sway Turbine, the prototype will be positioned on Norway's west coast near the town of Bergen, and be commercially available before 2015 (Sway Turbine, 2011). It is common, by contrast, for individual coal-fired, natural gas-fired, nuclear and conventional hydroelectric

Calculating Power Output From a Typical Three-Blade Wind Turbine

$$\text{Watts} = E \times (\pi/4) \times D^2 \times P$$

- E = .30 = a constant value for typical three-blade turbines
- $\pi/4$ = .785,398,163 = a constant value for typical three-blade turbines
- D = turbine rotor diameter, expressed in meters, squared
- P = power in "real" wind (held constant at 7 m/s)

Enercon E-53/800 kW
Watts = .30 x .785,398,163 x 52.9^2 x 401

Watts = .30 x .785,398,163 x 52.9^2 x 401
= .30 x .785,398,163 x 2798 x 401
= 264,365 W
=264 kW
x 8760
= 2,312,640 kWh/year or 2,313 MWh/year

Enercon E-82/2 MW
Watts = .30 x .785,398,163 x 82^2 x 401

Watts = .30 x .785,398,163 x 82^2 x 401
= .30 x .785,398,163 x 6724 x 401
= 635,306 W
= 635 kW
x 8760
= 5,562,600 kWh/year or 5,563 MWh/year

Enercon E-101/3 MW
Watts = .30 x .785,398,163 x 101^2 x 401

Watts = .30 x .785,398,163 x 101^2 x 401
= .30 x .785,398,163 x 10,201 x 401
= 963,825 W
= 964 kW
x 8760
= 8,444,640 kWh/year or 8,445 MWh/year

Enercon E-126/7.5 MW
Watts = .30 x .785,398,163 x 126^2 x 401

Watts = .30 x .785,398,163 x 126^2 x 401
= .30 x .785,398,163 x 15,876 x 401
= 1,500,018 W
= 1,500 kW
x 8760
= 13,140,000 kWh/year or 13,140 MWh/year

Calculation 8-3, Calculating Power Output from a Typical Three-Blade Wind Turbine

Turbine	Rotor Diameter (m)	Power Output (MWh/year)	Rotor Diameter Increase	Power Output Increase
E-53/800 kW	53	2,313	--	--
E-82/2 MW	82	5,563	55%	140%
E-101/3 MW	101	8,445	23%	52%
E-126/7.5 MW	126	13,140	25%	56%

Table 8-4, Larger Rotor Diameter Yields Greater Relative Increase in Power Output

Properly Siting the Enercon E-126/7.5MW Wind Turbine in Class 7 Winds

Enercon E-126/7.5 MW
Watts = .30 x .785,398,163 x 126² x 852

Watts = .30 x .785,398,163 x 126² x 852
= .30 x .785,398,163 x 15,876 x 852
= 3,187,071 W
= 3,187 kW
x 8760
= 27,918,120 kWh/year or 27,918 MWh/year

An Ideal Site for the E-126/7.5 MW
C.F. = actual generation/perfect generation

In "Perfect" Conditions:
7500 kilowatts x 24 hours x 365 days
= 65,700,000 kWh/year

In "Real" Conditions of Class 7 Wind:
Capacity factor = 27,918,120 / 65,700,000
= .4249 x 100 = 42%

Calculation 8-4, Properly Siting the Enercon E-126/7.5MW Wind Turbine in Class 7 Winds

power plants to carry nameplate capacities anywhere between 500 to 3,000 MW. Thus, while individual companies, schools or organizations may obtain plenty of electricity from a single wind turbine, electric utility companies need many dozens or even hundreds to adequately satisfy any appreciable demand from the electric grid.

The best place in the United States to find wind farms is Texas. **Texas has the most installed wind capacity (10,089 MW as of December 31, 2010) compared to any other state in America.** States that follow Texas, with their respective installed capacities, are Iowa (3,675 MW), California (3,253 MW), Washington (2,104 MW) and Minnesota (2,205 MW) (United States Department of Energy, 2011d). **The world's largest capacity wind farm is the Roscoe Wind Farm (RWF), located about 50 miles west of Abilene, Texas, and holds 627 turbines with a combined nameplate capacity of 781.5 MW** (Power-Technology.com, 2011). The landscape in west Texas is very flat, has few trees, and average winds that blow between 8.5 and 9 m/s. The rural area is sparsely populated and supported mainly by thousands of acres of cotton fields. Situated within these fields is the RWF, occupying the windiest parts of 4 separate counties (Nolan, Mitchell, Scurry and Fisher) and spanning almost 100,000 acres (O'Grady, 2009).

Notice that RWF is situated within the cotton fields. **Roscoe Wind Farm, as well as many others, are dual-use entities, allowing farmers to continue cultivating their land or grazing their cattle while electricity is generated by wind turbines spinning overhead.** Thus Roscoe's famers cultivate cotton, as well as earn between $5,000 and $15,000 of lease income per turbine per year, depending on the turbine size (Burnett, 2007). If, for example, a single farmer decides, therefore, to lease out enough space for 15 turbines, he stands to make at least $75,000 each year, in addition to whatever income is earned from cotton.

Many land-owners around Roscoe are pleased with the prospect of the local economy earning income from wind turbine leases. Nearby, however, at the Horse Hollow Wind Energy Center, local enthusiasm is less enthusiastic. **Horse Hollow Wind Energy Center, located about 30 miles southwest of Abilene, Texas, is the second largest capacity wind farm in the world, with 421 turbines and a combined nameplate capacity of 735.5 MW** (NextEra Energy Resources, 2011). Controversy began in 2005 when local landowners, displeased with the sight of large wind turbines dotting the landscape, filed a lawsuit against the project's owner, Florida Power and Light (now named NextEra Energy Resources). A Texas judge ruled that the landowners' nuisance claim that was based on the appearance of the turbines was insufficient grounds for a lawsuit. Plaintiffs then filed another suit, this time claiming the turbines' noise was a nuisance. After an 18 month trial, including hundreds of hours of decibel measurements both inside and outside local area

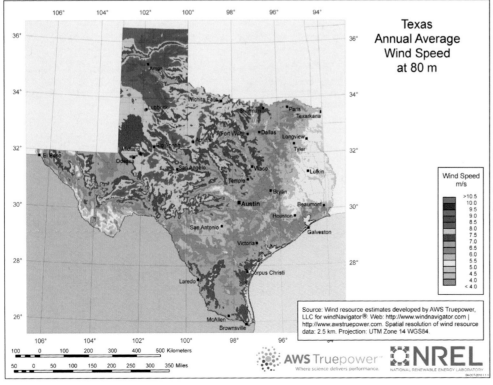

Figure 8-22, Wind Resource Map of Texas at 80 Meters, *Credit/Courtesy National Renewable Energy Laboratory*

Wind   221

homes, a jury found there was no measureable audio nuisance and awarded the plaintiffs nothing (Phadke, 2011).

The first offshore wind farm lease in United States federal waters was signed on October 6, 2010. **The Cape Wind Energy Project, located in Nantucket Sound on Horseshoe Shoal, is approximately 46 square miles and is expected to hold 130, 3.6 MW turbines, for a combined nameplate capacity of 468 MW.** The lease lasts 33 years and costs Cape Wind LLC $88,278.00 in rent prior to construction, and a 2% to 7% operating fee (depending on regional electricity sales) after elec-

trical generation begins (Bureau of Ocean Management, Regulation and Enforcement, 2011). The project is expected to satisfy about 75% of the electricity demand from Cape Cod and the Islands of Martha's Vineyard and Nantucket (Cape Wind, 2011).

Horseshoe Shoal is a shallow part of Nantucket Sound and was chosen by Cape Wind as the preferred wind farm site after considering the area's unique recreation, navigation and environmental issues. Even so, the project is fraught with local opposition. Save Our Sound: The Alliance to Protect Nantucket Sound, leads the opposition effort and claims that the proposed

Figure 8-23, Elbow Creek Wind Farm, Texas, *Credit/ Courtesy of Siemens*

Figure 8-24, A Turbine at the Wildorado Wind Farm, Texas, *Credit/Courtesy of Siemens*

wind farm will obstruct views, damage the environment and pose a hazard to shipping and fishing industries. Specifically worrying for Save Our Sound is the Cape Wind's large electrical service platform that will hold 40,000 gallons of oil and 1,000 gallons of diesel fuel, and be situated on a platform within the wind farm (Save Our Sound: Alliance to Protect Nantucket Sound, 2011). Save our Sound is of course quite disappointed by the recently-signed lease agreement, but vows to continue blocking the Cape Wind project by mounting continuing legal challenges and public protests.

The largest offshore wind farm in the world is the Thanet Offshore Wind Farm, located in Kent, United Kingdom, and holding 100, 3 MW turbines for a combined nameplate capacity of 300 MW. The farm is 13.5 square miles, with the closest turbine situated about 7.5 miles from the shoreline (Vattenfall, 2011). Since the turbines are quite tall—standing 377 feet above the sea's surface—the farm is visible from shore on a clear day, prompting mixed local opinions about the project. Some praise the government for pursuing sustainable energy resources, while others object to the turbines marring an otherwise unobstructed

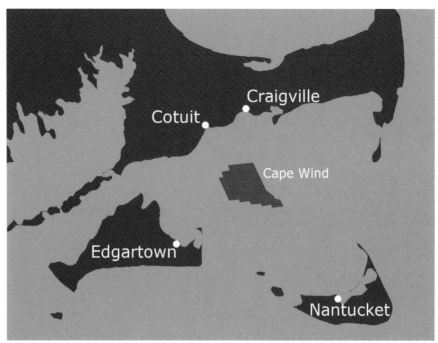

Figure 8-25, Map of Cape Wind Site and Surrounding Communities

Figure 8-26, Computer Simulated View of Cape Wind from Cotuit, *Credit/Courtesy of Cape Wind Associates*

Figure 8-28, Computer Simulated View of Cape Wind from Edgartown, *Credit/Courtesy of Cape Wind Associates*

Figure 8-27, Computer Simulated View of Cape Wind from Craigville, *Credit/Courtesy of Cape Wind Associates*

Figure 8-29, Computer Simulated View of Cape Wind from Nantucket, *Credit/Courtesy of Cape Wind Associates*

Figure 8-30, Model of Offshore Wind Farm Connection to Power Grid, *Credit/Courtesy of Siemens*

Figure 8-31, The Middelgrunden Offshore Wind Farm, Denmark, *Credit/Courtesy of Siemens*

Figure 8-32, The Lillgrund Offshore Wind Farm, Sweden, *Credit/Courtesy of Siemens*

ocean view (BBC News, 2010). Whatever their opinion about Thanet, United Kingdom residents are expecting massive expansions of offshore wind projects as the country works to comply with renewable energy targets established by the European Union in 2007. The House of Commons states the government's position quite plainly:

> Given the relative maturity of the wind sector, and the continuing construction of new wind capacity, we

believe that wind energy will make the greatest contribution to meeting our 2020 renewable energy targets. In order for the full potential of wind power to be realised, it is essential that the Government takes urgent steps to address operational barriers to its deployment.

(United Kingdom, House of Commons, 2008, p. 16)

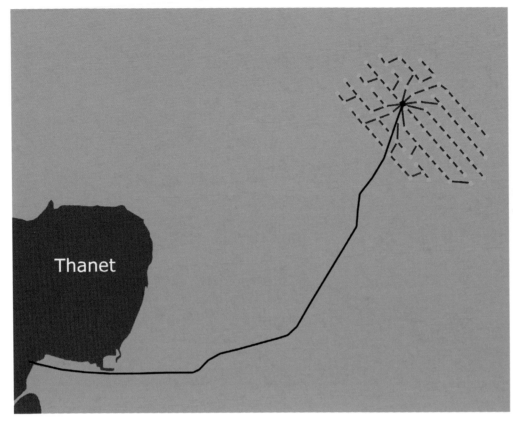

Figure 8-33, Thanet Offshore Wind Farm Map

Wind's Virtues and Vices

There is nothing quite like approaching a commercial wind turbine for the first time. From afar, one can see it is large, but this remote sense of scale transforms to a visceral awareness when standing beneath the massive blades that sweep through the sky. When hundreds of turbines are situated in a single wind farm, one truly appreciates the enormity of any single project. Called elegant, majestic, and even beautiful, the same turbines are often hated for transforming a previously unbroken vista with their incessant, spinning pinwheels. **An important challenge facing wind as an energy resource is the aesthetic transformation that wind turbines impart upon a landscape.**

There is simply no hiding a commercial wind turbine. In order to work properly it must be exposed to as much free airspace as possible, which necessarily makes it highly visible, sometimes for miles. Nearby property owners that advocate "green" energy often find themselves in the awkward position of fighting against one of the cleanest energy sources available. One potential solution is to require an extensive setback. **A setback is the distance from the base of a wind turbine to a property line, road, business, or residential structure.** 1,000 feet is *de facto* standard, but since no federal or state legislation exists regarding setback distance, local municipalities draft their

own standards, often with contentious community feedback. Recall that both Horse Hollow and Cape Wind projects have litigation histories, most of which stem from the projects' aesthetic impact.

Bird fatalities also plague wind as an energy source. Altamont Pass Wind Resource Area (APWRA), located in northern California near San Francisco, initially sparked this controversy. Bird fatalities in APWRA are a result of poor site planning and older-generation wind turbines that spin fast and are situated atop steel lattices. A four-year study commissioned by the California Energy Commission reports that the nearly 5,000 wind turbines in APWRA kill between 1,766 and 4,721 birds per year (California Energy Commission, 2004, p. 3). This number is high compared to other wind farms that are better-situated and use modern, slower-spinning turbines situated atop tubular monopoles.

As with most statistics, bird fatalities from wind turbines needs to be placed within context. **Of the approximately 500 million to 1 billion birds killed every year in the United States due to *all* anthropogenic sources (buildings, power lines, domestic cats, automobiles, pesticides, communication towers, wind turbines and airplanes), 28,500 are killed by wind turbines** (Erickson, Johnson, & Young, 2005, p. 1039). This works out to wind turbines being responsible for less than .01% of all bird fatalities in United States in any given year. Thus, while some birds do die as a result of wind turbines, the impact compared to other anthropogenic sources is quite minimal. The National Audubon Society weighs in on this issue and strongly supports wind power, provided that the turbines are properly "planned, sited and operated to minimize negative impacts on bird and wildlife populations" (National Audubon Society, Inc., 2011).

Bat fatalities from wind turbines have also recently gained attention. In fact, it is clear that bird fatalities are less a concern today than bat fatalities. Migratory bat species that roost in trees, in particular, account for over 75% of the bat fatalities from wind turbines. These deaths occur mostly during late summer and autumn, which is bat migration season, as well as during seasonal mating peaks (United States Geologic Survey, 2011). Concern gained national attention when a University of Maryland doctoral student found thousands of bat carcasses littering the ground around two wind farms in West Virginia (Blum, 2005). **Scientists are currently investigating possible causes of bat fatalities, with wind turbine height and cut-in speed being especially interesting variables.** For example, a 2007 study found that "bat fatalities increased exponentially as turbine height increased, with turbine towers 65 m or taller having the highest fatality rates" (Barclay, Baerwald, & Gruver, 2007, p. 384). This is especially important given that the wind industry is trending towards ever-higher turbines.

The bat-fatality problem inspired Bat Conservation International (BCI), the US Fish and Wildlife Service, the American Wind Energy Association (AWEA), and the United States National Renewable Energy Laboratory (NREL) to create Bats and Wind Energy Cooperative (BWEC). BWEC funds research and recommends solutions, often with wind industry help. A 2009 BWEC study, for example, used Iberdrola Renewables' Casselman Wind Power Project site in Pennsylvania as a research site. Scientists found bat fatalities dramatically fell when wind operators increased turbine cut-in speeds. Thus, in periods of low-wind, when bats are most likely to be flying about, there is less risk of them being struck by spinning blades. The energy loss from increasing the cut-in speed is small, between .3 and 1% of the annual energy production at the site (Bats and Wind Energy Cooperative, 2009).

Aerodynamic and mechanical noise caused by a wind turbine's blades and controller also poses a challenge for wind as an energy source. Turbine noise is often described as a mixture of whooshes, whines and thwumps, depending on turbine model, wind conditions and placement (Kintisch, 2010, p. 789). Volume is measured in units called decibels (dB), with 0 dB the

Figure 8-34, Scale Models of Wind Turbines, *Credit/ Courtesy of National Energy Education Development Program*

threshold of human hearing. Decibels are measured with an electronic device called a noise dosimeter (these are the same devices that were used during the Horse Hollow litigation and trial). Examples of dB ratings include: a humming refrigerator = 45 dB, normal conversation = 60 dB, heavy city traffic = 85 dB, (National Institutes of Health, 2011). **Modern, utility-scale turbines situated with a 350 meter (1,148 feet) setback produce a combined aerodynamic and mechanical noise profile between 35 and 45 dB** (American Wind Energy Association, 2010)**.**

While 35 to 45 dB is about the same level of background noise within a typical home, it is still perceived

Mortality Source	Annual mortality Estimate	Relative Share
Buildings	550 million	58.2 %
Power Lines	130 million	13.7 %
Cats	100 million	10.6 %
Automobiles	80 million	8.5 %
Pesticides	67 million	7.1 %
Communication Towers	4.5 million	0.5 %
Wind Turbines	28.5 thousand	<0.01 %
Airplanes	25 thousand	<0.01 %

Table 8-5,
Predicted Annual Avian Mortality from all Anthropogenic Sources

as an annoyance by many people living just outside a turbine's setback radius. Examples abound, with irritated and often angry people describing the noise as similar to a giant dishwasher or distant helicopter (Gunderson, 2009). As the demand for clean, renewable energy increases, more people are finding themselves living next to a proposed wind farm site. Inclusive community planning is essential to ensure the success of any wind farm project. Variables such as proposed setback, terrain, predicting noise levels, time of day or night that wind blows fastest, as well as the public's perception about wind turbines in a rural area, all need to be transparently considered with all stakeholders.

Figure 8-35, A Migratory Tree Bat, *Credit/Courtesy Paul Cryan, United States Geological Survey*

The wind power industry also struggles against public perception that it consumes vast amounts of government subsidies while other energy sectors are able to operate with little public financial assistance. This is simply wrong. **The wind power industry receives far fewer total dollars than most other electricity generating sectors.** The United States Government Accountability Office studied federal electricity subsidies between 2002 and 2007. It found that nuclear programs received the largest portion of public money ($6.2 billion), fossil fuel programs received the second largest portion ($3.1 billion), and all renewable programs received $1.4 billion (United States Government Accountability Office, 2007, pp. 2-3). More specifically, in 2007 wind research and development received $62.7 million, compared to solar ($203 million), hydrogen ($246 million) biomass ($254 million), coal ($573 million) and nuclear ($1.3 billion) (United States Government Accountability Office, 2007b, p. 21).

While fewer dollars flow to the wind industry, a balanced investigation must also consider the amount of public money spent *per unit of electricity production.* **The wind power industry receives some of the highest per MWh subsidies compared to other electricity generating sectors.** In 2007, for example, the wind industry received $23.37/MWh. This compares to $24.34/MWh for the solar industry, and $29.81/MWh for the refined coal industry. Compared to these three, the next most expensive electricity industry in terms of public subsidies was nuclear power, collecting $1.59/MWh. All the remaining electricity generating sectors receive less money/MWh (United States Energy Information Administration, 2008, p. 106).

Fuel Source	Type of Fuel	Total Subsidies ($ millions)
Non-Renewables	Coal	572.8
	Oil	3.6
	Natural Gas	16.1
	Nuclear	1,235.3
Renewables	Hydrogen	246.1
	Biomass	253.9
	Solar	202.6
	Wind	62.7
	Geothermal	6.4
	Hydropower	0

Table 8-6, Electricity-Related Subsidies in Total Dollars, FY 2007

One lesson here is to be careful when considering statistics, especially when they involve money. Both of the above methods of measuring public subsidies are valid. One measurement technique obviously works best for an anti-wind coalition, while the other works best for a pro-wind coalition. Our responsibility as informed energy consumers is to know what tools various associations might use to sway our opinion. Only then can we make the best, informed decisions about what energy resources we want to support.

Wind maps clearly illustrate that wind blows very strongly through a north/south belt of the United States, running from North Dakota through the Texas panhandle. This landscape consists of thousands of square miles of undulating plains buffeted by strong eastern winds flowing down the slopes of the Rocky Mountains. Loosely called America's Breadbasket, the High Plains, or the Midwest, it is also known as the Saudi Arabia of Wind. This wind-saturated corridor is in the heart of rural America, a region known for falling incomes, dwindling populations and failing towns (McGranahan, Cromartie, & Wojan, 2010). **Stopping and even reversing economic and population decline is happening in rural communities where farmers and ranchers shift their economic activities to include wind turbine leases on their land.**

Fuel	2007 Net Generation (billion kWh)	2007 Subsidies (millions of $)	Subsidy Support ($/Mwh)
Coal	1,946	854	0.44
Refined Coal	72	2,156	29.81
Natural Gas	919	227	0.25
Nuclear	794	1,267	1.59
Biomass	40	36	0.89
Geothermal	15	14	0.92
Hydroelectric	258	174	0.67
Solar	1	14	24.34
Wind	31	724	23.37
Landfill Gas	6	8	1.37
Municipal Solid Waste	9	1	0.13

Table 8-7, Electricity-Related Subsidies in Total Dollars/MWh, FY 2007

Sweetwater, Texas is well known in this respect, where wind farms are responsible for increasing property values in the district, from 35 million to 500 million, thereby providing the economic means for the construction of a new school (Gray, 2010). Curriculum expansion is also occurring at the Sweetwater branch of Texas State Technical College to satisfy demand for careers in wind energy (Texas State Technical College, 2010). These social and economic activities are what energy magnate T. Boone Pickens refers to when discussing how domestic energy sources can revitalize rural America. Part of his "Pickens Plan" calls for expanding America's installed wind capacity in order to satisfy 22% of the country's electricity demand. Not only will more wind capacity increase domestic energy security, it will also stimulate rural communities in rural America (PickensPlan, 2011).

Wind power has broad, popular support because it is inexhaustible, emission-free and diversifies a country's energy portfolio. Recall that renewable energy sources can be replenished in a short time span. With wind, this "short time span" is the next air pressure change, which is essentially immediately because air pressure is always changing. It is intermittent, however, sometimes wind simply does not blow enough for a turbine to cut-in. Thus, this clean and renewable energy must be balanced with other energy sources to provide reliable electricity to consumers.

Because it is intermittent, it is often wise to store wind-generated electricity rather than feeding it directly into a grid. The best known energy storage system is the battery, but the lead-acid type is generally not suited for utility-scale services because of its short life cycle and varying discharge over time. The sodium-sulfur battery (NAS, where the NA and S stand for sodium and sulfur) is a viable alternative. Japan's NGK Insulators, for example, is leading development in this area with its 34 MW NAS battery bank adjacent to a 51 MW wind farm in the Tohoku District of Japan (NGK Insulators Ltd., 2011). As the wind blows, turbines feed electricity into batteries, which in turn discharge into the grid during times of high demand. This electricity supply and delivery technique is called peak-shaving, and is an appropriate component of any expanding wind-energy sector.

Important Ideas:

- Wind is growing as an energy source partly because of the fact that it generates carbon-free electricity.

- Wind is growing as an energy source partly because of state tax incentives

- Wind is growing as an energy source partly because of renewable energy certificates (RECs).

- Renewable energy certificates (RECs) represent the environmental and social attributes of renewable energy sources, and are bought and sold apart from the physical commodity of electricity.

- A single REC is created for every MWh of electricity generated via renewable energy sources, including biomass, moving water, wind, geothermal and sunlight.

- RECs are retied—and therefore rendered no longer useable—when environmental or renewable energy claims are made specific to the tracking numbers associated with individual RECs.

- Buying and selling RECs takes place in REC exchanges, which are different entities from the regional tracking systems.

- Wind is growing as an energy source partly because of state legislatures adopting renewable energy portfolio standards (RPS).

- A renewable energy portfolio standard (RPS) is a law requiring utilities to supply a specific amount of electricity from renewable energy sources.

- Air moving down a pressure gradient is called wind.

- Wind-power density is how much energy is available in any square meter of wind and is calculated in Watts/m^2.

- Wind-power density in "real" wind is nearly double that of "perfect" wind.

- Wind Power Classes range from 1 to 7, with class 1 winds having poor resource potential, and class 7 winds having superb resource potential.

- Wind turbine towers are commonly made of tubular steel, with a base measuring several meters in diameter, and gently tapering at about 80 to 100 meters above the ground.

- A nacelle is the large, enclosed structure atop a wind turbine that houses the gears, generator, controller and brake.

- A rotor includes the turbine's hub and blades.

- Wind turbine blades are tapered similar to an airplane wing.

- Cut-in wind speed is the speed at which a turbine's blades begin to spin, and cut-out wind speed is the speed at which a turbine's blades are deliberately stopped.

- Wind turbines typically have capacity factors between 25-30%.

- No matter how well engineers craft the next generation of wind turbines, they will never exceed 59.3% efficiency—this is known as Betz's Law after the German physicist that first calculated the limit in the early twentieth century.

- Individual wind turbines have very small nameplate capacities compared to individual fossil fuel, nuclear, and conventional hydroelectric power plants.

- Texas has the most installed wind capacity (10,089 MW as of December 31, 2010) compared to any other state in America.

- The world's largest capacity wind farm is the Roscoe Wind Farm (RWF), located about 50 miles west of Abilene, Texas, and holds 627 turbines with a combined nameplate capacity of 781.5 MW.

- Roscoe Wind Farm, as well as many others, are dual-use entities, allowing farmers to continue cultivating their land or grazing their cattle while electricity is generated by wind turbines spinning overhead.

- Horse Hollow Wind Energy Center, located about 30 miles southwest of Abilene, Texas, is the second largest capacity wind farm in the world, with 421 turbines and a combined nameplate capacity of 735.5 MW.

- The Cape Wind Energy Project, located in Nantucket Sound on Horseshoe Shoal, is approximately 46 square miles and is expected to hold 130, 3.6 MW turbines, for a combined nameplate capacity of 468 MW.

- The largest offshore wind farm in the world is the Thanet Offshore Wind Farm, located in Kent, United Kingdom, and holding 100, 3 MW turbines for a combined nameplate capacity of 300 MW.

- An important challenge facing wind as an energy resource is the aesthetic transformation that wind turbines impart upon a landscape.

- A setback is the distance from the base of a wind turbine to a property line, road, or business or residential structure.

- Of the approximately 500 million to 1 billion birds killed every year in the United States due to *all* anthropogenic sources (buildings, power lines, domestic cats, automobiles, pesticides, communication towers, wind turbines and airplanes), 28,500 are killed by wind turbines.

- Scientists are currently investigating possible causes of bat fatalities, with wind turbine height and cut-in speed being especially interesting variables.

- Modern, utility-scale turbines situated with a 350 meter (1,148 feet) setback produce a combined aerodynamic and mechanical noise profile between 35 and 45 dB.

- The wind power industry receives far fewer total dollars than most other electricity generating sectors.

- The wind power industry receives some of the highest per MWh subsidies compared to other electricity generating sectors.

- Stopping and even reversing economic and population decline is happening in rural communities where farmers and ranchers shift their economic activities to include wind turbine leases on their land.

- Wind power has broad, popular support because it is inexhaustible, emission-free and diversifies a country's energy portfolio.

Works Cited

American Wind Energy Association. (2010, May). *Utility Scale Wind Energy and Sound*. Retrieved November 3, 2011, from The basics of sound: http://www.awea.org/_cs_upload/learnabout/publications/4138_3.pdf

Barclay, R. M., Baerwald, E., & Gruver, J. (2007). Variation in Bat and Bird Fatalities at Wind Energy Facilities: Assessing the Effects of Rotor Size and Tower Height. *Canadian Journal of Zoology, 85*, 381-387.

Bats and Wind Energy Cooperative. (2009, May 12). *Scientists Demonstrate Solution to Reduce Bat Deaths at Wind Turbines*. Retrieved November 3, 2011, from http://www.batsandwind.org/pdf/BWEC%20Curtailment%20Press%20Release%205-12-09.pdf

BBC News. (2010, September 23). *Largest offshore wind farm opens off Thanet in Kent*. Retrieved November 3, 2011, from The world's biggest offshore wind farm off Kent has been officially opened.: http://www.bbc.co.uk/news/uk-england-kent-11395964

Blum, J. (2005, January 1). Researchers Alarmed by Bat Deaths From Wind Turbines. *The Washington Post*, p. A01.

Bureau of Ocean Management, Regulation and Enforcement. (2011, April 20). *Cape Wind*. Retrieved November 3, 2011, from Project Overview: http://www.boemre.gov/offshore/renewableenergy/CapeWind.htm

Burnett, J. (2007, November 27). *National Public Radio*. Retrieved November 3, 2011, from Winds of Change Blow into Roscoe, Texas: http://www.npr.org/templates/story/story.php?storyId=16658695

California Energy Commission. (2004, August). *Developing Methods to Reduce Bird Mortality in the Altamont Pass Wind Resource Area*. Retrieved November 3, 2011, from Executive Summary: http://www.energy.ca.gov/reports/500-04-052/500-04-052_00_EXEC_SUM.PDF

Cape Wind. (2011). *Frequently Asked Questions*. Retrieved November 3, 2011, from How much electricity will Cape Wind provide?: http://www.capewind.org/FAQ-Category4-Cape+Wind+Basics-Parent0-myfaq-yes.htm#21

Danish Wind Industry Association. (2011). *Betz' Law*. Retrieved November 2, 2011, from The Ideal Braking of the Wind: http://wiki.windpower.org/index.php/Betz%27_law

Database of State Incentives for Renewables and Efficiency (DSIRE). (2011, June 3). *Nebraska, Incentives/Policies for Renewable Energy*. Retrieved November 2, 2011, from Sales and Use Tax Exemption for Community Wind Projects: http://www.dsireusa.org/incentives/incentive.cfm?Incentive_Code=NE10F&re=1&ee=0

Database of State Incentives for Renewables and Efficiency (DSIRE). (2011b, February 1). *Maine, Incentives/Policies for Renewable Energy*. Retrieved November 2, 2011, from Sales and Use Tax Refund for Qualified Community Wind Generators: http://www.dsireusa.org/incentives/incentive.cfm?Incentive_Code=ME12F&re=1&ee=0

Database of State Incentives for Renewables and Efficiency (DSIRE). (2011c, June 22). *Illinois, Incentives/Policies for Renewable Energy*. Retrieved November 2, 2011, from Commercial Wind Energy Property Valuation: http://www.dsireusa.org/incentives/incentive.cfm?Incentive_Code=IL25F&re=1&ee=0

Database of State Incentives for Renewables and Efficiency (DSIRE). (2011d, January 4). *Virginia, Incentives/Policies for Renewables & Efficiency*. Retrieved January 22, 2011, from Voluntary Renewable Energy Portfolio Goal: http://www.dsireusa.org/incentives/incentive.cfm?Incentive_Code=VA10R&state=VA&CurrentPageID=1&RE=1&EE=1

Database of State Incentives for Renewables and Efficiency (DSIRE). (2011e, August 4). *New Mexico, Incentives/Policies for Renewables & Efficiency*. Retrieved November 2, 2011, from Renewables Portfolio Standard: http://www.dsireusa.org/incentives/incentive.cfm?Incentive_Code=NM05R&state=NM&CurrentPageID=1

Database of State Incentives for Renewables and Efficiency (DSIRE). (2011f, September 1). *Illinois, Incentives/ Policies for Renewables & Efficiency*. Retrieved November 2, 2011, from Renewable Portfolio Standard: http://www.dsireusa.org/incentives/incentive.cfm?Incentive_Code=IL04R&state=IL&CurrentPageID=1&R E=1&EE=1

Enercon. (2011). *Products & Service*. Retrieved November 2, 2011, from Wind Turbines: http://www.enercon.de/ en-en/Windenergieanlagen.htm

Environmental Tracking Network of North America. (2011). *REC Questionsm and Answers*. Retrieved November 2, 2011, from How are RECs used?: http://www.etnna.org/images/PDFs/ETNNA-REC-QandA.pdf

Erickson, W. P., Johnson, G. D., & Young, D. P. (2005). *A Summary and Comparison of Bird Mortality from Anthropogenic Causes with an Emphasis on Collisions*. Retrieved November 3, 2011, from Table 2–Summary of predicted annual avian mortality, USDA Forest Service Gen. Tech. Rep. PSW-GTR-191: http://www.fs.fed. us/psw/publications/documents/psw_gtr191/Asilomar/pdfs/1029-1042.pdf

Gray, K. (2010, December 24). *Wind Industry Still Making Local Impact*. Retrieved November 4, 2011, from The Sweetwater Reporter: http://www.sweetwaterreporter.com/content/wind-industry-still-making-local-impact

Gunderson, D. (2009, August 4). *Minnesota Public Radio*. Retrieved November 3, 2011, from Wind Turbine Noise Concerns Prompt Investigation: http://minnesota.publicradio.org/display/web/2009/08/03/wind-turbine-noise/

Kintisch, E. (2010, August 13). Out of Site: Renewables like Wind Turbines are Spreading Fast, but can they Survive Complaints that they Mar Neighbohhoods and Threaten Wildlife? *Science, 329*, 788-789.

McGranahan, D., Cromartie, J., & Wojan, T. (2010, December). *The Two Faces of Rural Population Loss Through Outmigration*. Retrieved November 4, 2011, from Amber Waves: The Economics of Food, Farming, Natural Resources, and Rural America: http://www.ers.usda.gov/AmberWaves/december10/Features/ RuralPopulation.htm

National Audubon Society, Inc. (2011). *Policy Issues & Action / Energy*. Retrieved November 3, 2011, from Wind Power Overview: Audubon's Position on Wind Power: http://policy.audubon.org/wind-power-overview-0

National Institutes of Health. (2011, September 13). *National Institute on Deafness and Other Communication Disorders*. Retrieved November 3, 2011, from Noise-Induced Hearing Loss, What sounds cause NIHL?: http://www.nidcd.nih.gov/health/hearing/noise.html

National Weather Service. (2010, September 9). *JetStream - Online School for Weather*. Retrieved November 2, 2011, from The Atmosphere: http://www.srh.noaa.gov/jetstream/atmos/atmos_intro.htm

NextEra Energy Resources. (2011). *Horse Hollow I, II & III Wind Energy Center*. Retrieved November 3, 2011, from http://www.nexteraenergyresources.com/content/where/portfolio/pdf/horsehollow.pdf

NGK Insulators Ltd. (2011). *NAS Batteries*. Retrieved November 4, 2011, from http://www.ngk.co.jp/english/ products/power/nas/index.html

O'Grady, E. (2009, October 1). *Reuters, U.S. Edition*. Retrieved November 3, 2011, from E.ON completes world's largest wind farm in Texas: http://www.reuters.com/article/2009/10/01/wind-texas-idUSN3023624320091001

Phadke, R. (2011, September). *Horse Hollow*. Retrieved November 2, 2011, from Wind Energy - Visual Impacts and Public Perceptions: http://www.macalester.edu/windvisual/HorseHollow2p.pdf

PickensPlan. (2011). *The Plan*. Retrieved November 4, 2011, from The Pickens Plan: http://www.pickensplan. com/theplan/

Power-Technology.com. (2011). *The Website for the Power Industry*. Retrieved November 3, 2011, from Roscoe Wind Farm, Texas, USA: http://www.power-technology.com/projects/roscoe-wind-farm/

Save Our Sound: Alliance to Protect Nantucket Sound. (2011). *Cape Wind Threats*. Retrieved November 3, 2011, from The Environment: http://www.saveoursound.org/cape_wind_threats/environment/

Siemens Energy. (2010). *Wind Turbines*. Retrieved November 2, 2011, from Siemens Wind Turbine SWT-2.3-101: http://www.energy.siemens.com/fi/en/power-generation/renewables/wind-power/wind-turbines/swt-2-3-101.htm#content=Technical%20Specification

State of Illinois. (2007, August 28). *Illinois Power Agency Act, 20 ILCS 3855/175*. Retrieved November 2, 2011, from http://www.ilga.gov/legislation/ilcs/ilcs5.asp?ActID=2934&ChapAct=20%26nbsp%3BILCS%26nbsp%3B3855%2F&ChapterID=5&ChapterName=EXECUTIVE+BRANCH&ActName=Illinois+Power+Agency+Act%2E

State of New Mexico. (2007, August 30). *New Mexico Administrative Code, 17.9.572.7*. Retrieved November 2, 2011, from http://www.nmcpr.state.nm.us/NMAC/parts/title17/17.009.0572.htm

State of Virginia. (2007, April 11). *Code of Virginia, 56-585.2*. Retrieved November 2, 2011, from http://leg1.state.va.us/cgi-bin/legp504.exe?000+cod+56-585.2

Sway Turbine. (2011). *About Us*. Retrieved November 3, 2011, from http://www.swayturbine.no/?page=210

Texas State Technical College. (2010). *Science, Technology, Energineering & mathematics*. Retrieved November 4, 2011, from http://www.tstc.edu/careerprograms/

The Statue of Liberty-Ellis Island Foundation, Inc. (2011). *Fun Facts About The Statue of Liberty*. Retrieved February 4, 2011, from http://www.statueofliberty.org/Fun_Facts.html

United Kingdom, House of Commons. (2008, June 9). *Innovation, Universities,Science and Skills Committee*. Retrieved November 3, 2011, from Renewable Electricity-Generation Technologies: http://www.publications.parliament.uk/pa/cm200708/cmselect/cmdius/216/216.pdf

United States Department of Energy. (2010, August). *Energy Efficiency & Renewable Energy*. Retrieved November 2, 2011, from 2009 Wind Technologies Market Report, Figure 31. Average Cumulative Sample-Wide Capacity Factor by Calendar Year: http://eetd.lbl.gov/ea/emp/reports/lbnl-3716e.pdf

United States Department of Energy. (2011, October 19). *Energy Efficiency & Renewable Energy*. Retrieved October 31, 2011, from Wind Powering America: http://www.windpoweringamerica.gov/wind_installed_capacity.asp

United States Department of Energy. (2011b, July 22). *Energy Efficiency & Renewable Energy*. Retrieved November 2, 2011, from Green Power Markets, Renewable Energy Certificates (RECs), REC Marketers, Certificate Brokers/Exchanges: http://apps3.eere.energy.gov/greenpower/markets/certificates.shtml?page=2

United States Department of Energy. (2011c, November 2). *Energy 101: Wind Turbines*. Retrieved November 2, 2011, from Department of Energy video hosted on YouTube: http://youtu.be/tsZITSeQFR0

United States Department of Energy. (2011d, April 29). *Energy Efficiency & Renewable Energy*. Retrieved November 3, 2011, from Installed Wind Capacity by State: http://www.windpoweringamerica.gov/docs/installed_wind_capacity_by_state.xls

United States Energy Information Administration. (2008, April). *Independent Statistics and Analysis*. Retrieved November 3, 2011, from Federal Financial Interventions and Subsidies in Energy Markets 2007, Chapter 5. Subsidies per Unit of Production: http://www.eia.gov/oiaf/servicerpt/subsidy2/pdf/chap5.pdf

United States Energy Information Administration. (2011, October 28). *Independent Statistics and Analysis*. Retrieved October 31, 2011, from Renewable Energy Explained: http://www.eia.gov/energyexplained/index.cfm?page=renewable_home

United States Energy Information Administration. (2011b, October 19). *Independent Statistics and Analysis*. Retrieved October 31, 2011, from Electric Net Summer Capacity: Total (All Sectors), 1949-2010: http://www.eia.gov/totalenergy/data/annual/showtext.cfm?t=ptb0811a

United States Energy Information Administration. (2011c). *Independent Statistcs and Analysis*. Retrieved October 31, 2011, from International Energy Statistcs--Capacity, Wind: http://www.eia.gov/cfapps/ipdbproject/iedindex3.cfm?tid=2&pid=37&aid=7&cid=regions&syid=2008&eyid=2008&unit=MK

United States Environmental Protection Agency. (2008, July). *EPA's Green Power Partnership, Renewable Energy Certificates*. Retrieved November 2, 2011, from Who Owns a REC?: http://www.epa.gov/greenpower/documents/gpp_basics-recs.pdf

United States Environmental Protection Agency. (2010, October 7). *Combined Heat and Power Partnership*. Retrieved November 2, 2011, from Renewable Portfolio Standards Fact Sheet: http://www.epa.gov/chp/state-policy/renewable_fs.html

United States Environmental Protection Agency. (2011, June 2). *Green Power Partnership*. Retrieved October 31, 2011, from 2010 Award Winners, EPA Green Power Purchaser Awards, Green Power Partner of the Year: http://www.epa.gov/greenpower/awards/winners.htm

United States Environmental Protection Agency. (2011b, June 2). *Green Power Partnership*. Retrieved November 2, 2011, from Renewable Energy Certificates (RECs), What is a REC?: http://www.epa.gov/greenpower/gpmarket/rec.htm

United States Geologic Survey. (2011). *Fort Collins Science Center*. Retrieved November 3, 2011, from Bat Fatalities at Wind Turbines: Investigating the Causes and Consequences: http://www.fort.usgs.gov/batswindmills/

United States Government Accountability Office. (2007, October). *Federal Electricity Subsidies, Information of Research Funding, Tax Expenditures, And Other Activities That Support Electricity Production, GAO-08-102*. Retrieved November 3, 2011, from Results in Brief: http://www.gao.gov/new.items/d08102.pdf

United States Government Accountability Office. (2007b, October). *Federal Electricity Subsidies, Information of Research Funding, Tax Expenditures, And Other Activities That Support Electricity Production, GAO-08-102*. Retrieved November 3, 2011, from Research and Development, DOA Electricity-Related Funding Varied Widely Across Fules in FY 2007: http://www.gao.gov/new.items/d08102.pdf

United States National Renewable Energy Laboratory. (1986, October). *Wind Energy Resource Atlas of the United States*. Retrieved November 2, 2011, from http://rredc.nrel.gov/wind/pubs/atlas/

United States National Renewable Energy Laboratory. (2011, June 29). *Dynamic Maps, GIS Data, & Analysis Tools*. Retrieved November 2, 2011, from Wind Maps, United States Wind Resource Map: http://www.nrel.gov/gis/pdfs/windsmodel4pub1-1-9base200904enh.pdf

Vattenfall. (2011, October 28). *Thanet Offshore Wind Farm*. Retrieved November 3, 2011, from Facts: http://www.vattenfall.co.uk/en/thanet-offshore-wind-farm.htm

World Wind Energy Association. (2011, 31 October - 2 November). *World Wind Energy Report 2010*. Retrieved April 27, 2012, from 10th World Wind Energy Conference & Renewable Energy Exhibition: http://www.wwindea.org/home/images/stories/pdfs/worldwindenergyreport2010_s.pdf

Dry Steam Geothermal Power Plant at The Geysers in California, *Credit/Courtesy of Bureau of Land Management*

Chapter 9
Geothermal

Capturing Earth's Internal Heat

If the inside of the earth is hot enough to melt rock and keep it in a molten state, why are we not using more of this resource to quench humanity's energy demand? Surely the planet's internal temperature can boil water to drive steam turbines, which in turn can spin electrical generators. The lack of extensive geothermal installed capacity seems at first glance quite illogical, especially considering the increasing risks and costs associated with fossil fuels and our growing concern about climate change. The statistics are sobering. **In 2009, the United States led the world with only 3,421 MW of installed geothermal-electrical-generating capacity** (United States Energy Information Administration, 2011).

This figure constitutes the total capacity of all 222 geothermal generators in the country. Recall that a single coal-fired power plant can readily boast a nameplate capacity of 2,000 MW. International statistics of installed geothermal-electrical-generating capacity are equally diminutive: Philippines = 1,958 MW, Mexico = 965 kW, Indonesia = 933 kW, and Italy = 671 kW (United States Energy Information Administration, 2008). Consider these admittedly very small installed capacities to the fact the most recent assessment of geothermal-electrical-generating capacity in the United States (including identified, undiscovered and enhanced geothermal systems) is a whopping 556,890 MW (United States Geological Survey, 2008).

Geothermal's very small electrical-generating nameplate capacity predictably means that, when we turn on our lights, only a tiny fraction of the consumed kWh are generated from the earth's internal heat. In 2010 all renewable energy sources constituted about 8%, about 8 quadrillion Btu (8×10^{15}), of the total energy consumption in the United States, and of that total Btu, geothermal contributed about 3%, or 240 trillion Btu (2.4×10^{14}) (United States Energy Information Administration, 2011b). Examined internationally or nationally, the numbers beg explanation because, obviously, if one drills deeply enough, a vast quantity of energy is available anywhere within our planet. The United States Geological Survey makes a concise point in this regard:

> "Even if only 1 percent of the thermal energy contained within the uppermost 10 kilometers of our planet could be tapped, this amount would be 500 times that contained in all oil and gas resources of the world."

(Duffield & Sass, 2004, p. 1)

Country	Installed Geothermal Capacity (MW)
United States	3,421
Philippines	1,958
Mexico	965
Indonesia	933
Italy	671

Table 9-1, Geothermal-Electrical-Generating Capacity, World's Top Five Countries, 2008

As we'll find, however, simply drilling deeply enough to capture the earth's internal heat is fraught with imposing technological challenges. Before considering geothermal wells, however, we must arm ourselves with some knowledge of the interior structures and temperatures of our planet. Consider first that earth's diameter is 12,756 kilometers, or 7,926 miles (National Audubon Society, Inc., 2011). For simplicity's sake, let's just round up and say that the earth has an 8,000 mile diameter, and thus a trip to the very center is a 4,000 mile journey. Along the way, we first pass though the relatively rigid and very thin crust, which varies in depth but has an average thickness of about 19 miles (United States Geological Survey, 2011). The semi-molten mantle is next, with rock behaving like warm plastic instead of the familiar hard stones we

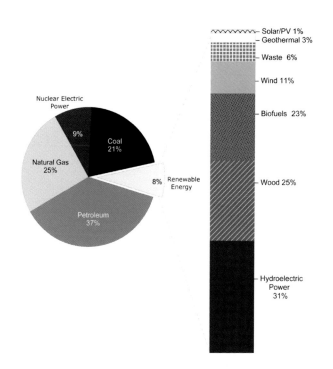

Figure 9-1, Geothermal's 3% Contribution to United States Renewable Energy Consumption, 2010, *Credit/Courtesy of United States Energy Information Administration*

experience on the earth's surface. The mantle is divided into upper and lower layers, and is in total about 1,800 miles thick (United States Department of Energy, 2007). Finally we arrive at the earth's core, which is also divided into two layers (a liquid nickel and iron outer core, and a solid iron inner core), and in total constitutes the remaining distance of just over 2,000 miles (National Aeronautics and Space Administration, 2009).

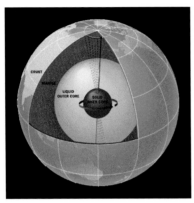

Figure 9-2, The Earth's Interior Structure, *Credit/ Courtesy of NASA*

Both pressure and temperature increase as one descends into the earth. Recall from Chapter 3 that the "oil window" lies between 1.24 and 3.73 miles beneath the surface, and that at these depths the temperature lies between 65 °C and 165 °C (149 °F and 329 °F). Descending deeper, say, to the 19 mile-deep mantle boundary, the temperature steadily rises to 400 °C (752 °F). Temperatures continue rising with depth until, at the center of the earth, estimated temperatures range between 5,000 °C and 7,000 °C (9,032 °F to 12,632 °F).

Geologists point to several reasons explaining the presence of this internal heat, including left-over thermal energy acquired as the planet coalesced from nebular material, as well as left-over thermal energy acquired when a large body impacted the earth and thus formed our moon. As the earth's solid iron core increases in size (albeit it very slowly), latent heat is also released into the mantle. Latent heat is the energy that is absorbed or released during a change of state. For example, when liquid water turns into solid water (ice), calories are released. Inside the earth—specifically at the inner and outer core boundary—liquid metal is slowly turning into solid metal, and this change of state releases heat (Buffett, 2003). By far the most important source of heat, however, comes from the decay of radioactive isotopes such as Potassium 40, Uranium 238 and 235, and Thorium 232. Recall from Chapter 5 that as radioactive isotopes spon-

taneously and naturally decay, they emit energy in the form of waves or particles. These waves or particles are energetic, and thus hot. **Radioactive isotopes, situated within the earth's mantle, account for about 90% of the earth's interior heat** (Anuta, 2006).

We are perhaps most familiar with the earth's internal heat when seeing a volcano erupt, visiting a hot spring, or watching Old Faithful geyser regularly spray hot water into the air. Volcanoes, hot springs and geysers exist in places where the earth's crust is compromised, being stretched too thinly, or subjected to melting due to subduction, diversion, or hot-spot plate tectonic forces. Wherever the earth's crust is stressed or cracked, mantle heat has a route to the surface. Any rocks, minerals and/or water in the way of that heat are subject to increasing temperatures. These clear displays of heat, however, often obscure the fact that earth's internal energy exists anywhere within our planet. The key for energy companies is to obtain access to this heat using today's technologies, while working within today's energy prices.

At a depth of 10 kilometers, nearly the entire continental United States has enough heat for geothermal-electrical-generation. This does not mean, however, that this energy is accessible. 10 kilometers is, after all, 10,000 meters or 6.21 miles deep. **Obtaining heat-energy from a depth of 10 kilometers is beyond present drilling technology and thus too expensive given today's energy-prices.** Fortunately, one need not drill so deeply to tap usable heat for geothermal-electrical-generation.

At about 4.5 kilometers (2.8 miles) deep, significant areas of the western United States become available for geothermal-electrical-generation. At this depth, portions of Oregon, Idaho, California, Nevada, Utah, Colorado, Arizona and New Mexico have temperatures in excess of 150 °C (302 °F). Locations with temperatures below 150 °C are usually classified as "low temperature" geothermal resources, and locations with temperatures above 150 °C are usually classified as "high temperature" geothermal resources. **High temperature (>150 °C) geothermal resources are the most important type for geothermal-electrical-generation** (Idaho National Laboratory, 2011).

In Chapter 7 we discussed how water circulates from the atmosphere to the surface and back in a movement called the hydrologic cycle. Some water in this cycle seeps into the ground and becomes groundwater, which is valued for drinking, agricultural, municipal and industrial use. Occasionally, and especially in areas where the earth's crust is compromised,

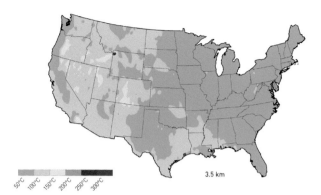

Figure 9-3, Average Temperature at 3.5 Kilometers, *Credit/Courtesy of United States Department of Energy, Geothermal Resource-Base Assessment*

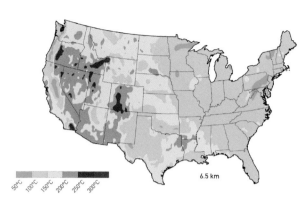

Figure 9-6, Average Temperature at 6.5 Kilometers, *Credit/Courtesy of United States Department of Energy, Geothermal Resource-Base Assessment*

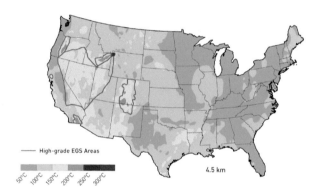

Figure 9-4, Average Temperature at 4.5 Kilometers, *Credit/Courtesy of United States Department of Energy, Geothermal Resource-Base Assessment*

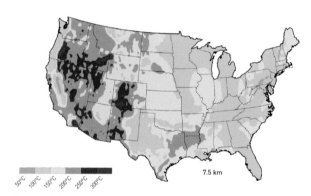

Figure 9-7, Average Temperature at 7.5 Kilometers, *Credit/Courtesy of United States Department of Energy, Geothermal Resource-Base Assessment*

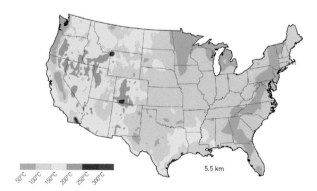

Figure 9-5, Average Temperature at 5.5 Kilometers, *Credit/Courtesy of United States Department of Energy, Geothermal Resource-Base Assessment*

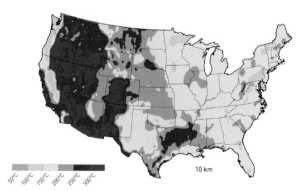

Figure 9-8, Average Temperature at 10 Kilometers, *Credit/Courtesy of United States Department of Energy, Geothermal Resource-Base Assessment*

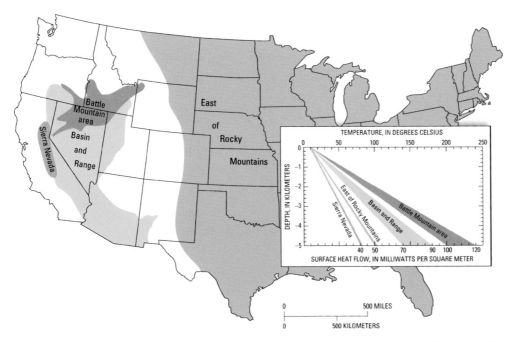

Figure 9-9, Geothermal Gradients Across the Continental United States, *Credit/Courtesy of United States Geological Survey*

groundwater penetrates several miles into the earth where temperatures exceed water's boiling point. **Hydrothermal energy resources are large reservoirs of hot groundwater water lying within a few miles of the earth's surface.** The geology of a naturally occurring hydrothermal resource is not unlike a petroleum resource in that permeable and porous rocks facilitate the movement and storage of a liquid—in this case water. Both hydrothermal and petroleum resources need impermeable or low-flow rock boundaries to trap the fluid (think of the caprock described in Chapter 3). Finally, both kinds of resources are tapped by drilling a production well. While petroleum wells yield liquid oil, geothermal wells yield hot water or, less-commonly, only steam. Each hydrothermal resource type requires a different kind of geothermal power plant.

Drilling Geothermal Wells for Electricity Production

Drilling into hydrothermal reservoirs for electricity production offers unique challenges to crews, engineers and geologists. While equipment and techniques are similar to those used in the oil and gas industry, geothermal wells: 1) are often more extensively fractured, 2) frequently contain especially abrasive rock, 3) usually have corrosive gasses, 4) exhibit high temperatures, and 5) require larger diameters. These conditions push the limits of today's drilling and casing technologies.

Recall from Chapters 3 and 4 that petroleum and methane resources are obtained when a crew strings together many dozens of 30 foot long drill pipe sections. Attached to the end of the string is a drill bit. Drilling mud is pushed down the string, through the bit, and up the annulus, carrying rock cuttings to the surface for analysis and dumping. This system is highly effective in clearing the hole of debris, thereby preventing the string from getting stuck.

Geothermal crews also (but not always) use mud to clear rock cuttings, and just as in petroleum and meth-

Figure 9-10, Geothermal Rotary Rig, *Credit/Courtesy of ThinkGeoEnergy.com*

Figure 9-11, Geothermal Drilling Crew at Work, *Credit/Courtesy of ThinkGeoEnergy.com*

ane operations, drilling proceeds as long as the mud flows properly. Problems arise when a well penetrates fractures, however, because at these junctures the higher-pressure drilling mud is literally pushed into these low-pressure underground cracks. Fractures are often several centimeters wide, and while petroleum and/or methane drilling crews would likely seal off a fractured zone with cement, geothermal wells actually need these fractures to remain open. Large fractures are desirable in geothermal wells because they make a well productive (Finger & Blankenship, 2010, p. 15). The challenge with geothermal drilling is to not lose too much mud into the fractures, while continuing to drill to the desired depth. Inevitably, some mud is pushed into fractures because geothermal drilling crews are reluctant to use "lost circulation materials." **Lost circulation materials are fluids, foams or solids used to plug fractures and thereby avoid losing drilling mud.** Lost circulation materials are typically ineffective in bridging wide gaps, or (in the case of cement) permanent and therefore unacceptable in a geothermal well.

Recall from Chapter 3 that sandstone is a sedimentary rock that, due to its porosity and permeability, is especially effective at holding petroleum deposits. Sandstone, as the name implies, is made up of small sand particles, as well as various cementing agents. Sand is primarily the mineral quartz, which has a Mohs Hardness of 7. **Mohs Hardness Scale ranks minerals from 1 to 10, with harder, higher-ranked minerals able to scratch softer, lower-ranked minerals.** Oil and gas drilling operations frequently use tricone drill bits (although these are often replaced with drag bits) to

penetrate sandstone. Tricone bits have three cones that rotate and thereby pulverize and chip away the rock.

Many geothermal drilling crews use the same tricone drill bits commonly seen in the oil and gas industry (Raymond & Grossman, 2006, p. 11), but they often penetrate into hard rock such as granite. Granite is an igneous rock that is composed of large, interlocking quartz and feldspar minerals (feldspar has a Mohs hardness of 6). The minerals' combined hardness, as well as the fact their crystals are large and interlocking (granite it is often described as having a "crystalline" structure), make granite a much harder rock than sandstone. Tricone drill bits quickly degrade when encountering granite because the quartz and feldspar crystals are highly abrasive. Geothermal drilling crews thus often struggle with a painfully slow rate of penetration (ROP)—as slow as 3 to 5 feet per hour (United States Department of Energy, 2011)—compared to oil and gas drilling crews because of slow penetration and frequent tripping (pulling the string up and out of the well) to replace worn bits (Augustine, Tester, Anderson, Petty, & Livesay, 2006, p. 13). For comparison, consider that a high average ROP, commonly seen when drilling with drag bits through rocks softer than granite, is around 80 feet per hour (Wise, Mansure, & Blankenship, 2005, p. 5). Tripping the string and replacing bits adds time and equipment costs to geothermal drilling operations.

One possible solution to the problem of excessive bit wear is to use fixed polycrystalline-diamond-compact cutters on non-rotating drag bits. **Polycrystalline-diamond-compact (PDC) cutters are thin layers of synthetic diamonds bonded to tungsten carbide disks.** These disks are then bonded to a bit which, unlike

Figure 9-12, Tricone Drill Bit at the Krafla Geothermal Power Plant, Iceland, *Credit/Courtesy of ThinkGeoEnergy.com*

Figure 9-13, Drag Bit with Polycrystalline-Diamond-Compact (PDC) Cutters, *Credit/Courtesy of Sandia National Laboratories*

a roller cone, shears rock instead of chipping it. The diamond edge of any cutter is exceptionally hard and when combined with the bit's shearing action performs well in some (but not all) rock formations. PDC technology is only gradually being adopted by the geothermal industry because of the challenges unique to geothermal wells. Geothermal engineers are seeking solutions to the following problems with PDC cutters: 1) they are known to fail at temperatures above 350 ^{0}C, and 2) they have a low impact strength that leads to blunting and shattering (United States Department of Energy, 2009, p. 1). While these are serious issues, the fact that PDC bits have no moving parts makes them appealing in some geothermal drilling operations. Too often, roller cone bearing-seals succumb to thermal decomposition in high-temperature environments, which ultimately leads to bearing failure (Shakhovsky, Dick, Carter, & Jacobs, 2011).

Figure 9-14, Drag Bit with PDC Cutters at a Geothermal Drilling Rig, *Credit/Courtesy of ThinkGeoEnergy. com*

The earth's crust contains solid rocks and minerals, liquid water, and gases. Two especially common gases are carbon dioxide (CO_2) and hydrogen sulfide (H_2S). Visitors to hot springs do not notice the CO_2 (it is colorless and odorless), but they most definitely perceive the sharp, rotten egg odor of H_2S, even in such minute quantities as 4.7 parts per billion (ppb). While the smell of H_2S is indeed foul, geothermal drilling operators worry about it for reasons other than olfactory unpleasantness. This gas—specifically the hydrogen from dissociated H_2S molecules—renders the strongest metals brittle and susceptible to cracking and failure. **Hydrogen embrittlement occurs when high-strength steel is subjected to mechanical loading in the presence of hydrogen gas, and thereby becomes fractured, brittle, and less resistant to breakage.** When H_2S encounters high-strength steel alloy in geothermal drilling environments, a hydrogen atom can detach from a hydrogen sulfide molecule and works its way into a tiny crack in the metal's surface. When a second hydrogen atom enters that same crack, a hydrogen molecule is created that then exerts mechanical pressure, widening the crack and thereby allowing more hydrogen atoms into the growing fissure.

Since geothermal drilling operations often penetrate very hard rock, it seems quite sensible to assume that engineers would choose the very hardest metals for bits, drill pipe and casing. However, the risk of hydrogen embrittlement renders the hardest steel alloys a poor choice for geothermal operations. This is a Catch-22 if ever one existed: harder rock requires harder metals to cut through, yet the hardest metals become brittle and likely to fail in geothermal environments (Finger & Blankenship, 2010, p. 14). Thus, lower-strength steels are selected for geothermal drilling and casing equipment because these are less susceptible to hydrogen embrittlement. This raises costs as this equipment suffers from mechanical wear in the especially abrasive underground environments of geothermal wells.

As if combating rock fractures, abrasive environments, and corrosive gases were not enough, geothermal drilling operators must also struggle with the very resource they hope to recover—heat. Recall that at 4.5 kilometers deep temperatures are in excess of 150 °C (302 °F). Down-hole electronic components, such as those allowing engineers to steer the drill string or log drilling information, reliably operate at temperatures up to 200 °C. Most geothermal wells, however, have temperatures in excess of 300 °C, which quickly degrades unprotected electronics. A bandage kind of solution is to encase down-hole electronics in a Dew-

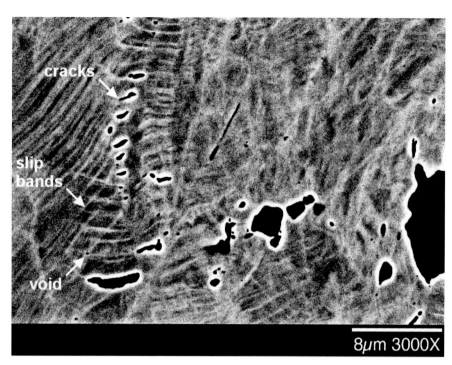

Figure 9-15, A Close-Up of Stainless Steel Exposed to Hydrogen Gas, *Credit/Courtesy of Sandia National Laboratories*

ar flask. **A Dewar flask is a double-walled, vacuum-sealed vessel that dramatically slows heat transfer** (you may actually use a version of a Dewar flask to hold your morning coffee). Dewar-encasing electronics (termed "Dewared") buys geothermal drilling operators about 10 extra hours of component life, at which point the component must be pulled back up to the surface and replaced, typically at a cost of $8,000 to $10,000 (Sandia National Laboratories, 1998).

Using Dewar flasks is well established in the geothermal industry, but a more practical and cost effective solution is to develop electronic components capable of withstanding extremely high temperatures. Just such work has been conducted and recently commercialized, with silicon-on-insulator (SOI) and silicon-carbide (SiC) technology leading the way (United States Department of Energy, 2010, pp. 43-45). **Today commercially-available SOI-based semiconductors operate indefinitely at 300 °C, and SiC-based semiconductors operate at temperatures up to 500 °C** (Normann, Henfling, & Chavira, 2005, p. 1). Today's geothermal operators realize significant savings from these electronics that are able to withstand very high-temperatures.

A final challenge for the geothermal industry is the fact that wells—and thus casing diameters—must be very wide. Geothermal wells require more steel and cement than oil and gas wells of comparable length. If casings are too narrow, fluid production is restricted, the well under-produces and thus fails to be profitable (King, Freeston, & Winmill, 1995, p. 241). **Large-diameter wells are essential in the geothermal industry because the production fluid—hot water or steam—has a low inherent value.** To get some idea of the width of these wells, imagine standing next to a hole in the ground *nearly 3 feet in diameter* (yes, you could literally fall into this hole!). 90 centimeters (over 35 inches) easily constitutes a typical "surface casing" width for geothermal wells. **Surface casing is casing pipe that reaches the earth's surface and is the widest-diameter of all casing pipe.** As drilling proceeds, successively smaller-diameter casing strings are cemented into place. At production zones casing diameters are still quite wide, typically measuring between 20 and 34 centimeters (7.87 inches to 13.39 inches) (Finger & Blankenship, 2010, p. 26). Casing pipe itself often measures about 2 centimeters thick.

Thus geothermal crews must string together and cement into place sections of very wide and very thick

Figure 9-16, A Dewared Geothermal Logging Tool, *Credit/Courtesy of United States Department of Energy*

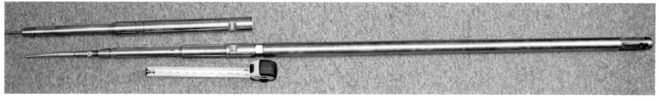

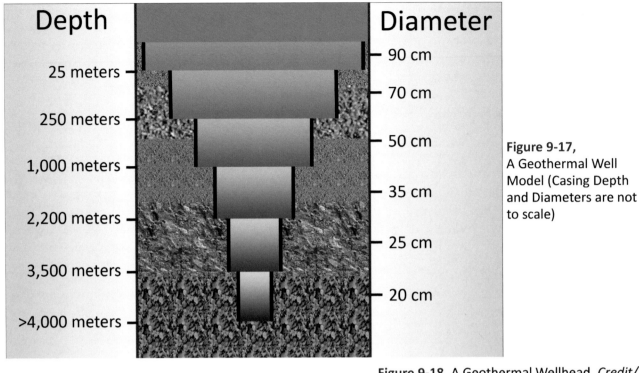

Depth	Diameter
	90 cm
25 meters	70 cm
250 meters	50 cm
1,000 meters	35 cm
2,200 meters	25 cm
3,500 meters	20 cm
>4,000 meters	

Figure 9-17, A Geothermal Well Model (Casing Depth and Diameters are not to scale)

Figure 9-18, A Geothermal Wellhead, *Credit/ Courtesy of Nicole Reed, National Renewable Energy Laboratory*

casing pipe, with a single well often having three to five different, steadily decreasing-diameter casing strings. Casing and cementing efforts thus account for about 30%-35% of the total price of any geothermal well, and thus reducing the number of casing strings can significantly reduce development costs (United States Department of Energy, 2007b, p. 5).

Generating Electricity

Not all geothermal power plants produce electricity in the same manner. Since underground environmental conditions vary from location to location, so too does the pressure, temperature, size and state of hydrothermal resources. Thus one geothermal field might produce only steam, another might produce only very hot water, while a third might produce only moderately hot water. In the geothermal industry these are called 1) vapor-dominated (or dry-steam) hydrothermal systems, 2) hot-water (or flash-steam) hydrothermal systems, and 3) binary-cycle hydrothermal systems (Duffield & Sass, 2004, p. 11). While all three constitute hydrothermal resources, each requires a site-specific electricity generation technique.

Vapor-dominated (or dry-steam) hydrothermal systems use steam directly from a geothermal well to drive a steam turbine generator. While these are the simplest, oldest and most-preferred systems used to capture the earth's internal heat to generate electricity, they are also the rarest. In vapor-dominated hydrothermal resources, a highly permeable reservoir rock is bordered and capped by highly impermeable non-reservoir rock. Due to this impermeable non-reservoir rock, such systems do not readily recharge with infiltrating rainwater or groundwater (Kaya & O'Sullivan, 2010, p. 1). Thus, using these systems without condensed-steam or surface-water injection, eventually lowers reservoir pressure and renders them tapped-out.

The largest and most successful vapor-dominated hydrothermal system in the world is The Geysers, located about 70 miles northeast of San Francisco, with a nameplate capacity of 725 MW. The Geysers first came online in 1960 with an 11 MW facility and was the first commercial geothermal power plant in the United States. The oil crisis and high energy prices of the late 1970s and early 1980s, coupled with legislation incentivizing electricity generation by independent power producers, led to rapid capacity expansion at the site. This meant that the same reservoir was repeatedly tapped by new wells, which eventu-

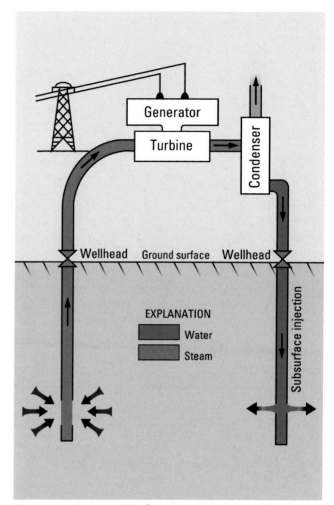

Figure 9-19, A Model of a Vapor-Dominated (or Dry-Steam) Hydrothermal System, *Credit/Courtesy of United States Geological Survey*

ally boosted the site's nameplate capacity to 2,043 MW in 1989 (Calpine, 2011). In 1990, however, it was clear that the site was overdeveloped because pressure and steam were decreasing throughout the field, thereby yielding steadily decreasing kWh of generated electricity. Since the reservoir had no natural way of recharging, continued electricity production depended upon injecting condensed steam to maintain reservoir volume and pressure. The practical limit of condensed-steam injection was reached, however, so over the course of the next thirteen years a network of underground pipelines was constructed to carry treated wastewater from nearby cities to the site in order to supplement existing injection efforts (Sanyal & Enedy, 2011, pp. 4-7). This novel effort proved so successful that The Geyser's output is now declining by only 1-2% a year, and Calpine (the site's operator) is now developing new capacity expansion.

Hot-water (or flash-steam) hydrothermal systems convert hot water from a geothermal well into steam

to drive a steam turbine generator. While not as desirable as vapor-dominated systems, hot-water systems are the most common electricity-producing geothermal systems on earth (United States Department of Energy, 2011b). Water temperature in these systems is usually greater than 182 °C (360 °F) (Idaho National Laboratory, 2011), which sounds improbably given our common knowledge that water boils into steam at a much lower temperature of 100 °C (212 °F). How can liquid water exist at such a high temperature?

The key to understanding this apparent contradiction is to consider two issues, pressure and boiling. Recall from chapters 3 and 8 that air pressure is the force of atmospheric gases pressing down, up and side-to-side on every square inch of everything. Also recall that the average amount of pressing (1 atm) is 14.7 pounds for every square inch of surface. Now consider the act of boiling water. If someone asks you "Why does water boil?" the most common answer is "Because it gets hot." This seemingly obvious response is only half-correct. Certainly, heating water is a common step to get it to change state from liquid to vapor, but heating is merely a *mechanism* to induce state change, not the *reason* for state change. Water boils not because we add energy, but because the energy we add changes the water's vapor pressure. Thus, if the "average amount of pressing" is less than 14.7 pounds (say, at the top of a mountain), comparatively less energy

is required to boil water. Conversely, if the "average amount of pressing" is greater than 14.7 pounds (say, deep underground), comparatively more energy is required to boil water. In hot-water hydrothermal reservoirs then, liquid water exists at temperatures greater than 182 ^{0}C *because it is not hot enough to boil.*

Hot-water (or flash-steam) hydrothermal production wells induce hot, high pressure liquid water to flow up to the earth's surface. Since a steam turbine awaits, this liquid water must change state before it proves useful for electricity production. Since the water's temperature is already very high, one need only reduce the ambient pressure to induce state change. This is accomplished in a flash chamber. **A flash chamber quickly lowers the ambient pressure surrounding hot geothermal water, thereby inducing state change from liquid to steam.** The steam is then routed to a steam turbine generator. The water that does not convert to steam inside the flash chamber is collected along with the condensed steam and injected back into the hydrothermal reservoir. As with vapor-dominated hydrothermal systems, fluid injection extends a reservoir's working life by maintaining productive volume and pressure levels.

Binary-cycle hydrothermal systems use moderately hot water from a geothermal well to vaporize a working fluid, which then drives a steam turbine generator. These systems operate with water temperature less that 182 ^{0}C (360 ^{0}F) (Idaho National Laboratory, 2011). While the water is not hot enough to flash into significant quantities of steam, its heat is captured in a heat exchanger to vaporize a secondary "working fluid" (sometimes called a "heat-transfer fluid"). The geothermal water and working fluids are circulated inside separate, closed loops of pipe and therefore never mix. A common working fluid is isopentane (C_5H_{12}), a volatile (highly flammable & easily-evaporated) hydrocarbon condensate found in oil and gas drilling operations. Isopentane has a very low normal (1 atm) boiling point, just 27.7 °C (81.86 °F). Thus, when liquid C_5H_{12} is piped into one side of a heat exchanger, it encounters a metal barrier that is over 100 °C hotter than its boiling point. The isopentane instantly vaporizes into high-pressure steam, which is then routed to drive a steam turbine generator. After passing through the turbine, the isopentane is condensed back into liquid, only to be immediately piped back into the heat exchanger for vaporization.

The obvious advantage of using C_5H_{12} as a working fluid is its low normal boiling point. It is flammable and toxic however, and safety concerns arise when it is used in systems that may be near population centers. Thus engineers and chemists continually search for other heat-transfer fluids in the hopes of finding one that has a similarly low boiling point, is environ-

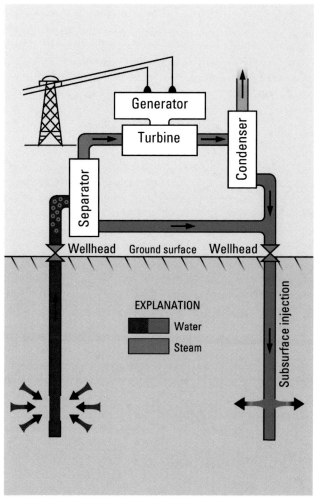

Figure 9-21, A Model of a Hot-Water (or Flash-Steam) Hydrothermal System, *Credit/Courtesy of United States Geological Survey*

Figure 9-22, A Hot-Water (Flash-Steam) Geothermal Power Plant in Southern California, *Credit/ Courtesy of Warren Gretz, National Renewable Energy Laboratory*

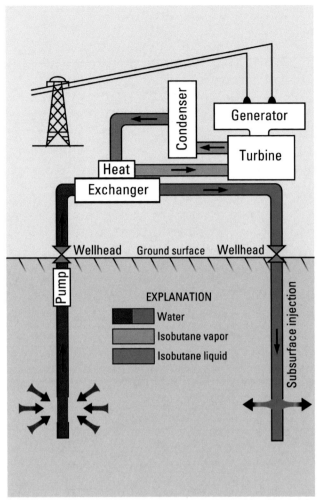

Figure 9-23, A Model of a Binary Cycle Hydrothermal System, *Credit/Courtesy of United States Geological Survey*

mentally and biologically benign, and is an efficient heat carrier (United States Department of Energy, 2011c). One such commercialized and trademarked product that has demonstrated experimental success is Solkatherm, made by Solvay Chemicals in the United Kingdom (Bombarda & Gaia, 2006). Solkatherm is non-flammable, non-toxic and boils at 36.7 °C (just 9 °C higher than isopentane) (Solvay Chemicals, 2011).

Enhanced Geothermal Systems (EGS)

Hydrothermal resources must contain commercially-viable quantities of heat, fluid, pressure and fractures. Unfortunately, naturally-occurring systems containing all these variables are relatively rare. Mining the earth's heat is still possible, however, using a technique once called "Hot-Dry-Rock" technology (HDR), and now referred to as Enhanced Geothermal Systems. **Enhanced Geothermal Systems (EGS) are engineered hydrothermal reservoirs created in otherwise dry, impermeable, crystalline rock.** EGS require production and injection wells situated within "stimulated" rock. Stimulated simply means fractured, which is normally accomplished by pumping high pressure water into deep crystalline rock until it literally breaks into an interconnected, networking series of cracks. An injection well then pumps water into the reservoir where it is naturally heated by the hot rocks. A production well draws the hot water through the reservoir and up to the surface where it is used to spin a

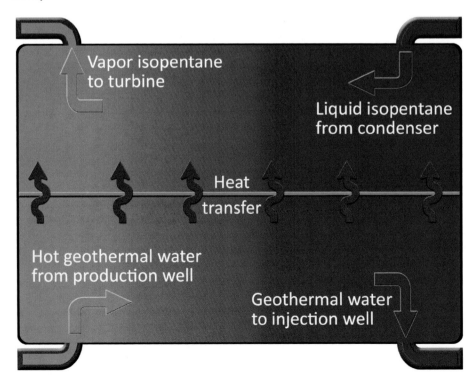

Figure 9-24, A Model of a Heat Exchanger in a Binary Cycle Hydrothermal System

Figure 9-25, Part of the Binary Cycle Geothermal System in the Rocky Mountain Testing Center, Wyoming, *Credit/Courtesy of Patrick Laney, National Renewable Energy Laboratory*

steam-turbine. After condensing, the same water is injected back into the reservoir in a continuous, circulating loop.

Theoretically just about anywhere on the earth can host an EGS. In practice however, specific sites are more feasible than others. Engineers in the United States have been studying EGS since 1974 when the first experimental HDR site was created at Fenton Hill, New Mexico. Fenton Hill proved that 1) engineered hydrothermal reservoirs can be created and 2) these systems can produce electricity (Massachusetts Institute of Technology, 2006, pp. 7-13). What Fenton Hill did not prove was that these accomplishments could be achieved on a large-enough scale, and within economically-reasonable boundaries, to provide significant quantities of electricity for the commercial marketplace.

As one might imagine, it is essential that EGS production and injection wells are connected underground via the series of manufactured fractures so that water can easily flow from one well to another. This is quite difficult to attain because wells can be over 4,000 meters deep, rock fractures in unpredictable patterns, and the only way to "see" underground fractures is with acoustic technology that has limited scope and accuracy. **The flow rate in naturally-occurring commercial hydrothermal reservoirs is about 80 kg/second at 200 °C, while the best-performing EGS reservoir has a flow rate of 25 kg/second** (United States Department of Energy, 2008, p. 6). Thus EGS flow rates must significantly improve before commercial viability is an option.

In addition to the challenge of improving flow rates, EGS projects must fracture and sufficiently interconnect many cubic kilometers of rock. **For every 100 MW of nameplate capacity, approximately 5 km^3 of subsurface rock must be properly stimulated in EGS projects** (Massachusetts Institute of Technology, 2006b, p. 17). Since a linear relationship between plant capacity and reservoir size is expected to exist beyond 100 MW, we need only scale-up to attain some idea of required reservoir size. Thus a 500 MW EGS power plant needs 25 km^3 of stimulated rock, and a 1,000 MW power plant about 50 km^3. To envision a 50 km^3 reservoir, consider a block of rock measuring 3.68

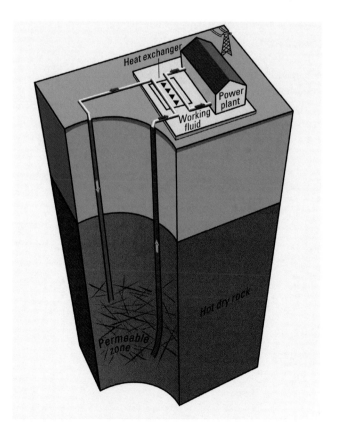

Figure 9-26, A Model of an Enhanced Geothermal System (EGS), *Credit/Courtesy of United States Geological Survey*

Figure 9-27, EGS Research at Fenton Hill, New Mexico, *Credit/Courtesy of Sandia National Laboratories, National Renewable Energy Laboratory*

km long, 3.68 km wide, and 3.68 km tall. Fracturing this block of rock at a depth of about 4,000 kilometers, in a predictable fashion with adequate permeability, is an enormous challenge.

Even with the challenges, however, the prospect of capturing essentially unlimited internal heat from inside the earth is attracting many government and commercial entities to EGS technology (The Economist, 2010, pp. 18-20). Australia is an especially active country for EGS start-up companies—dozens have appeared on various stock exchanges in the last decade. Australia's current EGS leader is Geodynamics Limited, a company currently drilling in South Australia, New South Wales and Queensland. A 25 MW EGS power plant is expected to come on-line soon, and be followed by over 500 MW of installed EGS technology throughout Australia by 2018 (Geodynamics Limited, 2011).

Direct-Use Geothermal

Capturing the earth's internal heat is not restricted to places having hydrothermal energy resources with temperatures in excess of 150 °C. While generally unsuitable for electrical-generation, geothermal sites with temperatures below 150 °C are still valuable.

Direct-use geothermal applications use low temperature hydrothermal water (between 20 °C and 150 °C) to heat buildings, warm greenhouses, farm fish, process food and warm swimming and spa facilities. Employing direct-use geothermal energy can save homeowners and commercial facilities an enormous amount of money, sometimes up to 80% of the energy costs compared to using traditional fossil fuels (United States Department of Energy, 2004, p. 3). Couple these savings with the fact that there are nearly zero emissions from geothermal resources, and a powerful incentive exists to exploit these energy sources whenever possible. The United States Department of Energy has identified over 10,000 existing and potential direct-use resources in the western United States alone (United States Department of Energy, 2009b).

When individual buildings employ direct-use geothermal, one of the more efficient and cost-savings techniques involves a district heating arrangement. **District geothermal heating involves pumping hot water or steam from a central location to recipient buildings through a network of underground, insulated flow and return pipes.** Usually most of the returned fluid is injected back into the ground in order to sustain the hydrothermal resource. The city of Boise, Idaho has the largest direct use geothermal heating network in the United States (City of Boise, 2011). Four independent district heating systems operate within the city, routing about 775 million gallons of hydrothermal water per year to 200 homes and 85 government buildings and businesses (Idaho Office of Energy Resources, 2010). Idaho's state capitol building is one of these government buildings, and is unique in the United States because it is the only capitol heated with geothermal power.

While notable, Boise's district geothermal heating pales in size compared to Reykjavik, Iceland. The homes of about 204,000 people—nearly the entire

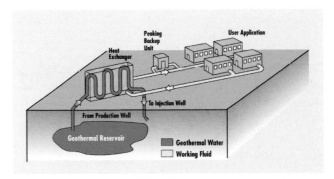

Figure 9-28, Model of District Geothermal Heating, *Credit/Courtesy of United States Department of Energy*

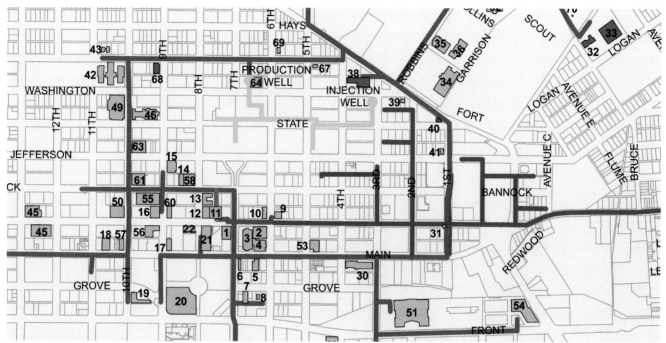

urban population of the country—are heated with district geothermal heating supplied by Reykjavik Energy, the largest district-heating company in Iceland (Orkustofnun, 2010, p. 21). The company draws hot water from low temperature hydrothermal reservoirs that exist near the city, as well as high temperature resources from farther afield, which are then routed to the city via insulated pipelines. District geothermal heating has eliminated coal and nearly eliminated oil as heating fuels, both of which cause air pollution and must be imported.

As well as tapping into its low temperature geothermal reservoirs for district heating, Iceland generates about 75% of its electricity from hydroelectric plants,

Figure 9-29, Map of Geothermal Systems in Boise City, *Credit/Courtesy of City of Boise*

and about 25% from geothermal-electrical-generation (Orkustofnun, 2011). Successfully tapping these cheap and clean indigenous energy resources make Iceland the cleanest country in the world, according to its 2010 Environmental Performance Index (EPI) score. EPI scoring is a collaborative effort by the Yale Center for Environmental Law and Policy and the Center for International Earth Science Information Network at Columbia University (Environmenal Performance Index, 2010).

Figure 9-30, Reykjavik, Iceland, Capital of the Cleanest Country in the World, *Credit/Courtesy of Andreas Tille, Wikimedia*

Figure 9-31, Geothermally Heated Greenhouse, *Credit/ Courtesy of National Renewable Energy Laboratory*

Figure 9-32, Geothermally Heated Aquaculture Facility, *Credit/Courtesy of National Renewable Energy Laboratory*

Figure 9-33, Geothermally Heated Cascading Fish Raceways, *Credit/Courtesy of National Renewable Energy Laboratory*

In addition to heating homes and buildings, direct-use geothermal is often observed in greenhouses, fish farms and food processing facilities. Several dozens of direct-use geothermal heated greenhouses and fish farms are found throughout the western United States. The greenhouses cultivate vegetables, houseplants, tree seedlings and flowers (Witcher, Lund, & Seawright, 2002), while the fish farms are known for growing bass, catfish, and especially tilapia (AmeriCulture Inc., 2011). In both greenhouses and aquaculture facilities, hot water is often piped into the buildings where it warms the air via radiant heating. At some aquaculture locations the hot geothermal water is mixed with cold surface water in order to make an ideal water temperature to support fish and other animals (United States Department of Energy, 2004, p. 14). Food processing facilities, by contrast, use geothermal heat primarily to dry and thus preserve vegetables, including onions, potatoes, mushroom, beans and tomatoes. These facilities exist throughout the world, including the United States, Mexico, Iceland, Tunisia, India, Philippines and Indonesia (Geo-Heat Center Bulletin Comprehensive Index, 2011).

Geothermal Heat Pumps

The earth's internal energy is exploitable even if direct-use geothermal is impossible and the location is unsuitable for generating electricity with geothermal energy sources. The earth's temperature is relatively constant within a meter or two beneath the surface, ranging from 45 °F to 75 °F (7 °C to 24 °C) depending upon a location's latitude and soil composition (United States Department of Energy, 2011d). You intuitively know this if you have a basement or if you have ever ventured into a cave. No matter the temperature above ground—blistering sun or deep-freeze—the basement or cave maintains a steady temperature. This thermal consistency is itself an energy resource, and exploited by means of a geothermal heat pump (GHP), also sometimes called ground source heat pump (GSHP). **A geothermal heat pump (GHP) runs on electricity; during the winter, it warms buildings by moving heat from the earth and depositing it into the building, and during the summer it cools buildings by moving heat from the building and depositing it into the earth.** Unlike typical heating, ventilating and air conditioning (HVAC) systems, GHPs create comfortable interior climates simply by transferring heat from one location to another (Geothermal Exchange Organization, 2011).

A typical GHP system consists of three parts: 1) the pump, 2) the heat exchanger and 3) a building's ductwork. The pump simply moves fluid, usually water and environmentally-benign antifreeze. **A GHP heat exchanger is a series of pipes, often called a loop, that is buried in the ground near a building and through which the GHP fluid circulates** (National Renewable Energy Laboratory, 2011). Most GHP heat exchangers are closed, meaning the same fluid indefinitely circulates through the piping and never leaves the system. Loops are situated either horizontally (where adjacent land is available for trenching) or vertically (where ad-

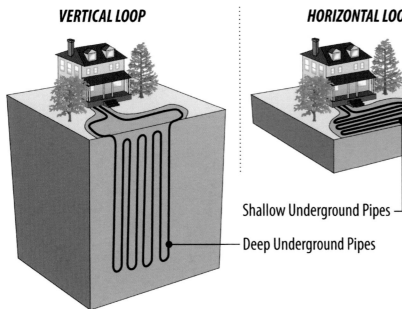

VERTICAL LOOP   **HORIZONTAL LOOP**

Shallow Underground Pipes ⌐

⌐ Deep Underground Pipes

Figure 9-34, Models of Vertical & Horizontal Closed-Loop GHP Heat Exchangers, *Credit/Courtesy of National Energy Education Development Program*

jacent land cannot be trenched, for a variety of reasons). Sometimes a site may be adjacent to a pond or lake, in which case the piping can be run underground, down into the water, and situated deeply enough to prevent freezing (United States Department of Energy, 2011e).

GHPs are most economically installed during construction when the ground's surface is already disturbed and all planning and permitting activities are complete. They are also, however, routinely retrofitted into existing structures after people seek out energy savings and more environmentally-friendly energy consumption patterns. Since a typical GHP system uses 25%–50% less electricity than conventional heating or cooling systems (United States Department of Energy, 2011f), as well as being more efficient at heating and cooling a building, the energy savings add up each month and eventually pay for the system, usually within two to ten years (United States Department of Energy, 2011g). In 2009 a total of 115,442 geothermal heat pumps were shipped throughout the United States and to international destinations (United States Energy Information Administration, 2010). Ohio, Pennsylvania and Illinois were the top three recipient states of GHPs in 2009 (United States Energy Information Administration, 2010b).

Figure 9-35, Installing a Horizontally-Situated Closed Loop Heat Exchanger, *Credit/Courtesy of Joseph Cinefro*

Figure 9-36, Supply and Return Pipes in a Closed Loop Heat Exchanger, *Credit/Courtesy of Joseph Cinefro*

Figure 9-37, A Geothermal Circulation Pump, *Credit/ Courtesy of Joseph Cinefro*

Geothermal's Virtues and Vices

Geothermal-electrical-generation produces clean energy. A geothermal power plant "emits about 1 percent of the sulfur dioxide (SO_2), less than 1 percent of the nitrous oxides (NO_x), and 5 percent of the carbon dioxide (CO_2)" compared to a coal-fired power plant of equivalent size (Duffield & Sass, 2004, p. 26). Couple these clean profiles with the fact that geothermal-electrical-generation produces virtually no particulate matter. Very little gas, as well as near-zero ash emissions, makes these plants highly desirable in a world with growing energy demands and environmental concerns. Citizens of Reykjavik know these lessons well, having learned them far back in the last century. Iceland is of course unique, situated atop the volcanically active mid-Atlantic ridge with easily-tapped hydrothermal reservoirs. Other countries are not nearly so geologically fortunate and thus must tap much more challenging terrain to access geothermal energy. This of course increases costs and sometimes makes existing energy sources that are much dirtier, such as coal, more attractive.

Geothermal-electrical-generation uses relatively little land. This is an important issue as federal, state and local governments balance our growing energy demands with our concurrent reluctance to develop

Figure 9-38, The Inside of a Closed Loop Geothermal Heat Pump, *Credit/Courtesy of Joseph Cinefro*

Power Generation System	Land Use M²/MW
110 MW geothermal flash plant (excluding wells)	1,260
47 MW solar thermal plant (Mojave Desert, CA)	28,000
10 MW solar PV plant (Southwestern US)	66,000
125 MW Dutch Hill/Cohocton Wind Farm (New York)	69,929

Table 9-2, Comparison of Power Generation Land Requirements

undisturbed terrain. Once piping is established, the land on top of a hydrothermal reservoir is still available for agriculture, livestock, wilderness recreation or other projects. The same cannot be said for large solar thermal or solar photovoltaic energy projects, and while wind farms are conducive to agriculture and livestock farming, each turbine has setback restrictions that can inhibit other kinds of land use.

Recall from Chapter Two that baseload power plants operate continuously and thus constantly produce electricity. **Geothermal power plants generate baseload electricity.** This virtue means that geothermal-generated electricity is reliable and thus a preferred renewable energy source for electricity dispatchers. By contrast, consider sunlight and wind. While also clean and renewable, these energy sources are intermittent and thus pose a much greater supply and distribution challenge for electricity companies.

In addition to generating baseload power, geothermal plants have a very impressive capacity factor. Recall from Chapter 1 that capacity factor is the percent of time a generating system operates at maximum capacity. **Geothermal power plants typically have a 90% capacity factor, which is the highest capacity factor of any renewable energy generating technology** (United States Energy Information Administration, 2010c, p. 161).

While environmentally benign and highly reliable, geothermal-electrical-generation does have some drawbacks. Recall that hydrogen sulfide (H_2S) is a common gas found within the earth's crust, and that it causes embrittlement challenges during geothermal drilling operations. This gas also plagues geothermal operators after drilling ceases. At some locations, considerable quantities of H_2S are carried up with the steam during normal operations. The hydrogen sulfide is re-

leased into the air after the steam condenses, much to the annoyance of anyone downwind from the power plant. The stinky, rotten egg odor can be overwhelmingly pungent, even at very low concentrations. A bad smell is only part of the problem, however. A high concentration (500-1,000 ppm) of H_2S is poisonous and can render victims unable to breathe—a conditioned called respiratory paralysis (Agency for Toxic Substances and Disease Registry, 2010). At lower concentrations (20-50 ppm) H_2S is an eye irritant that can lead to the whites of the eyes becoming inflamed, swollen and red—a condition called conjunctivitis and commonly known as "pink eye". **Hydrogen sulfide (H_2S) concentrations are sometimes high enough to make its abatement a priority for operators of geothermal-electrical-generation plants.**

Hydrogen sulfide mitigation can be accomplished chemically, for instance by wet scrubbing the H_2S out of the steam using a solution of sodium hydroxide (NaOH) (commonly called lye or caustic soda) and an oxidizing agent such as bleach (NaClO) (Arnórsson, 2004, p. 318). This method produces salts, water and sulfur, but is relatively expensive compared to the products it produces and is thus not a preferred mitigation method. A simpler and cheaper method is to inject the H_2S back into the hydrothermal reservoir. This practice is being tested by Reykjavík Energy at Hellisheiði, a power plant that annually releases 13,000 tons of hydrogen sulfide into the atmosphere, and is located only 20 kilometers southeast of the city of Reykjavik (Gunnarsson, Sigfússon, Stefánsson, Arnórsson, Scott, & Gunnlaugsson, 2011).

Another important issue confronting geothermal-electrical-generation projects is induced seismicity associated with the development of enhanced geothermal systems (EGS). Recall that EGS projects stimulate a hydrothermal reservoir by pumping high pressure water into deep crystalline rock until it literally breaks into an interconnected, networking series of cracks. When rock breaks, it creates earthquakes, in this case the induced earthquakes are called "microseismic events". They are acoustically tracked by geologists to determine the size and length of each fracture, and thus the size and three-dimensional area of the created hydrothermal reservoir (Massachusetts Institute of Technology, 2006c, pp. 9-10). As the name suggests, microseismic events are quite small, typically no greater than magnitude 2. The energy released from magnitude 3 or lower earthquakes is so small that such events typically go unnoticed except by a few people "under especially favorable conditions" (United States Geological Survey, 2010).

This fact, however, is often cold-comfort to people who have never before experienced an earthquake and then suddenly feel an induced one caused by a nearby energy company. People worry about these small events being a catalyst for much larger ones, and if public outcry is loud and sustained, it can easily derail a geothermal energy project. Just such worries were experienced by the crews at the EGS pilot site of Soultz-sous-Forêts, in France. Induced seismicity between 2000 and 2005 produced many thousands of microseismic events, the largest measuring magnitude 2.9. Due to local geologic conditions, many of the events were felt by the nearby population, sparking complaints, critical editorials, insurance claims and increasing general opposition to the project (Cuenot, 2010). Public worries were assuaged, however, and today, despite many more microseismic events occurring during circulation tests between 2005 and 2010, a completed 1.5 MW EGS power plant now stands at the site (Cuenot, Frogneux, Dorbath, & Calo', 2011).

Important Ideas:

- In 2009, the United States led the world with only 3,421 MW of installed geothermal-electrical-generating capacity.

- Radioactive isotopes, situated within the earth's mantle, account for about 90% of the earth's interior heat.

- At a depth of 10 kilometers, nearly the entire continental United States has enough heat for geothermal-electrical-generation.

- Obtaining heat-energy from a depth of 10 kilometers is beyond present drilling technology and thus too expensive given today's energy-prices.

- High temperature (>150 °C) geothermal resources are the most important type for geothermal-electrical-generation.

- Hydrothermal energy resources are large reservoirs of hot groundwater water lying within a few miles of the earth's surface.

- Lost circulation materials are fluids, foams or solids used to plug fractures and thereby avoid losing drilling mud.

- Mohs Hardness Scale ranks minerals from 1 to 10, with harder, higher-ranked minerals able to scratch softer, lower-ranked minerals.

- Polycrystalline-diamond-compact (PDC) cutters are thin layers of synthetic diamonds bonded to tungsten carbide disks.

- Hydrogen embrittlement occurs when high-strength steel is subjected to mechanical loading in the presence of hydrogen gas, and thereby becomes fractured, brittle, and less resistant to breakage.

- A Dewar flask is a double-walled, vacuum-sealed vessel that dramatically slows heat transfer.

- Today commercially-available SOI-based semiconductors operate indefinitely at 300 °C, and SiC-based semiconductors operate at temperatures up to 500 °C.

- Large-diameter wells are essential in the geothermal industry because the production fluid—hot water or steam—has a low inherent value.

- Surface casing is casing pipe that reaches the earth's surface and is the widest-diameter of all casing pipe.

- Vapor-dominated (or dry-steam) hydrothermal systems use steam directly from a geothermal well to drive a steam turbine generator.

- The largest and most successful vapor-dominated hydrothermal system in the world is The Geysers, located about 70 miles northeast of San Francisco, with a nameplate capacity of 725 MW.

- Hot-water (or flash-steam) hydrothermal systems convert hot water from a geothermal well into steam to drive a steam turbine generator.

- A flash chamber quickly lowers the ambient pressure surrounding hot geothermal water, thereby inducing state change from liquid to steam.

- Binary-cycle hydrothermal systems use moderately hot water from a geothermal well to vaporize a working fluid, which then drives a steam turbine generator.

- Enhanced Geothermal Systems (abbreviated "EGS") are engineered hydrothermal reservoirs created in otherwise dry, impermeable, crystalline rock.

- The flow rate in naturally-occurring commercial hydrothermal reservoirs is about 80 kg/second at 200 °C, while the best-performing EGS reservoir has a flow rate of 25 kg/second.

- For every 100 MW of nameplate capacity, approximately 5 km^3 of subsurface rock must be properly stimulated in EGS projects.

- Direct-use geothermal applications use low temperature hydrothermal water (between 20 °C and 150 °C) to heat buildings, warm greenhouses, farm fish, process food and warm swimming and spa facilities.

- District geothermal heating involves pumping hot water or steam from a central location to recipient buildings through a network of underground, insulated flow and return pipes.

- A geothermal heat pump (GHP) runs on electricity; during the winter, it warms buildings by moving heat from the earth and depositing it into the building, and during the summer it cools buildings by moving heat from the building and depositing it into the earth.

- A GHP heat exchanger is a series of pipes, often called a loop, that is buried in the ground near a building and through which the GHP fluid circulates.

- Geothermal-electrical-generation produces clean energy.

- Geothermal-electrical-generation uses relatively little land.

- Geothermal power plants generate baseload electricity.

- Geothermal power plants typically have a 90% capacity factor, which is the highest capacity factor of any renewable energy generating technology.

- Hydrogen sulfide (H_2S) concentrations are sometimes high enough to make its abatement a priority for operators of geothermal-electrical-generation plants.

- Another important issue confronting geothermal-electrical-generation projects is induced seismicity associated with the development of enhanced geothermal systems (EGS).

Works Cited

Agency for Toxic Substances and Disease Registry. (2010, march 25). *Public Health Assessments & Health Consultations*. Retrieved November 8, 2011, from Puna Geothermal Venture, Pahoa, Hawaii County, Hawaii: http://www.atsdr.cdc.gov/hac/pha/pha.asp?docid=1036&pg=0

AmeriCulture Inc. (2011). *About Us*. Retrieved November 7, 2011, from http://www.americulture.com/

Anuta, J. (2006, March 30). *Probing Question: What heats the earth's core?* Retrieved November 7, 2011, from PhysOrg.com: http://www.physorg.com/news62952904.html

Arnórsson, S. (2004). Environmental Impact of Geothermal Energy Utilization. In R. Giere, & P. Stille, *Energy, Waste and the Environment: A Geochemical Perspective* (pp. 297-336). Bath, UK: The Geological Society.

Augustine, C., Tester, J. W., Anderson, B., Petty, S., & Livesay, B. (2006). A Comparison of Geothermal with Oil and Gas Drilling Costs. *Thirty-First Workshop on Geothermal Reservoir Engineering* (p. 13). Stanford: Stanford University.

Bombarda, P., & Gaia, M. (2006). Geothermal Binary Plants Utilising an Innovative Non-Flammable Azeotropic Mixture as a Working Fluid. *Proceedings, 28th New Zealand Geothermal Workshop.* Aukland: New Zealand Geothermal Workshop.

Buffett, B. A. (2003). The Thermal State of Earth's Core. *Science, 229*(5613), 1675-1676.

Calpine. (2011). *History - The Geysers.* Retrieved November 7, 2011, from http://www.geysers.com/history.htm

City of Boise. (2011). *Geothermal Heating District.* Retrieved November 7, 2011, from Geothermal Heating in Boise: http://www.cityof-boise.org/Departments/Public_Works/Services/Geothermal/index.aspx

Cuenot. (2010, November 15-17). *Induced Microseismic Activity at the Soultz-sous-Forêts EGS site: Main scientific Results Obtained in Different Experimental Conditions.* Retrieved November 8, 2011, from European Center for Geodynamics and Seismology: ftp://ftp.ecgs.lu/public/publications/ecgs-induced/PRESENTATIONS/Cuenot_TALK.pdf

Cuenot, N., Frogneux, M., Dorbath, C., & Calo', M. (2011). Induced Microseismic Activity During Recent Circulation Tests at the EGS Site of Soultz-sous-Forêts (France). *Proceedings, Thirty-Sixth Workshop on Geothermal Reservoir Engineering.* Stanford, California: Thirty-Sixth Workshop on Geothermal Reservoir Engineering.

Duffield, W. A., & Sass, J. H. (2004). *Geothermal Energy—Clean Power From the Earth's Heat, Circular 1249.* Retrieved November 7, 2011, from United States Geological Survey: http://pubs.usgs.gov/circ/2004/c1249/c1249.pdf

Environmenal Performance Index. (2010). *Country Scores.* Retrieved November 7, 2011, from EPI Scores 100-85: http://www.epi.yale.edu/Countries

Finger, J., & Blankenship, D. (2010, December). *Handbook of Best Practices for Geothermal Drilling.* Retrieved November 7, 2011, from Sandia National Laboratories: http://www1.eere.energy.gov/geothermal/pdfs/drillinghandbook.pdf

Geodynamics Limited. (2011). *Power From the Earth.* Retrieved November 7, 2011, from About Geodynamics: http://www.geodynamics.com.au/IRM/content/starthere.html

Geo-Heat Center Bulletin Comprehensive Index. (2011, June 20). *Agribusiness.* Retrieved November 7, 2011, from http://geoheat.oit.edu/ghcindex.htm

Geothermal Exchange Organization. (2011). *Comparing HVAC Systems.* Retrieved November 8, 2011, from Geothermal Heat Pump Systems: http://www.geoexchange.org/index.php?option=com_content&view=article&id=52:comparing-hvac-systems&catid=375:geothermal-hvac&Itemid=32

Gunnarsson, I., Sigfússon, B., Stefánsson, A., Arnórsson, S., Scott, S. W., & Gunnlaugsson, E. (2011). Injection of H2S from Hellisheiði Power Plant, Iceland. *Proceedings, Thirty-Sixth Workshop on Geothermal Reservoir Engineering.* Stanford, California: Stanford Geothermal Workshop.

Idaho National Laboratory. (2011). *Production of Geothermal Energy.* Retrieved November 7, 2011, from What is Geothermal Energy?: https://inlportal.inl.gov/portal/server.pt/community/geothermal/422/what_is_geothermal_energy_

Idaho Office of Energy Resources. (2010, February 9). *District Heating.* Retrieved November 7, 2011, from http://www.energy.idaho.gov/renewableenergy/district_heating.htm

Kaya, E., & O'Sullivan, M. (2010). Modelling of Injection into Vapour-Dominated Geothermal Systems. *Proceedings World Geothermal Congress.* Bali, Indonesia: World Geothermal Congress.

King, T. R., Freeston, D. H., & Winmill, R. L. (1995). A Case Study of Wide Diameter Casing for Geothermal Systems. *17th New Zealand Geothermal Workshop* (pp. 241-244). Auckland: New Zealand Geothermal Workshop.

Massachusetts Institute of Technology. (2006). *The Future of Geothermal Energy: Impact of Enhanced Geothermal Systems (EGS) on the United States in the 21st Century: Chapter 4: Review of EGS and Related Technology – Status and Achievements.* Retrieved November 7, 2011, from United States Department of Energy, Energy Efficiency & Renewable Energy, Geothermal Technologies Program: http://www1.eere.energy.gov/geothermal/pdfs/egs_chapter_4.pdf

Massachusetts Institute of Technology. (2006b). *The Future of Geothermal Energy: Impact of Enhanced Geothermal Systems (EGS) on the United States in the 21st Century: Chapter 1: Synopsis and Executive Summary, Table 1.2 Estimated land area and subsurface reservoir volumes needed for EGS development.* Retrieved November 7, 2011, from United States Department of Energy, Energy Efficiency & Renewable Energy, Geothermal Technologies Program: http://geothermal.inel.gov/publications/future_of_geothermal_energy.pdf

Massachusetts Institute of Technology. (2006c). *The Future of Geothermal Energy: Impact of Enhanced Geothermal Systems (EGS) on the United States in the 21st Century: Chapter 8, Environmental Impacts,Attributes, and Feasibility Criteria.* Retrieved November 8, 2011, from United States Department of Energy, Energy Efficiency & Renewable Energy, Geothermal Technologies Program: http://www1.eere.energy.gov/geothermal/pdfs/egs_chapter_8.pdf

National Aeronautics and Space Administration. (2009, December 22). *Jet Propulsion Laboratory, California Institute of Technology.* Retrieved November 7, 2011, from As the World Churns, Peeling Back the Onion: http://www.jpl.nasa.gov/news/features.cfm?feature=2420

National Aeronautics and Space Administration. (2011). *Goddard Space Flight Center.* Retrieved November 7, 2011, from High Energy Astrophysics Science Archive Research Center: http://heasarc.nasa.gov/docs/cosmic/earth_info.html

National Renewable Energy Laboratory. (2011, March 8). *Learning About Renewable Energy.* Retrieved November 8, 2011, from Geothermal Heat Pumps: http://www.nrel.gov/learning/re_geo_heat_pumps.html

Normann, R. A., Henfling, J. A., & Chavira, D. J. (2005). Recent Advancements in High-Temperature, High-Reliability Electronics Will Alter Geothermal Exploration. *Proceedings World Geothermal Congress.* Antalya: World Geothermal Congress.

Orkustofnun. (2010, February). *National Energy Authority.* Retrieved November 7, 2011, from Geothermal Development and Research in Iceland: http://www.nea.is/media/utgafa/GD_loka.pdf

Orkustofnun. (2011). *National Energy Authority.* Retrieved November 7, 2011, from Hydro Power and Generation of Electricity Using Geothermal Energy: http://www.nea.is/hydro and http://www.nea.is/geothermal/electricity-generation/nr/76

Raymond, D. W., & Grossman, J. W. (2006). *Development and Testing of a Mudjet-Augmented PDC Bit.* Albuquerque: Sandia National Laboratories.

Sandia National Laboratories. (1998, April 29). *Hot research at Sandia may make producing electricity from geothermal energy more cost competitive.* Retrieved November 7, 2011, from http://www.sandia.gov/media/geother.htm

Sanyal, S. K., & Enedy, S. L. (2011). Fifty Years of Power Generation at The Geysers Geothermal Field, California--Lessons Learned. *Proceedings, Thirty-Sixth Workshop on Geothermal Reservoir Engineering.* Stanford, California: Thirty-Sixth Workshop on Geothermal Reservoir Engineering.

Shakhovsky, D., Dick, A., Carter, G., & Jacobs, M. (2011, January). *World Oil Online.* Retrieved April 13, 2011, from Drilling Technology, New Rubber Parts Improve Roller Cone Performance at High Temperatures: http://www.worldoil.com/New-rubber-parts-improve-roller-cone-bit-performance-at-high-temperatures-January-2011.html

Solvay Chemicals. (2011). *Hydrofluorocarbons HFC.* Retrieved November 7, 2011, from Solkatherm® SES36: http://www.solvaychemicals.com/EN/products/Fluor/Hydrofluorocarbons_HFC/SolkathermSES36.aspx

The Economist. (2010, September 4-10). Hot Rocks and High Hopes. *The Economist*, 18-20.

United States Department of Energy. (2004, August). *Energy Efficiency & Renewable Energy.* Retrieved November 7, 2011, from Geothermal Technologies Program, Direct Use: http://www.nrel.gov/docs/fy04osti/36316.pdf

United States Department of Energy. (2007, June 25). *Argonne National Laboratory.* Retrieved November 7, 2011, from A New Picture of Earth's Lower Mantle?: http://www.aps.anl.gov/Science/Highlights/2007/20070625.htm

United States Department of Energy. (2007b, October 16). *Enhanced Geothermal Systems Wellfield Construction Workshop.* Retrieved November 7, 2011, from Summary Report Draft: http://www1.eere.energy.gov/geothermal/pdfs/well_construction.pdf

United States Department of Energy. (2008). *Energy Efficiency & Renewable Energy.* Retrieved November 7, 2011, from An Evaluation of Enhanced Geothermal Systems Technology, Geothermal Technologies Program, Chapter 3: Assessment of Assumptions in the MIT Study: http://www1.eere.energy.gov/geothermal/pdfs/evaluation_egs_tech_2008.pdf

United States Department of Energy. (2009, March). *Energy Efficiency & Renewable Energy, Industrial Technologies Program.* Retrieved November 7, 2011, from Ultratough, Thermally Stable Polycrystalline Diamond/Silicon Carbide Nanocomposites for Drill Bits: http://www1.eere.energy.gov/industry/nanomanufacturing/pdfs/nanocomposites_drill_bits.pdf

United States Department of Energy. (2009b, March 20). *Energy Efficiency & Renewable Energy, Geothermal Technologies Program.* Retrieved November 7, 2011, from Direct Use of Geothermal Energy, The Direct-Use Resource: http://www1.eere.energy.gov/geothermal/directuse.html

United States Department of Energy. (2010). *Energy Efficiency & Renewable Energy, Geothermal Technologies Program.* Retrieved November 7, 2011, from A History of Geothermal Energy Research and Development in the United States, Drilling 1976-2006: http://www1.eere.energy.gov/geothermal/pdfs/geothermal_history_2_drilling.pdf

United States Department of Energy. (2011). *National Energy Technology Laboratory (NETL)*. Retrieved November 7, 2011, from Exploration & Production Technologies, Advanced Drilling - Advanced Tools & Methods - Advanced Drill Bit Technologies: http://www.netl.doe.gov/technologies/oil-gas/EP_Technologies/AdvancedDrilling/AdvancedToolsMethods/AdvancedDrillBits/AdvDrillBits.html

United States Department of Energy. (2011b, March 8). *National Renewable Energy Laboratory*. Retrieved November 7, 2011, from Geothermal Electricity Production: http://www.nrel.gov/learning/re_geo_elec_production.html

United States Department of Energy. (2011c, September 27). *Energy Efficiency & Renewable Energy*. Retrieved November 7, 2011, from Geothermal Technologies Program; High-Potential Working Fluids for Next Generation Binary Cycle Geothermal Power Plants: http://www4.eere.energy.gov/geothermal/projects/177

United States Department of Energy. (2011d, February 9). *Energy Efficiency & Renewable Energy*. Retrieved November 8, 2011, from Geothermal Heat Pumps: http://www.energysavers.gov/your_home/space_heating_cooling/index.cfm/mytopic=12640

United States Department of Energy. (2011e, February 9). *Energy Efficiency & Renewable Energy*. Retrieved November 8, 2011, from Types of Geothermal Heat Pump Systems: http://www.energysavers.gov/your_home/space_heating_cooling/index.cfm/mytopic=12650

United States Department of Energy. (2011f, February 9). *Energy Efficiency & Renewable Energy*. Retrieved November 8, 2011, from Benefits of Geothermal Heat Pump Systems: http://www.energysavers.gov/your_home/space_heating_cooling/index.cfm/mytopic=12660

United States Department of Energy. (2011g, February 9). *Energy Efficiency & Renewable Energy*. Retrieved November 8, 2011, from Selecting and Installing a Geothermal Heat Pump System, Economics of Geothermal Heat Pumps: http://www.energysavers.gov/your_home/space_heating_cooling/index.cfm/mytopic=12670

United States Energy Information Administration. (2008). *Independent Statistics and Analysis*. Retrieved November 7, 2011, from International Energy Statistics: http://tonto.eia.gov/cfapps/ipdbproject/iedindex3.cfm?tid=2&pid=35&aid=7&cid=&syid=2008&eyid=2008&unit=MK

United States Energy Information Administration. (2010, December). *Independent Statistics and Analysis*. Retrieved November 8, 2011, from Geothermal Heat Pump Manufacturing Activities, 2009: http://www.eia.doe.gov/cneaf/solar.renewables/page/ghpsurvey/ghpssurvey.html#_ftn1

United States Energy Information Administration. (2010b, December). *Independent Statistics and Analysis*. Retrieved November 8, 2011, from Geothermal Heat Pump Manufacturing Activities, 2009, Table 4.6 Geothermal Heat Pump Shipments by Destination, 2008 and 2009: http://www.eia.doe.gov/cneaf/solar.renewables/page/ghpsurvey/table4_6.pdf

United States Energy Information Administration. (2010c, April 9). *Independent Statistics and Analysis*. Retrieved November 8, 2011, from Assumptions to the Annual Energy Outlook 2010, Renewable Fuels Module, Table 13.2. Capacity Factors1 for Renewable Energy Generating Technologies in Three Cases: http://www.eia.gov/oiaf/aeo/assumption/pdf/0554(2010).pdf

United States Energy Information Administration. (2011, January 4). *Independent Statistics and Analysis*. Retrieved November 7, 2011, from Existing Capacity by Energy Source, 2009: http://www.eia.doe.gov/cneaf/electricity/epa/epat1p2.html

United States Energy Information Administration. (2011b, June). *Independent Statistics and Analysis*. Retrieved November 7, 2011, from U.S. Energy Consumption by Energy Source, 2010: http://www.eia.gov/energyexplained/index.cfm?page=renewable_home

United States Geological Survey. (2008). *Assessment of Moderate- and High-Temperature Geothermal Resources of the United States*. Retrieved November 7, 2011, from Fact Sheet 2008–3082: http://pubs.usgs.gov/fs/2008/3082/pdf/fs2008-3082.pdf

United States Geological Survey. (2010, February 22). *Earthquake Hazards Program*. Retrieved November 8, 2011, from Magnitude / Intensity Comparison: http://earthquake.usgs.gov/learn/topics/mag_vs_int.php

United States Geological Survey. (2011, January 14). *The Interior of the Earth, by Eugene C. Robertson*. Retrieved November 7, 2011, from Data on the Earth's Interior: http://pubs.usgs.gov/gip/interior/

Wise, J. L., Mansure, A. J., & Blankenship, D. A. (2005). Hard-Rock Field Performance of Drag Bits and a Downhole Diagnostics-While-Drilling (DWD) Tool. *Proceedings World Geothermal Congress*. Antalya: World Geothermal Congress.

Witcher, J. C., Lund, J. W., & Seawright, D. E. (2002, December). *Lightning Dock KGRA, New Mexico's Largest Geothermal Greenhouse, Largest Aquaculture Facility, and First Binary Electrical Power Plant*. Retrieved November 7, 2011, from Geo-Heat Center, Oregon Institute of Technology: http://geoheat.oit.edu/bulletin/bull23-4/art8.pdf

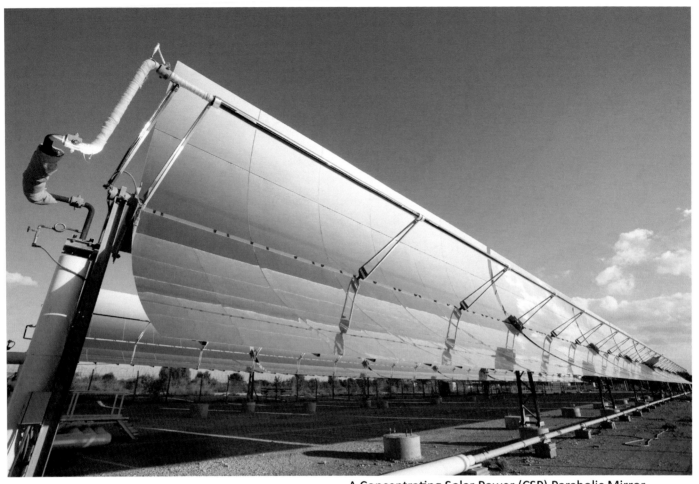

A Concentrating Solar Power (CSP) Parabolic Mirror,
Credit/Courtesy of Siemens

Chapter 10
Sunlight

Sunlight's Energy

"More energy from the sun falls on the earth in one hour than is used by everyone in the world in one year" (National Renewable Energy Laboratory, 2009). Latitude, seasons, weather and elevation all play a significant role in the amount of sunlight the earth's surface receives at any given moment. Even so, the numbers are impressive. For instance, averaged over one year, over the entire planet, every square meter of the earth absorbs about 4.2 kilowatt hours of energy every day ($4.2 kWh/m^2/day$) (Union of Concerned Scientists, 2010). Since this is an average, we can expect middle latitude locations to receive less, say 3.4 $kWh/m^2/day$, and lower latitude locations to receive more, say, 5.1 $kWh/m^2/day$. Places receiving the most sunlight energy (>6.5 $kWh/m^2/day$) are desert locations with little or no cloud cover.

The above facts are made all the more impressive when considering that sunlight is limitless and pollution free. Sunlight is the ultimate renewable energy resource. Solar energy will be available to humanity as long as the sun keeps shining, about 5 billion more years. With such accolades it is thus sobering to see the paltry installed solar electrical generating capacity of the United States (and most every other countries). **In 2010, the United States had only 888 MW of installed solar electrical generating capacity, the least installed capacity of any energy resource** (United States Energy Information Administration, 2011)**.** This means that for yet another year, sunlight constitutes only about 1% of all the renewable energy consumed in the United States, and far less than 1% of all the energy consumed in the United States. This solar deficiency is changing however, as 26,771 MW of utility-scale solar projects are currently under construction or under development in the United States as of late June, 2011 (Solar Energy Industries Association, 2011).

Sunlight is energy—specifically electromagnetic radiation—that can travel through a vacuum. This means

Figure 10-1, National Solar Photovoltaics (PV) Resource Potential, KWh/m²/Day, *Credit/Courtesy of National Renewable Energy Laboratory*

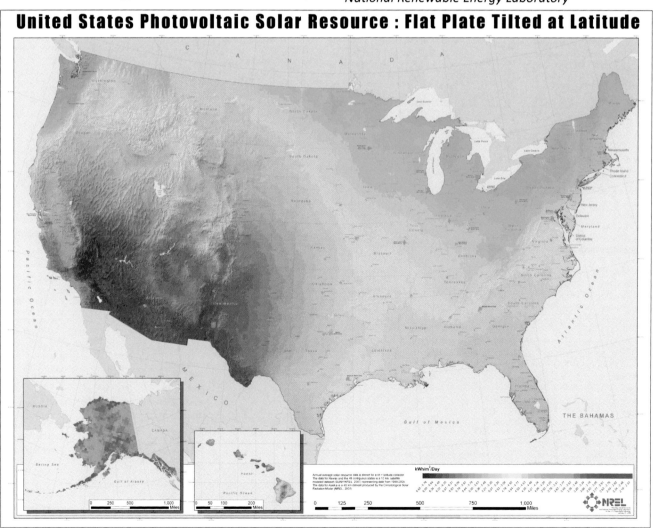

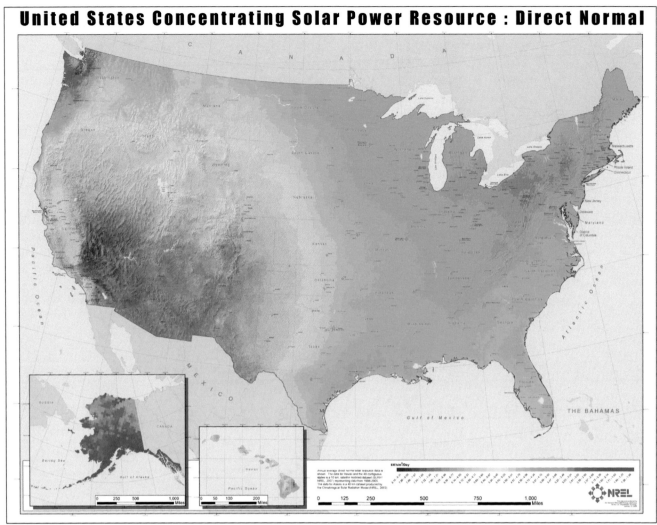

Figure 10-2, National Concentrating Solar Power (CSP) Resource Potential, KWh/m^2/Day, *Credit/ Courtesy of National Renewable Energy Laboratory*

that sunlight does not need a medium, like air or water or rock, to travel through in order to get from point A to point B. This auspicious property of sunlight means that half of the earth is always bathed in sunlight, thereby facilitating photosynthesis and thus enabling nearly all forms of life on our planet. Other kinds of energy, say thermal energy or sound waves, do need something between points A and B in order to propagate. Thus the tagline from the 1979 science fiction thriller *Alien*, "In space, no one can hear you scream" is in fact correct. Sound waves need something to travel through, but in the vacuum of space there is nothing, so there is no sound.

Electromagnetic radiation is oscillating electrical and magnetic fields that have the peculiar ability to be correctly described as both a particle (a photon) and a wave. This dual nature facilitates contextual discussions about the properties of different kinds of light, as well as how they influence objects. Thus you might see a comet's tail being created from photons hitting the comet's nucleus, but you communicate the expe-

rience by calling your friend on your mobile phone using radio waves. Both photons and radio waves are light—oscillating electrical and magnetic fields—while the context of each experience is different.

Electromagnetic radiation is readily portrayed on the electromagnetic spectrum. **The electromagnetic spectrum illustrates the entire range of electromagnetic radiation, which includes gamma rays, X-rays, ultraviolet rays, visible light, infrared light, microwaves and radio waves.** Keep in mind that these are all forms of light, but they differ in terms of their wavelength and frequency. **A wavelength is the distance between successive crests of a wave.** Electromagnetic wavelengths are often measured (and thus classified) in units called nanometers (nm). Recall from Chapter 5 that a nanometer is one billionth of a meter (10^{-9} of a meter). To visualize 1 nm, imagine taking a meter stick and slicing it into 1,000,000,000 equal slices—each of these slices is a nanometer.

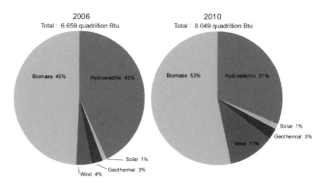

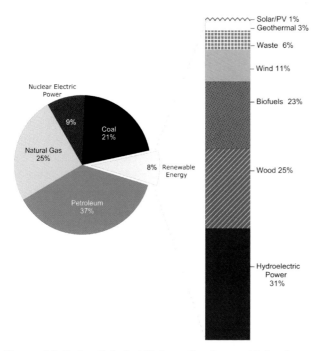

Figure 10-3, Sunlight's 1% Contribution to United States Renewable Energy Consumption, 2010, *Credit/Courtesy of United States Energy Information Administration*

Figure 10-4, United States Energy Renewable Consumption by Energy Source, 2006 Compared to 2010, *Credit/Courtesy of United States Energy Information Administration*

Recall also from Chapter 5 that within the visible portion of the electromagnetic spectrum, the color red has a wavelength of around 700 nm, while the color violet light has a wavelength of around 400 nm. **Most of the electromagnetic radiation emitted from the sun is between 500 and 600 nm, which our eyes see as greens and yellows** (Wall, 2011). Light with shorter wavelengths (such as violet) contains more energy than light with longer wavelengths (such as red) because shorter wavelength light has a higher frequency (Burke & Sarlina, 2011). Higher frequency light simply means more wavelengths pass a point in one second compared to lower frequency light (Goddard Space Flight Center, 2010). **All forms of electromagnetic radiation have 1) no mass and 2) travel in a vacuum at the universe's speed limit—the speed of light—of about 300,000 kilometers/second (186,000 miles/second).**

On a clear day, at noon, about 1,000 Watts/m² of solar electromagnetic radiation penetrates our atmosphere and makes it to the earth's surface (Union of Concerned Scientists, 2010). This radiation is primarily composed of visible and infrared wavelengths. The earth's land and water surfaces absorb this energy differently, depending on soil type and water composition. Fortunately, we are able to capture some of this energy, using both simple and complex technologies, before it is absorbed by the earth. Thus these raw Watts can be turned into useful kilowatt-hours. An example of very simple solar technology is a large black basin full of water you are likely to see situated on people's rooftops in very sunny, arid parts of the world. The black plastic absorbs the solar radiation, converts it into thermal radiation, which then heats water inside the container for use in the home. Much more complex solar technologies, by contrast, absorb and then convert solar radiation into electricity, either directly—as with photovoltaic systems—or indirectly—as with concentrating solar power systems.

Photovoltaics

Photovoltaic (PV) technology employs silicon-based semiconductors to convert sunlight into electricity

Figure 10-5, The Electromagnetic Spectrum, *Credit/Courtesy of NASA*

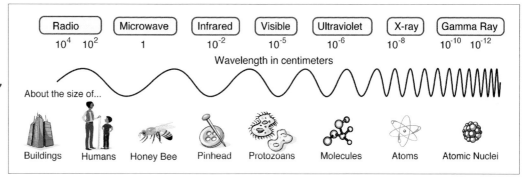

Figure 10-6, A Model of a Silicon Atom

without any emissions, moving parts or noise. Silicon (Si) is the 14th element on the periodic table, meaning it has 14 protons in its nucleus and 14 electrons in its electron cloud. The electron cloud of an atom is normally described as having levels, or shells, into which only a fixed number of electrons may fit at any single moment. The first shell of silicon can only hold two electrons, the second can only hold eight, and while the third shell can also hold eight, the atom itself has only 4 remaining electrons, so this last shell is half full (2 + 8 + 4 = 14). Silicon *can* hold eight electrons in its outer shell but *is* holding only four. PV technology exploits silicon's half-full outer shell, as well as the fact that it is a metal and thus, in concentrated samples, its electrons are free to drift about within the sample's crystalline lattice (recall that this concept was touched

on in Chapter 1 when discussing copper's propensity to carry an electric current).

Pure silicon is a poor electrical conductor. Compared to copper, one could even say silicon is "semi" good at conducting electricity—thus a "semiconductor". **A semiconductor is a solid substance that conducts small, precise electric currents, and whose conductivity can be manipulated with the addition of impurities.** Silicon is the best known and most established semiconducting material. While many dozens of other semiconductor materials also exist, silicon dominates modern electronic circuitry, including photovoltaic cells (United States Department of Energy, 2011). Semiconductor "impurities" common in the PV industry are phosphorus (P, element number 15) and boron (B, element number 5). **Adding impurities to a semiconducting substance is called "doping".**

Remember that silicon *can* hold eight electrons in its outer shell but *is* holding only four. Phosphorus, by contrast, *can* hold eight electrons in its outer shell but *is* holding only five. Doping a layer of silicon with phosphorus creates the compound (SiP) that contains a single, weakly-bonded electron in the compound's outermost electron shell. Eight electrons fill the SiP compound's penultimate electron shell, and a ninth electron hangs out all by itself in the ultimate shell. This kind of material is called an n-type (or negative) semiconductor. **N-type (or negative) semiconductors have an abundance of electrons.**

Doping a layer of silicon with boron, by contrast, creates the compound (SiB). Boron's three outermost electrons share the space with silicon's four outermost electrons. The compound's ultimate electron shell thus contains only seven electrons. There is

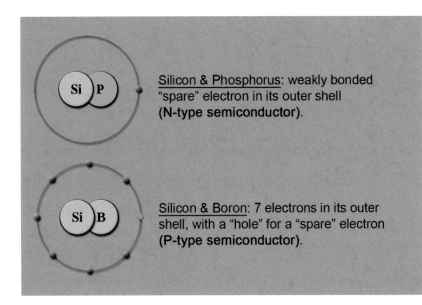

Silicon & Phosphorus: weakly bonded "spare" electron in its outer shell (N-type semiconductor).

Silicon & Boron: 7 electrons in its outer shell, with a "hole" for a "spare" electron (P-type semiconductor).

Figure 10-7,
SiP (a N-Type Semiconductor) and SiB (a P-Type Semiconductor) Compounds

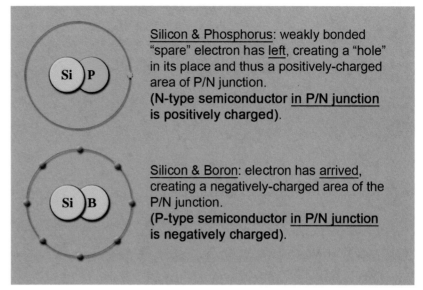

Silicon & Phosphorus: weakly bonded "spare" electron has <u>left</u>, creating a "hole" in its place and thus a positively-charged area of P/N junction.
(**N-type semiconductor** <u>in P/N junction</u> **is positively charged**).

Silicon & Boron: electron has <u>arrived</u>, creating a negatively-charged area of the P/N junction.
(**P-type semiconductor** <u>in P/N junction</u> **is negatively charged**).

Figure 10-8, The P-N Junction (Note the Ionic Charge Within the Junction)

enough space, therefore, —a single "hole"—in the compound's outermost electron shell for one additional electron. This kind of material is called a p-type (or positive) semiconductor. **P-type (or positive) semiconductors have an abundance of holes.**

Left to themselves, the n-type and p-type semiconductor layers do nothing. Sandwiching them together, however, unleashes a short-lived flood of electron activity. At the junction where the semiconductors meet, electrons rush from the n-type semiconductor (remember, n-types have an abundance of electrons) to the p-type semiconductor (remember, p-types have an abundance of holes just right for the incoming electrons). In a split second the junction between the two semiconductors is electrically stabilized, with no additional electron movement. This finished boundary is called the p-n junction. **The p-n junction is the stable electric boundary between n-type and p-type semiconductors.**

Note that the p-n junction now contains ions that were created as a result of this electron exchange. **Ions are charged atoms and/or molecules that are created by either losing or gaining electrons.** Thus, *within* the p-n junction, the n-type semiconductor has a net positive charge (because it lost electrons), and the p-type semiconductor has a net negative charge (because it gained electrons) (United States Department of Energy, 2011b). *Outside* the p-n junction, however, the SiP and SiB compounds remain unchanged, retaining their respective abundance of electrons and abundance of holes. Admittedly, this can be challenging to visualize since we are dealing with very thin layers of semiconducting materials.

The semiconductor is now ready for action. When sunlight strikes the semiconductor, some of the wavelengths go right through the material, some are reflected, and some are absorbed. The absorbed wavelengths are critical because these have the right amount of energy to knock electrons loose from the silicon atoms in both the n-type and p-type layers, and thereby set the stage for an electric current to commence.

The electrons swimming about in the semiconductor's n-type layer are naturally attracted to the semiconductor's p-type layer, but are blocked at the electrically stable p-n junction. The only way to get to the p-type layer is via a closed, external electric circuit (a wire) that physically connects the two layers. Thus, if provided a route, electrons will flow from the n-type layer, through a wire and a load, and into the p-type layer. While this is happening, electrons swimming about in the semiconductor's p-type layer that approach the p-n junction are swept into and through the junction. These very electrons then enter the external circuit because they too are attracted back to the p-type layer. **The stable electric boundary of the p-n junction allows electrons to flow in one direction only, from the p-type layer into the n-type layer.**

Recall that when sunlight strikes the silicon in a PV cell, some wavelengths go through it, some are reflected, and some are absorbed. Also recall that the absorbed wavelengths are the ones that do the work—that is, they knock silicon's electrons loose and allow them to become current carriers. **Silicon's electrons are knocked loose when it absorbs sunlight with wavelengths measuring 1.11 micrometers** (μm). To visualize 1.11 μm, imagine taking a meter stick and slicing it into 1,000,000 equal slices—each of these slices is a micrometer. Take one of these slices, and a bit more than a tenth of another slice, and you have 1.11 μm.

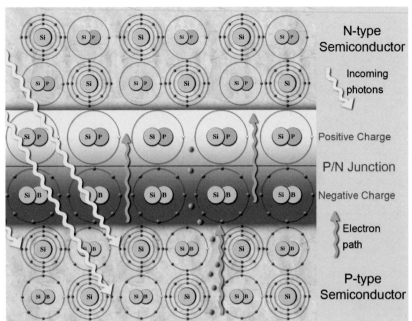

Figure 10-9, A Model of a Working Photovoltaic Semiconductor

Light with a wavelength of 1.11 μm is in the infrared portion of the electromagnetic spectrum, which is invisible to us because our eyes are only able to see light in the visible portion of the electromagnetic spectrum. When this particular wavelength of sunlight hits one of silicon's electrons, the electron acquires just the right amount of energy—1.18 electron volts (eV—a very tiny amount of kinetic energy)—to render it impossible to remain attached to its silicon atom (Low, Kreider, Pulsifer, Jones, & Gilani, 2008, p. 32). This is called silicon's bandgap energy. **Bandgap energy is the amount of energy required to free an electron from its atomic bond.**

Silicon therefore obviously captures only a tiny slice of the entire electromagnetic spectrum. Other materials, however, have bandgap energies different from silicon, and many of these can also be made into semiconductors. By using elements such as indium (In, element number 49), gallium (Ga, element number

31), manganese (Mn, element number 25), tellurium (Te, element number 52), and cadmium (Cd, element number 48), solar engineers work to capture ever-greater portions of the electromagnetic spectrum. Often this is accomplished in a single photovoltaic cell by literally stacking different semiconducting materials. **A multijunction solar cell consists of semiconductors constructed from different elements—thus**

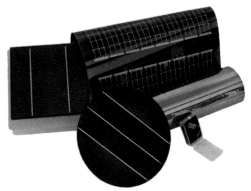

Figure 10-10, Old and New Generation Photovoltaic Cells, *Credit/Courtesy of National Renewable Energy Laboratory*

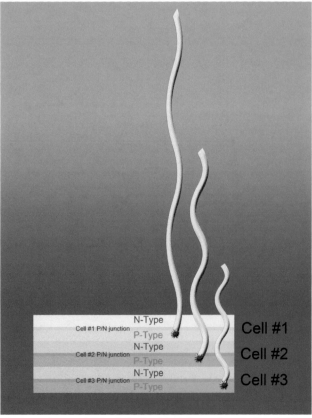

Figure 10-11, A Model of a Multijunction Solar Cell

Figure 10-12, Installing Photovoltaic Modules, *Credit/Courtesy of United States Department of Energy*

having different bandgap energies—that are stacked on top of each other. Literally, then, a multijunction solar cell has "multiple" p-n junctions. The top layer absorbs longer wavelengths of sunlight, and then each respectively-lower layer absorbs increasingly shorter wavelengths of energy (United States Department of Energy, 2011c).

The fundamental building block of PV systems is a cell. A photovoltaic cell is a small semiconductor typically measuring no larger than four inches across, and having a capacity no larger than 2 Watts (United States Department of Energy, 2011d). Since 2 Watts is a very small capacity, multiple cells are interconnected together to form modules, and modules are interconnected to form arrays. These interconnections boost a PV system's capacity by the product of all the combined cells. Thus, if a cell's capacity is 2 Watts, and a module is built using 72 cells, the module's capacity is 144 Watts (2 x 72 = 144 Watts). Then, if 10 of these

modules are interconnected, the resulting array capacity is 1,440 Watts (10 x 144 = 1,440 Watts). Finally, if 100 of these arrays are interconnected in one operational photovoltaic system, the entire photovoltaic system capacity is 144,000 Watts, or simply 144 kW (144 x 100 = 144,000). A photovoltaic system includes everything that is needed to satisfy a particular electricity demand, including the PV cells, modules, arrays, sunlight tracking gear, and any inverters (inverters convert direct current to alternating current electricity).

The energy industry categorizes photovoltaic systems as flat plate, thin film or concentrator. Flat plate systems are the most common and consist of modules or arrays of flat plates where individual PV cells are neatly arranged and securely encapsulated on a rigid substrate. A transparent, thin cover of glass or plastic protects the cells from weathering and dust (United States Department of Energy, 2011e). Thin film sys-

Figure 10-13, A Photovoltaic Array, *Credit/Courtesy of United States Department of Energy*

Figure 10-14, The 20 MW Photovoltaic System Near Beneixama, Spain, *Credit/ Courtesy of Siemens*

tems, by contrasts, are created by depositing semiconducting materials on a molecular scale and thereby reducing the thickness of a semiconductor to no more than 10 μm, which is hundreds of times thinner than traditional PV cells (United States Department of Energy, 2011f). Thin film PV systems have the advantage of having no individual PV cells since the film can be created *en masse*, made to nearly any scale and conform to nearly any shape.

Concentrator photovoltaic systems take advantage of the fact that PV cells are able to capture more of their essential bandgap energy wavelengths when bathed in concentrated sunlight (United States Department of Energy, 2011g). By focusing a lot of sunlight on a very small PV cell, a tremendous increase in efficiency is obtained. Table 10-1 illustrates this fact, with concentrator PV systems approaching 40% efficiency—that is, able to convert almost 40% of the incoming particular bandgap energy wavelengths into electricity. Concentrator systems need comparatively less space than flat plate and thin film systems, but have technical challenges associated with high temperatures and concentrating optics.

Figure 10-15, Flat Plate Photovoltaics on Sun-Tracking Panels, *Credit/Courtesy of Siemens*

Figure 10-16, A Thin Film Photovoltaic System at the U.S. Army Fort Carson Base, *Credit/Courtesy of United States Department of Energy*

In 2009, flat plate photovoltaic systems dominated the PV marketplace, with 984,160 peak kilowatts of semiconductor materials shipped. Thin film systems followed with 266,547 peak kilowatts shipped. Concentrator systems trailed, with 31,852 peak kilowatts shipped. These figures are about 30% greater than those in 2008 (United States Energy Information Administration, 2011b), thus suggesting to many photovoltaic advocates that the industry is entering an aggressive growth stage. PV optimism is also apparent at power-generation conventions where industry professionals gather to discuss the state of the industry. In 2011, for example, at *Electric Power 2011*, participants in the Solar Power and Photovoltaics session enthusiastically reported on the following PV facts:

- 1.9 GW of large-scale PV projects are now under construction

- 12 PV power plants having a nameplate capacity >50 MW have been completed in the past year

- 10 PV power plants are currently under construction, each having a nameplate capacity >100 MW

Figure 10-17, Thin-Film Photovoltaics on Solar House, *Credit/Courtesy of United States Department of Energy*

- 22 GW of PV projects are currently being planned

- 2,508 MW of PV capacity is expected in the United States by 2014

(Neville, 2011)

Table 10-1, Average Photovoltaic Cell/Module Energy Conversion Efficiencies, 2009

PV Type	Average Efficiency
silicon (single-crystal)	20%
silicon (cast)	14%
silicon (ribbon)	13%
Thin-film (amorphous silicon)	8%
Thin-film other (cadmium telluride & copper indium gallium selenide)	12%
Concentrator	38%

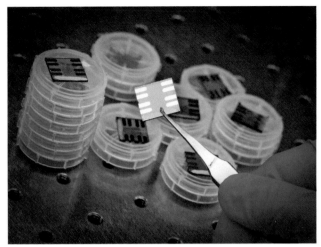

Figure 10-18, Concentrator Photovoltaic Cells, *Credit/ Courtesy of United States Department of Energy*

The world's largest capacity PV power plant just recently came online. **As of July 2011, the largest photovoltaic power plant in the world is the 80 MW Sarnia Solar Project in Ontario, Canada, occupying 950 acres (about 1.5 miles^2) and using 1.3 million thin film panels** (Enbridge, 2010).

Concentrating Solar Power (CSP)

Concentrating solar power (CSP) systems use mirrors to focus sunlight onto receivers and thereby convert sunlight into thermal energy, which is then used to generate electricity via steam turbines or heat engines. Utility-scale CSP systems use sun-tracking software, mirrors with very high reflectivity (the latest CSP mirrors reflect about 94% of light), and increasingly, thermal storage with molten salt to provide baseload electricity generation long after the sun sets. As with photovoltaics, considerable enthusiasm exists within the CSP industry, and many analysts expect significant

Figure 10-19, A Concentrator Photovoltaic Unit, *Credit/Courtesy of United States Department of Energy*

expansion in the near future. In the southwestern United States, for example, nearly 9,000 MW of new CSP projects are set to begin construction over the coming several years (Solar Today, 2010, p. 32). Moreover, as of July, 2011, the Bureau of Land Management approved 3,566 MW of new CSP projects on public land, including one 1,000 MW project—the Blythe Solar Power Project—that will rank as the world's largest CSP plant when completed in 2016 (Bureau of Land Management, 2011).

Utility-scale electricity producers have at their disposal three different kinds of CSP technologies: 1) linear, 2) dish, and 3) power towers. Linear systems consist of numerous parallel rows of sunlight collectors aligned in a north/south fashion. As the sun rises in the east, tracks across the sky, and sets in the west, the system's tracking gears continually adjust the angle of the collectors and thereby maximizes the amount of sunlight they receive throughout a day. Linear CSP technology includes the parabolic trough and Fresnel reflector systems.

Figure 10-20, Sarnia Solar Project, Array View, *Credit/ Courtesy of Enbridge, Inc.*

Figure 10-21, Sarnia Solar Project, Aerial View, *Credit/Courtesy of Enbridge, Inc.*

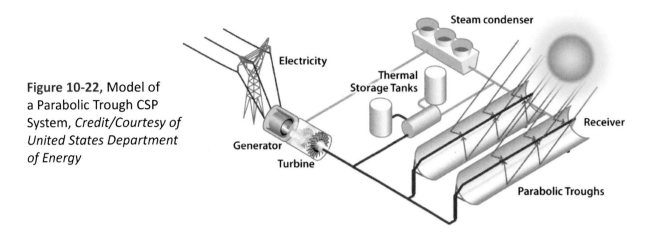

Figure 10-22, Model of a Parabolic Trough CSP System, *Credit/Courtesy of United States Department of Energy*

Figure 10-23, Close-Up View of a Hot Receiver Tube at Focus of Parabola, *Credit/Courtesy of United States Department of Energy*

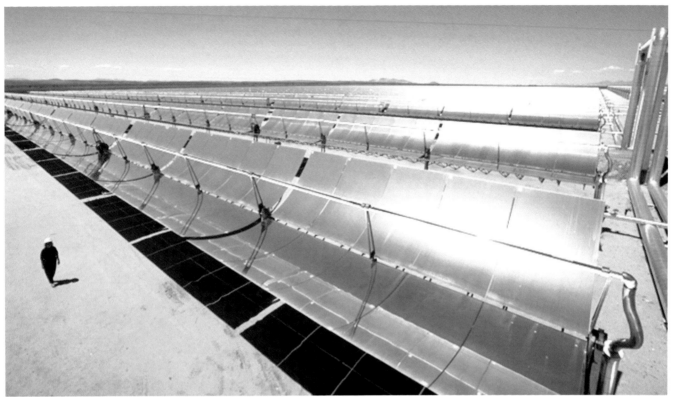

Figure 10-24, Numerous Parabolic Troughs at SEGS,
Credit/Courtesy of Sandia National Laboratories

Figure 10-25, SEGS Workers Installing Receiver Tube on a Parabolic Trough,
Credit/Courtesy of Sandia National Laboratories

Figure 10-26, Cleaning the Parabola Mirrors at SEGS, *Credit/Courtesy of Sandia National Laboratories*

Parabolic trough CSP systems use very long, U-shaped mirrors to focus sunlight on a receiver tube situated at the focus of the parabola and running the entire length of the trough. This technology has a proven work record in southern California, having produced utility-scale electricity there since the 1980s. The scale is quite large, with troughs often measuring 15 feet wide and running 25 feet long, dozens of which are linked together to form one long trough over 1,000 feet long. The receiver tubes circulate a heat transfer fluid—often synthetic oil—that that absorbs heat well and can reach temperatures as high as 700 °F (Solar Energy Industries Association, 2010). After being heated, the transfer fluid circulates through a heat exchanger where it boils water into steam, which in turn

Figure 10-27, Building a Parabolic Trough at the Lebrija Solar Thermal Plant in Spain, *Credit/Courtesy of Siemens*

Figure 10-28, A Solar Thermal Receiver Tube, *Credit/Courtesy of Siemens*

Figure 10-29, Manufacturing Solar Thermal Receiver Tubes, *Credit/Courtesy of Siemens*

Figure 10-30, Ariel View of a Portion of SEGS, *Credit/ Courtesy of Sandia National Laboratories*

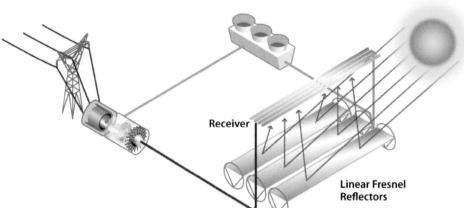

Figure 10-31, Model of Compact Linear Fresnel Reflector CSP System, *Credit/Courtesy of United States Department of Energy*

Receiver

Linear Fresnel Reflectors

spins a turbine to generate electricity. **As of July, 2011, the world's largest parabolic trough CSP system is the Solar Electric Generating Systems (SEGS), located in California's Mojave Desert, with a nameplate capacity of 310 MW and covering 1,500 acres (about 2.3 miles^2) of desert with over 900,000 mirrors** (NextEra Energy Resources, 2011).

An alternative to the trough CSP system is the Fresnel reflector. **Compact Linear Fresnel Reflector (CLFR) CSP systems use long, parallel rows of flat mirrors to focus sunlight on a receiver tube situated at the** **focal point of the mirrors and running their entire length.** Like parabolic troughs, the scale of CLFR systems is large, with mirrors measuring 6 to 8 feet wide and running 50 feet long, dozens of which are linked together to form one long mirror over 1,000 feet long . Also similar to parabolic troughs, CLFR receiver tubes can circulate synthetic oil as a heat transfer fluid, but the best-known Kimberlina Solar Power Plant simply

Figure 10-32, Close-Up View of a Hot Receiver Tube at Focus of Compact Linear Fresnel Reflector, *Credit/ Courtesy of Ausra*

Figure 10-33, Receiver Tubes Situated in A-Frames at Kimberlina Solar Power Plant, *Credit/Courtesy of Ausra*

Figure 10-34, Walking Next to Fresnel Mirrors, *Credit/Courtesy of Ausra*

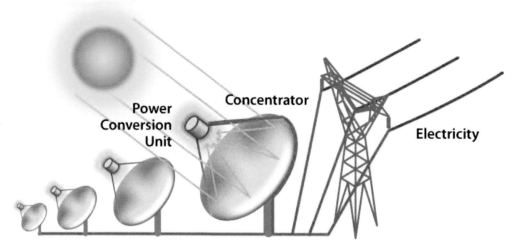

Figure 10-35, Model of a Dish CSP System, *Credit/ Courtesy of United States Department of Energy*

uses water and turns it directly to superheated steam with a temperature of about 750 °F (RenewableEnergyWorld.com, 2011). **As of July, 2011, the world's largest Compact Linear Fresnel Reflector CSP system is the Kimberlina Solar Power Plant, located in Bakersfield, California, with a nameplate capacity of 5 MW and covering 12 acres (about 0.02 miles^2) of desert** (National Renewable Energy Laboratory, 2009b).

Important differences set CLFR CSP systems apart from parabolic troughs. First, the flat mirrors are cheaper to manufacture, and are more rigid and sturdy, than parabolas. Second, these systems use less land area than parabola systems, thereby gleaning 1.5 to 3 times more power per acre compared to competing designs (Areva, 2011). Third, CLFR technicians can "stow" CLFR mirrors upside-down by rotating them, thereby protecting them from weather hazards such as hail storms (RenewableEnergyWorld.com, 2011). These three advantages can make CLFR technology a competitive alternative to parabolic trough systems. However, since CLFR systems are still quite young compared to the parabolic trough's 30+ year working record, more years will pass before the we see widespread CLFR adoption.

Dish CSP systems consist of large parabolic dishes lined with mirrors that focus sunlight on a single power conversion unit situated at the dish's focal point. The best known and most technologically advanced dish design is the SunCatcher, designed and engineered by Stirling Energy Systems Inc. under a partnership with Sandia National Lab and the Department of Energy (Stirling Energy Systems, Inc., 2011). Each SunCatcher measures 38 feet in diameter and has a nameplate capacity of 25 kW (Tessera Solar, 2010), which by itself is not a lot, but when linked with others—dozens, hundreds, or even thousands—

a utility-scale power plant is created. **As of July, 2011, the only dish CSP power plant in the United States is the 1.5 MW Maricopa Solar Plant, located in Peoria, Arizona, containing 60 linked SunCatchers** (National Renewable Energy Laboratory, 2011). This admittedly small power plant is a reference facility, designed to prove the technology and thereby open new opportunities for much larger power plants in the near future. The Bureau of Land Management, in fact, recently approved plans for two such facilities—the 709 MW Imperial Valley Solar Project, and the 663 MW Calico Solar Energy Project, both of which are located in California (Bureau of Land Management, 2011). For a sense the scalability of dish technology, consider that the Imperial Valley Solar Project will contain about 42,000 SunCatchers located on 10 miles^2 of land (United States Department of Energy, 2010).

Unlike trough and CLFR technologies, dish CSP systems produce electricity using heat engines, the most common one being the Stirling. Each SunCatcher has its own engine (this is the "power conversion unit" of the

Figure 10-36, CSP Dish Technology on Test at Sandia National Laboratories in 2008, *Credit/Courtesy of United States Department of Energy*

Figure 10-37, SunCatchers on Test at Sandia National Laboratories in 2009, *Credit/Courtesy of United States Department of Energy*

dish) situated at its focal point. There is no steam line, no molten salt, no synthetic oil and no long pipelines of working fluid in a dish CSP power plant. Instead, as each dish tracks the sun across the sky, each dish's engine generates electricity which is then fed into the grid. **Stirling engines operate with a fixed amount of working gas (often hydrogen) that is sealed inside the engine and is alternately heated (which induces expansion) and cooled (which induces contraction) by external sources.**

The hot, expanding gas inside a Stirling engine drives the engine's power piston. As the power piston moves, the hot gas flows to the cool part of the engine where it contracts. Contraction reduces pressure inside the engine, thereby enabling the power piston to return to its starting point. As the power piston returns, the cooled gas flows back, where it is again heated and thus again expands (California Energy Commission,

2008, pp. 2-7). This cycling rotates the engine's crankshaft, which in turn spins a generator. Stirling engines operate continuously until the external heating and cooling sources are interrupted.

Dish systems operate at the highest temperatures of all CSP systems, with the best heat absorber surfaces currently reaching temperatures of about 1,472 °F (Andraka, 2008). Stirling engines do not consume water, have no internal combustion and therefore no exhaust, can operate autonomously or in modules, and are very quiet. The Stirling engines installed on SunCatchers currently hold the world record for solar-energy-to-electricity conversion efficiency of 31.25% (Sandia National Laboratory, 2008).

Figure 10-38, SunCatchers at the 1.5 MW Maricopa Solar Plant in Peoria, Arizona, *Credit/Courtesy of United States Department of Energy*

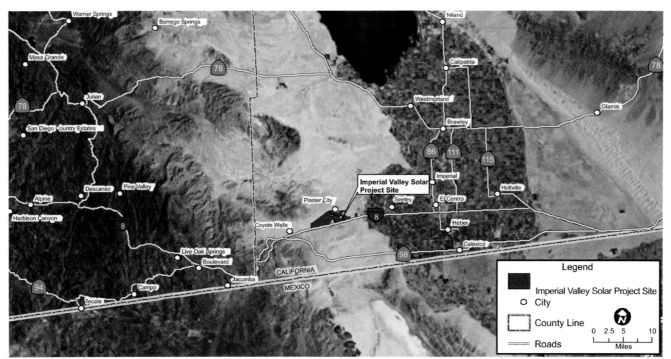

Figure 10-39, Map of the 709 MW Imperial Valley Solar Project Site, *Credit/Courtesy of Bureau of Land Management*

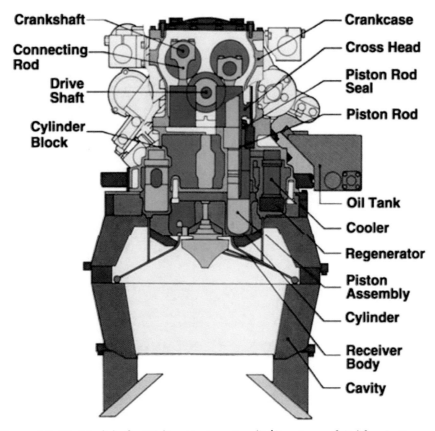

Figure 10-40, Model of a Stirling Engine, *Credit/Courtesy of California Energy Commission*

Power towers are a third CSP technology available to utility-scale electricity producers. **Power tower CSP systems use thousands of flat mirrors called heliostats to reflect sunlight onto a single, tall thermal receiver.** The thermal receiver literally glows, even at noon on cloudless days, with internal heat transfer fluids (water, oil or molten salts) reaching temperatures of 1,050 °F (Gould, 2008). The latest power towers under development or under construction (there are 10 total as of August, 2011) consist of thermal receivers measuring between 400 and 550 feet tall, heliostat fields covering between 2.5 and 5.5 miles^2, and total mirror areas measuring nearly 1 mile^2 (National Renewable Energy Laboratory, 2011b).

Heliostat dimensions vary, with common smaller dimensions measuring 4.68 by 2.95 meters, for a total mirror area of 13.8 m^2 (4.68 x 2.95 = 13.8), and common larger dimensions measuring 9.35 by 7.37 meters, for a total mirror area of 68.9 m^2 (9.35 x 7.37 = 68.9) (Kusek, 2011, p. 16). Nearly 70 m^2 of mirrors is indeed a large surface area, but solar power researchers have also investigated the costs and benefits associated with "large-area heliostats", those with surfaces measuring between 147 and 200 m^2. These are true giants, with each side measuring between 12 and 15 meters long, depending on the model (Strachan & Houser, 1993).

Economies of scale reduce costs, so as one might imagine, larger heliostats, as well as a large number of them, are often preferred by power producers because they literally cost less per meter^2 than smaller ones. This must be balanced, however, with

Figure 10-41, Glowing Hot Thermal Receiver on a Dish CSP System, *Credit/Courtesy of Sandia National Laboratories*

Figure 10-42, Model of a Power Tower CSP System, *Credit/Courtesy of United States Department of Energy*

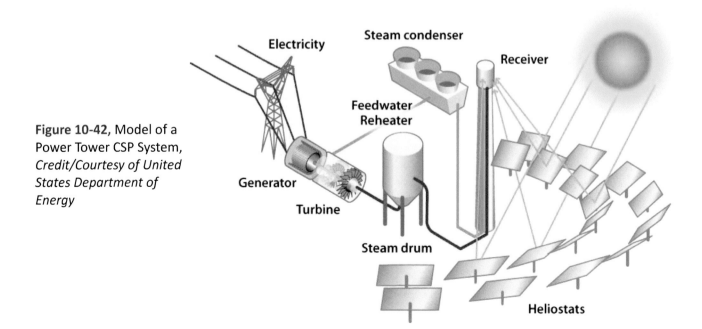

Figure 10-43, The Solar Two CSP Power Tower System, *Credit/Courtesy of United States Department of Energy*

Figure 10-44, A Glowing Hot Power Tower, *Credit/Courtesy of United States Department of Energy*

Figure 10-45, Solar Two Tower With Mirrors Focused, *Credit/Courtesy of Sandia National Laboratories*

Figure 10-46, Solar Engineer Standing Beneath a Heliostat Field, *Credit/Courtesy of Sandia National Laboratories*

Figure 10-47, A Large-Area Heliostat, *Credit/Courtesy of Sandia National Laboratories*

additional cost considerations such as the individual drives that maneuver each heliostat throughout the day, the mirror support structure, as well as wiring, foundation and pedestal sizes, because larger systems require more, and more robust (and thus more expensive) parts and materials (Kolb G. , 2008, p. 7). In 2006, a single heliostat cost between $126.00/m² and $164.00/m², depending on production rate (Kolb, et al., 2007, p. 21). Thus a 68.9 m² heliostat would cost between $8,681.40 and $11,299.6, depending on a project's scale. Since the largest power tower projects currently under development contain over 100,000 heliostats, funding requirements approach and even exceed 1 billion dollars.

As of August, 2011, the world's largest power tower CSP power plant is the 20 MW Planta Solar 20 (PS 20), located near Seville, Spain, with its 165 meter tall receiving tower and 1,255 heliostats (National Renewable Energy Laboratory, 2009c). 20 MW is a relatively small nameplate capacity for a utility-scale power plant, and this, coupled with the fact that it came online in 2009, illustrates the relative youth of this particular type of CSP system. The success of PS 20, as well as the now decommissioned power tower research facilities of Solar One and Solar Two (these

Figure 10-49, Planta Solar 20, Near Seville, Spain, *Credit/Courtesy of Wikimedia*

Figure 10-48, A Heliostat at the Solar Two Research Facility, *Credit/Courtesy of Sandia National Laboratories*

were located near Barstow, California), set the stage for much larger projects currently under development or construction.

The largest capacity power tower CSP project currently under construction is the 392 MW Ivanpah Solar Electric Generating System (ISEGS), located in southeastern California, with 170,000 heliostats covering nearly 5.5 miles² of Ivanpah Dry Lake (Bright-Source, 2010). ISEGS will begin electricity production in 2013. It consists of three interconnected power tower plants, each with its own receiving tower using water as the heat transfer fluid. Natural gas fired turbines kick on to supply electricity after the sun goes down, or on cloudy days when sunlight intensity is insufficient (California Energy Commission, 2011).

One advantage that the latest CSP systems have over photovoltaics is their ability to store thermal energy and thus generate electricity throughout the night, during the off-peak hours (Mehos, 2008, p. 332). The worn aphorism that "solar power is great until the sun sets" is finally being put to rest with this technology. **Thermal storage enables concentrating solar power plants to provide baseload, as well as dispatchable, power to an electrical grid.** Spain currently leads the world in commercial CSP plants with thermal-storage systems (Porteous, 2011, p. 15). Andasol-1 (AS-1) and Andasol-2 (AS-2), both parabolic trough systems, have enough thermal storage to run their turbines "for up to 7.5 hours at full load" without the sun shining (National Renewable Energy Laboratory, 2011c). Spain's recently completed Gemasolar Thermosolar Plant (a power tower), by contrast, has enough thermal stor-

Figure 10-50, A Model Rendering of Ivanpah Solar Power Complex, Developed by BrightSource Energy, *Credit/Courtesy of BrightSource Energy, Inc.*

age capacity to run its turbines a full 15 hours without the sun shining (National Renewable Energy Laboratory, 2009d).

The most established and commercialized CSP thermal storage technique is a two-tank, indirect system (this is how Spain's AS-1 and AS-2 work) (Hopwood, 2009). This technology commonly uses oil as the heat-transfer fluid and molten salt as the thermal-storage fluid. As the name suggests, the two fluids are stored in separate tanks. They never directly mix, but instead continually swap heat through a heat exchanger. Oil is heated within the receiver tubes situated at the focus of CSP mirrors. It then flows to a heat exchanger containing molten salt. The salt takes the oil's heat, thus cooling the oil, necessitating another pass through the CSP receiver tubes. As this happens, the molten salt is pumped to a second heat exchanger where it flashes water to steam, which in turn drives a steam turbine (United States Department of Energy, 2011h).

Figure 10-51, A Model of a Two-Tank, Indirect Thermal Storage System

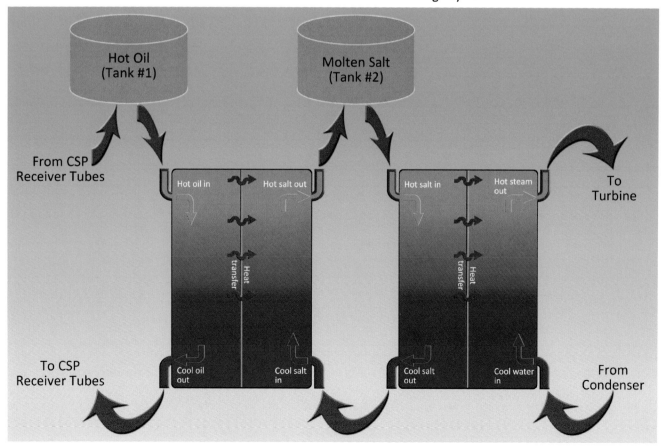

Plant Type	Capacity Factor (%)	Total System Levelized Cost (2009 $/MWh) for 2016
Conventional Combined Cycle	87	66.1
Advanced Nuclear	90	113.9
Conventional Coal	85	94.8
Offshore Wind	34	243.2
Solar PV	25	210.7
Solar Thermal	18	311.8

Table 10-2, Estimated Levelized Cost of New Generation Resources, 2016

Sunlight's Virtues and Vices

Both photovoltaic and concentrating solar power systems are burdened by excessive costs that place them at a market disadvantage compared to other energy sources (Solar Energy Industries Association, 2011b). This is true when comparing PV and CSP to the non-renewables, as well as to nearly all of the renewables. In order to judge the overall financial competitiveness of different generating technologies, the United States Energy Information Administration recently completed a study forecasting the average levelized costs of various generating technologies brought online by 2016. Of the 16 different generating technologies examined, CSP modeled as the most expensive, at $311.80/MWh, and PV the third most expensive, at $210.70/MWh. For perspective, consider that the same study found conventional coal pricing at $94.80/MWh, and conventional combined cycle natural gas pricing at $66.10/MWh (United States Energy Information Administration, 2011c).

While expensive, prices for utility-scale PV solar technologies have dropped in recent years. In 2005 the average cost of a single PV cell was $2.17/Watt, while in 2009 it was $1.27/Watt. PV modules followed suit, pricing at $3.19/Watt in 2005, and $2.79 in 2009 (United States Energy Information Administration, 2011d). Unfortunately, CSP technologies are pricing in the opposite direction. For example, in 2008 the average price for solar thermal collectors was $4.80/ft^2, while in 2009 it jumped to $7.01/ft^2 (United States Energy Information Administration, 2011e).

Knowing that price has always plagued the solar industry, The United States Department of Energy (DOE) is now encouraging price-reduction research for the PV sector of the solar industry. In its SunShot initiative, photovoltaic installation and permitting costs (non-technical barriers) are especially targeted for price reduction, as well as costs associated with cell technology and manufacturing. Non-technical barriers include digitizing local permitting processes and developing rigorous and standardized uniform codes and standards (United States Department of Energy, 2011i). The hope is that by reducing soft cost while simultaneously improving PV cell technology, the United States can cut the price of utility-scale PV energy systems by 75% and thereby regain its leadership in the global PV marketplace by 2020 (United States Department of Energy, 2011j). SunShot-funded research takes place in national laboratories, universities, industry, and federal, state and local government offices across the country and measures in the hundreds of millions of dollars for a wide variety of projects.

The DOE effort is worth it, considering that "all the energy stored in Earth's reserves of coal, oil, and natural gas is matched by the energy from just 20 days of sunshine" (Union of Concerned Scientists, 2010). Of course, such a statement must be tempered by the fact that sunlight is intermittent, varying from location to location depending on the time of day, the season and the weather. In addition, relatively large land areas are needed to capture enough solar energy to feed a utility grid. For example, consider two plants in California, SEGS (310 MW), which occupies 2.3 miles^2, and Diablo Canyon (2,240 MW), which occupies just over 1 mile^2. While comparing solar and nuclear power plants is problematic on many levels, the land use comparison is revealing—with less than half the land area, Diablo Canyon delivers over seven times more electricity to California's grid.

Fuel costs $0.00 at solar energy facilities. There are no trainloads, barges, truckloads, pipelines, or fuel processing costs associated with solar fuel. It literally comes from the sky, every day, for free. This is coupled with the fact that the most intense sunlight coincides with the most demand. Consider a typical afternoon during the summertime in a heavily-populated, arid part of the United Sates. Between 9:00 AM and 5:00 PM, thousands of air conditioners will draw thousands of MWhs worth of electricity from the grid. This is exactly the time period when solar energy facilities produce the most power. **Solar energy facilities produce the most electricity during periods of peak electricity demand.** Overcoming PV and CSP facility costs, therefore, makes a lot of sense for national and local governments around the world and they seek to ensure energy security and stability for growing populations of energy-hungry people.

While our solar discussion has thus far concentrated on utility-scale projects, small-scale residential applications are also possible with this energy source. Both active and passive solar energy systems are options for homeowners. **Active solar energy systems use pumps or fans to circulate heated fluids or air between solar collectors and a home, while passive systems use no electrical or mechanical devices to provide heat and lighting to a structure.** A set of PV modules installed on a home's rooftop is an example of an active system because electricity is generated from the sun and moves into the home. Active solar water heaters are also (as the name implies), active systems. These systems include a rooftop solar collector through which a heat-transfer fluid is pumped. The fluid absorbs solar energy and then circulates into a home's hot water heater where its heat is absorbed by the hot water heater's water supply (United States Department of Energy, 2011k). Since a circulating pump is needed for this system, it is an active solar energy system.

Passive solar energy systems are simpler, require little to no maintenance, and are most appropriately (though not exclusively) designed and planned during the pre-construction phase of a home. Most significantly, passive systems make use of a building's orientation to the sun. In the northern hemisphere,

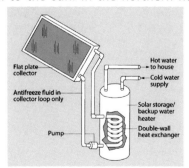

Figure 10-52, A Model of an Active, Closed Loop Solar Water Heater, *Credit/Courtesy of United States Department of Energy*

higher than 23.5° N latitude, the sun always shines somewhere in the southern skies. During the summer months the sun tracks across the sky at a higher altitude than during the winter months. With large windows and thoughtful rooftop designs, buildings with their broadest side situated on an east-west axis can take advantage of this altitude difference to supplement interior heating and lighting requirements. One straightforward passive heating design is the installation of dark tiles on floors of sun-drenched, south facing rooms. During the daytime, these tiles absorb solar energy, which is then released into the home during the evening when the temperature drops. Another application includes a sunspace, in which interior air is heated in a manner similar to a greenhouse, and then circulated through the home with clever ventilation placement (National Renewable Energy Laboratory, 2009e).

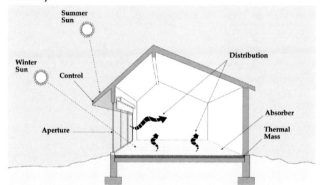

Figure 10-53, A Model of a Passive Solar Design, *Credit/Courtesy of National Renewable Energy Laboratory*

Important Ideas:

- More energy from the sun falls on the earth in one hour than is used by everyone in the world in one year.

- In 2010, the United States had only 888 MW of installed solar electrical generating capacity, the least installed capacity of any energy resource.

- The electromagnetic spectrum illustrates the entire range of electromagnetic radiation, which includes gamma rays, X-rays, ultraviolet rays, visible light, infrared light, microwaves and radio waves.

- A wavelength is the distance between successive crests of a wave.

- Most of the electromagnetic radiation emitted from the sun is between 500 and 600 nm, which our eyes see as greens and yellows.

- All forms of electromagnetic radiation have 1) no mass and 2) travel in a vacuum at the universe's speed limit—the speed of light—of about 300,000 kilometers/second (186,000 miles/second).

- On a clear day, at noon, about 1,000 Watts/m² of solar electromagnetic radiation penetrates our atmosphere and makes it to the earth's surface.

- Photovoltaic (PV) technology employs silicon-based semiconductors to convert sunlight into electricity without any emissions, moving parts or noise.

- A semiconductor is a solid substance that conducts small, precise electric currents, and whose conductivity can be manipulated with the addition of impurities.

- Adding impurities to a semiconducting substance is called "doping".

- The p-n junction is the stable electric boundary between n-type and p-type semiconductors.

- Ions are charged atoms and/or molecules that are created by either losing or gaining electrons.

- The stable electric boundary of the p-n junction allows electrons to flow in one direction only, from the p-type layer into the n-type layer.

- Silicon's electrons are knocked loose when it absorbs sunlight with wavelengths measuring 1.11 micrometers.

- Bandgap energy is the amount of energy required to free an electron from its atomic bond.

- A multijunction solar cell consists of semiconductors constructed from different elements—thus having different bandgap energies—that are stacked on top of each other.

- A photovoltaic cell is a small semiconductor typically measuring no larger than four inches across, and having a capacity no larger than two Watts.

- A photovoltaic system includes everything that is needed to satisfy a particular electricity demand, including the PV cells, modules, arrays, sunlight tracking gear, and any inverters (inverters convert direct current to alternating current electricity).

- The energy industry categorizes photovoltaic systems as flat plate, concentrator or thin film.

- As of July 2011, the largest photovoltaic power plant in the world is the 80 MW Sarnia Solar Project in Ontario, Canada, occupying 950 acres (about 1.5 miles²) and using 1.3 million thin film panels.

- Concentrating solar power (CSP) systems use mirrors to focus sunlight onto receivers and thereby convert sunlight into thermal energy, which is then used to generate electricity via steam turbines or heat engines.

- Parabolic trough CSP systems use very long, U-shaped mirrors to focus sunlight on a receiver tube situated at the focus of the parabola and running the entire length of the trough.

- As of July, 2011, the world's largest parabolic trough CSP system is the Solar Electric Generating Systems (SEGS), located in California's Mojave Desert, with a nameplate capacity of 310 MW and covering 1,500 acres (about 2.3 miles²) of desert with over 900,000 mirrors.

- Dish CSP systems consist of large parabolic dishes lined with mirrors that focus sunlight on a single power conversion unit situated at the dish's focal point.

- As of July, 2011, the only dish CSP power plant in the United States is the 1.5 MW Maricopa Solar Plant, located in Peoria, Arizona, containing 60 linked SunCatchers.

- Stirling engines operate with a fixed amount of working gas (often hydrogen) that is sealed inside the engine and is alternately heated (which induces expansion) and cooled (which induces contraction) by external sources.

- Power tower CSP systems use thousands of flat mirrors called heliostats to reflect sunlight onto a single, tall thermal receiver.

- As of August, 2011, the world's largest power tower CSP power plant is the 20 MW Planta Solar 20 (PS 20), located near Seville, Spain, with its 165 meter tall receiving tower and 1,255 heliostats.

- The largest capacity power tower CSP project currently under construction is the 392 MW Ivanpah Solar Electric Generating System (IS-EGS), located in southeastern California, with 170,000 heliostats covering nearly 5.5 miles[2] of Ivanpah Dry Lake.

- Thermal storage enables concentrating solar power plants to provide baseload, as well as dispatchable, power to an electrical grid.

- Both photovoltaic and concentrating solar power systems are burdened by excessive costs that place them at a market disadvantage compared to other energy sources.

- Fuel costs $0.00 at solar energy facilities.

- Solar energy facilities produce the most electricity during periods of peak electricity demand.

- Active solar energy systems use pumps or fans to circulate heated fluids or air between solar collectors and a home, while passive systems use no electrical or mechanical devices to provide heat and lighting to a structure.

Works Cited

Andraka, C. E. (2008). Cost/Performance Tradeoffs for Reflectors used in Solar Concentrating Dish Systems, ES2008-54048. *Proceedings of Energy Sustainability, 2008.* Jacksonville, Florida: Energy Sustainability.

Areva. (2011). *Industry-Leading Competitive Solar Solutions.* Retrieved July 23, 2011, from http://www.areva.com/EN/global-offer-725/concentrated-solar-power-renewable-energies-solutions.html

BrightSource. (2010). *Ivanpah Solar Electric Generating System.* Retrieved August 4, 2011, from http://ivanpahsolar.com/

Bureau of Land Management. (2011, March 31). *Approved Renewable Energy Projects.* Retrieved July 22, 2011, from Solar Energy Projects: http://www.blm.gov/wo/st/en/prog/energy/renewable_energy/approved_projects.html

Burke, S., & Sarlina, T. (2011). *Fermilab Science Education Office.* Retrieved July 4, 2011, from Light and Color: http://ed.fnal.gov/trc_new/demos/present/Light_Color_TJ-Susan.ppt

California Energy Commission. (2008, June 18). *Imperial Valley Solar (Formerly called SES Solar Two Project).* Retrieved July 27, 2011, from Power Plant Licensing Case, SES Solar Two AFC Volume 2 and 3, Master Appendix B: http://www.energy.ca.gov/sitingcases/solartwo/documents/applicant/afc/volume_02+03/MASTER_Appendix%20B.pdf

California Energy Commission. (2011, March 11). *Ivanpah Solar Electric Generating System.* Retrieved August 4, 2011, from Process Description: http://www.energy.ca.gov/sitingcases/ivanpah/index.html

Enbridge. (2010, October 4). *Enbridge and First Solar Complete the Largest Photovoltaic Facility in the World.* Retrieved July 19, 2011, from http://www.enbridge.com/MediaCentre/News.aspx?yearTab=en2010&id=1329131

Goddard Space Flight Center. (2010, February 3). *Electromagnetic Spectrum - Introduction .* Retrieved July 4, 2011, from A Radio Wave is not a Gamma-Ray, a Microwave is not an X-ray ... or is it?: http://imagine.gsfc.nasa.gov/docs/science/know_l1/emspectrum.html

Gould, B. (2008, April 23). *Molten Salt Power Towers.* Retrieved August 3, 2011, from DOE Program Review: http://www1.eere.energy.gov/solar/review_meeting/pdfs/prm2008_gould_solarreserve.pdf

Hopwood, D. (2009, September 8). *Engineering and Technology Magazine.* Retrieved August 5, 2011, from The Hot New Ticket: http://eandt.theiet.org/magazine/2009/15/new-hot-ticket.cfm

Kolb, G. (2008). *CSP Advanced Systems – Advanced Heliostats, Sandia National Laboratories.* Retrieved August 4, 2011, from 2008 Solar Annual Review Meeting: http://www1.eere.energy.gov/solar/review_meeting/pdfs/prm2008_kolb_sandia.pdf

Kolb, G. J., Jones, S. A., Donnelly, M. W., Gorman, D., Thomas, R., Davenport, R., et al. (2007, June). *Heliostat Cost Reduction Study, Sandia National Laboratories.* Retrieved August 3, 2011, from Sandia Report, SAND2007-3293: http://prod.sandia.gov/techlib/access-control.cgi/2007/073293.pdf

Kusek, S. (2011, May 17). *Low Cost Heliostat Development.* Retrieved August 3, 2011, from DOE –CSP, Program Review: http://www1.eere.energy.gov/solar/pdfs/csp_pr2011_hitek_services.pdf

Low, J. J., Kreider, M. L., Pulsifer, D. P., Jones, A. S., & Gilani, T. H. (2008). Band Gap Energy in Silicon. *American Journal of Undergraduate Research, 7*(1), 27-32.

Mehos, M. (2008). Concentrating Solar Power. In D. Hafemeister, B. Levi, M. Levine, & P. Schwartz (Ed.), *AIP Conference Proceedings Volume 1044, Physics of Sustainable Energy: Using Energy Efficiently and Producing It Renewably* (pp. 331-339). Berkeley, California: American Institute of Physics.

National Renewable Energy Laboratory. (2009, October 7). *Learning About Renewable Energy.* Retrieved September 9, 2011, from Solar Energy Basics: http://www.nrel.gov/learning/re_solar.html

National Renewable Energy Laboratory. (2009b, May 11). *Concentrating Solar Power Projects.* Retrieved July 23, 2011, from Kimberlina Solar Thermal Power Plant: http://www.nrel.gov/csp/solarpaces/project_detail.cfm/projectID=37

National Renewable Energy Laboratory. (2009c, April 21). *Concentrating Solar Power Projects.* Retrieved August 4, 2011, from Power Tower Projects, Planta Solar 20: http://www.nrel.gov/csp/solarpaces/project_detail.cfm/projectID=39

National Renewable Energy Laboratory. (2009d, April 17). *Concentrating Solar Power Projects.* Retrieved August 5, 2011, from Gemasolar Thermosolar Plant: http://www.nrel.gov/csp/solarpaces/project_detail.cfm/projectID=40

National Renewable Energy Laboratory. (2009e, September 29). *Learning About Renewable Energy.* Retrieved September 9, 2011, from Passive Solar: http://www.nrel.gov/learning/re_passive_solar.html

National Renewable Energy Laboratory. (2011, July 25). *Concentrating Solar Power Projects.* Retrieved July 26, 2011, from Maricopa Solar Project: http://www.nrel.gov/csp/solarpaces/project_detail.cfm/projectID=58

National Renewable Energy Laboratory. (2011b, May 13). *Concentrating Solar Power Projects.* Retrieved August 3, 2011, from Power Tower Projects: http://www.nrel.gov/csp/solarpaces/power_tower.cfm

National Renewable Energy Laboratory. (2011c, January 20). *Concentrating Solar Power Projects.* Retrieved August 5, 2011, from Andasol-1 & Andasol-2: http://www.nrel.gov/csp/solarpaces/project_detail.cfm/projectID=3 and http://www.nrel.gov/csp/solarpaces/project_detail.cfm/projectID=4

Neville, A. (2011, July 1). *Power: Business and Technology for the Global Generartion Industry.* Retrieved July 19, 2011, from Sunny Days Ahead for Solar: http://www.powermag.com/issues/features/3807.html

NextEra Energy Resources. (2011). *Solar Electric Generating Systems.* Retrieved July 22, 2011, from http://www.nexteraenergyresources.com/pdf_redesign/segs.pdf

Porteous, J. (2011, May 4). *Progress Towards Baseload Solar Thermal Power.* Retrieved August 5, 2011, from Ecos Magazine: http://www.ecosmagazine.com/?paper=EC10095

RenewableEnergyWorld.com. (2011, March 4). *Power Engineering.* Retrieved July 22, 2011, from Tech Tour: Kimberlina Solar Power Plant: http://www.renewableenergyworld.com/rea/video/tech-tour-kimberlina-solar-power-plant

Sandia National Laboratory. (2008, February 12). *News Releases.* Retrieved July 27, 2011, from Sandia, Stirling Energy Systems set new world record for solar-to-grid conversion efficiency: https://share.sandia.gov/news/resources/releases/2008/solargrid.html

Solar Energy Industries Association. (2010, March 18). *Factsheets.* Retrieved July 22, 2011, from Concentrating Solar Power: Utility-Scale Solutions for Pollution-Free Electricity: http://www.seia.org/galleries/FactSheets/Factsheet_CSP.pdf

Solar Energy Industries Association. (2011, June 28). *Utility-Scale Solar Projects in the United States: Operating, Under Construction, or Under Development.* Retrieved July 27, 2011, from http://www.seia.org/galleries/pdf/Major%20Solar%20Projects.pdf

Solar Energy Industries Association. (2011b). *About Solar Energy.* Retrieved August 16, 2011, from http://www.seia.org/cs/about_solar_energy

Solar Today. (2010, May). *Bold, Decisive Times for Concentrating Solar Power.* Retrieved July 22, 2011, from http://www.solartoday-digital.org/solartoday/201005?pg=32&pm=2&fs=1#pg32

Stirling Energy Systems, Inc. (2011). *Technology.* Retrieved July 26, 2011, from http://www.stirlingenergy.com/technology.htm

Strachan, J. W., & Houser, R. M. (1993, February). *Testing and Evaluation of Large-Area Heliostats for Solar Thermal Applications.* Retrieved August 4, 2011, from Sandia National Laboratories Technical Library, SAND92-1381 UC-235: http://www.sandia.gov/solar/CSP_papers/Tower/SAND92_1381_heliostat_testing.pdf

Tessera Solar. (2010, January 22). *Tessera Solar and Stirling Energy Systems Unveil World's First Commercial-Scale SunCatcher Plant, Maricopa Solar, with Utility Partner Salt River Project.* Retrieved July 26, 2011, from http://tesserasolar.com/north-america/pdf/2010_01_22.pdf

Union of Concerned Scientists. (2010). *Clean Energy*. Retrieved September 9, 2011, from How Solar Energy Works, The Solar Resource: http://www.ucsusa.org/clean_energy/technology_and_impacts/energy_technologies/how-solar-energy-works.html

United States Department of Energy. (2010, February 24). *Energy Efficiency & Renewable Energy*. Retrieved July 26, 2011, from Environmental Assessment Issued for 750-Megawatt Solar Two Project: http://apps1.eere.energy.gov/news/news_detail.cfm/news_id=15817

United States Department of Energy. (2011, May 31). *Energy Efficiency & Renewable Energy*. Retrieved July 13, 2011, from Photovoltaic Cells, Crystalline Silicon Cells: http://www.eere.energy.gov/basics/renewable_energy/pv_cells.html

United States Department of Energy. (2011b, May 31). *Energy Efficiency & Renewable Energy*. Retrieved July 13, 2011, from Semiconductors and the Built-In Electric Field for Crystalline Silicon Photovoltaic Cells, PV Semiconductors: http://www.eere.energy.gov/basics/renewable_energy/semiconductors.html

United States Department of Energy. (2011c, May 31). *Energy Efficiency & Renewable Energy*. Retrieved July 17, 2011, from Photovoltaic Cell Structures, Multijunction Devices: http://www.eere.energy.gov/basics/renewable_energy/pv_cell_structures.html

United States Department of Energy. (2011d, May 31). *Energy Efficiency & Renewable Energy*. Retrieved July 14, 2011, from Photovoltaic Systems: http://www.eere.energy.gov/basics/renewable_energy/pv_systems.html

United States Department of Energy. (2011e, May 31). *Energy Efficiency & Renewable Energy*. Retrieved July 18, 2011, from Flat-Plate Photovoltaic Systems: http://www.eere.energy.gov/basics/renewable_energy/flat_plate_pv_systems.html

United States Department of Energy. (2011f, May 31). *Energy Efficiency & Renewable Energy*. Retrieved July 18, 2011, from Polycrystalline Thin Film Used in Photovoltaics: http://www.eere.energy.gov/basics/renewable_energy/polycrystalline_thin_film.html

United States Department of Energy. (2011g, May 31). *Energy Efficiency & Renewable Energy*. Retrieved July 18, 2011, from Concentrator Photovoltaic Systems: http://www.eere.energy.gov/basics/renewable_energy/concentrator_pv_systems.html

United States Department of Energy. (2011h, May 31). *Energy Efficiency & Renewable Energy*. Retrieved August 5, 2011, from Thermal Storage Systems for Concentrating Solar Power, Two-Tank Indirect System: http://www.eere.energy.gov/basics/renewable_energy/thermal_storage.html

United States Department of Energy. (2011i, August 9). *SunShot Initiative*. Retrieved August 14, 2011, from Reaching the Goal: http://www1.eere.energy.gov/solar/sunshot/about.html#background

United States Department of Energy. (2011j, February 4). *Energy Efficiency & Renewable Energy*. Retrieved August 14, 2011, from DOE Pursues SunShot Initiative to Achieve Cost Competitive Solar Energy by 2020: http://apps1.eere.energy.gov/news/daily.cfm/hp_news_id=288

United States Department of Energy. (2011k, February 9). *Energy Efficiency & renewable Energy*. Retrieved August 16, 2011, from Solar Water Heaters, How They work: http://www.energysavers.gov/your_home/water_heating/index.cfm/mytopic=12850

United States Energy Information Administration. (2011, June). *Renewable Energy Consumption and Electricity Preliminary Statistics 2010*. Retrieved July 3, 2011, from Table 8. Total renewable net summer capacity by energy source and State, 2010: http://www.eia.gov/renewable/annual/preliminary/pdf/table8.pdf

United States Energy Information Administration. (2011b, January). *Solar Photovoltaic Cell/Module Manufacturing Activities 2009*. Retrieved September 9, 2011, from Figure 3.2 Photovoltaic Cell and Module Shipments by Type, 2005-2009: http://www.eia.gov/cneaf/solar.renewables/page/solarreport/solarpv.html

United States Energy Information Administration. (2011c, April 26). *Independent Statistics and Analysis*. Retrieved August 14, 2011, from Levelized Cost of New Generation Resourcesin the Annual Energy Outlook 2011: http://www.eia.gov/oiaf/aeo/electricity_generation.html

United States Energy Information Administration. (2011d, January). *Independent Statistics and Analysis*. Retrieved August 14, 2011, from Solar Photovoltaic Cell/Module Manufacturing Activities 2009: http://www.eia.gov/cneaf/solar.renewables/page/solarreport/solarpv.html

United States Energy Information Administration. (2011e, January). *Independent Statistics and Analysis*. Retrieved August 14, 2011, from Solar Thermal Collector Shipments by Type, Quantity, Revenue, and Average Price : http://www.eia.gov/cneaf/solar.renewables/page/solarreport/table2_12.html

Wall, M. (2011). *Idaho Naitonal Laboratory, INL, ISUTeam on Nanoparticle Production Breakthrough*. Retrieved July 17, 2011, from A Better Solar Cell: https://inlportal.inl.gov/portal/server.pt?open=514&objID=1269&mode=2&featurestory=DA_524323

Index